SECOND EDITION

Industrial Chemical Exposure

Guidelines for Biological Monitoring

SECOND EDITION

Industrial Chemical Exposure

Guidelines for Biological Monitoring

Robert R. Lauwerys
Perrine Hoet

LEWIS PUBLISHERS
Boca Raton   Ann Arbor   London   Tokyo

Library of Congress Cataloging-in-Publication Data

Industrial chemical exposure: guidelines for biological monitoring /
  Robert R. Lauwerys and Perrine Hoet. — 2nd ed.
      p. cm.
  Includes bibliographical references and index.
  ISBN 0-87371-650-7
  1. Biological monitoring.   2. Industrial toxicology.
I. Lauwerys, Robert R.   II. Hoet, Perrine.
RA1223.B54I53   1993
615.9′02—dc20   92-42736
                                                        CIP

PRINTED IN THE UNITED STATES OF AMERICA
    3 4 5 6 7 8 9 0
Printed on acid-free paper

PREFACE TO THE FIRST EDITION

Traditionally, the prevention of excessive exposure to chemicals in industry has been approached by setting standards for the concentration of pollutants in ambient air. For a long time, air monitoring has constituted the major means of assessing workers for exposure to chemicals in industry. This monitoring method considers only exposure by the pulmonary route and, even for chemicals which mainly enter the organism with the inspired air, it does not indicate the true uptake of the exposed workers. These shortcomings have stimulated the search for biological methods of evaluating individual exposure. Such methods had already been in practice for many years but only for a limited number of chemicals (e.g., lead, carbon monoxide). During the last 20 years, more biological methods have been proposed and this field is still rapidly expanding. The data regarding these tests are scattered in various scientific journals; it was useful, therefore, to summarize them.

The objective of this book is not to review in detail all the toxicokinetic data related to industrial chemicals, but to summarize the practical applications of these studies. My aim is to offer a practical guide to those individuals, including occupational physicians, industrial hygienists, and clinical chemists, concerned with the assessment of exposure to industrial chemicals.

R. R. Lauwerys, 1983

PREFACE TO
THE SECOND EDITION

During the last 10 years, significant progress has been made in the field of biological monitoring of exposure to industrial chemicals. The toxicokinetics of several substances in humans has been further clarified, which has permitted the proposal of new biological markers of exposure. For a few chemicals, at least, attempts have been made to better assess the relationship between the biological marker of exposure and the risk of adverse effects.

The field of biological monitoring in occupational hygiene has now clearly emerged from infancy. Several official agencies and/or private organizations have endorsed this approach for health risk assessment and some of them have proposed reference values as guidelines for the evaluation of potential health hazards in industry.

It was necessary to update our previous review on this topic. The present publication aims at offering a practical guide for those concerned with health risk assessment in occupational settings. It does not intend to summarize all the published experimental toxicokinetic studies on industrial chemicals but focuses on the biological markers which might be of practical use for assessing exposure to industrial chemicals.

R. R. Lauwerys and P. Hoet, 1993

THE AUTHORS

Robert R. Lauwerys is Professor of occupational medicine and industrial toxicology at the Catholic University of Louvain, Belgium. He received an M.D. and an M.I.H. degree from that university in 1962 and 1965 and an M.Sc. and D.Sc. in toxicology from Harvard University in 1966 and 1968, respectively. After completing one year of post-doctoral research at the M.R.C. Toxicology Research Unit, Carshalton, U.K., he returned to Belgium to set up a research unit in occupational medicine and industrial toxicology at the medical school of the Catholic University of Louvain. He heads a multidisciplinary research team whose activities concern the metabolism and the mechanism of action of industrial chemicals with the aim of developing biological methods for the assessment of their internal dose and the early detection of adverse effects. His group also performed epidemiologic studies on workers and in the general population to assess the health risk resulting from long-term exposure to several industrial and environmental pollutants.

Professor Lauwerys teaches courses on industrial toxicology and occupational diseases and has frequently lectured on these topics at international meetings. He has written a book in French (*Toxicologie Industrielle et Intoxications Professionnelles*) which is now in its third edition, and has contributed to about 20 books. He has been the author and co-author of approximately 300 research articles.

Perrine Hoet received an M.D., an M.I.H., and an M.Sc. in toxicology from the Catholic University of Louvain, Belgium in 1984, 1987, and 1989, respectively. She is research associate in the occupational medicine and industrial toxicology unit from the same university and is in charge of the outpatient clinic in industrial toxicology. She participates as an expert for the Commission of the European Communities in the development of criteria for the recognition of occupational diseases. She is also medical consultant to the occupational disease compensation board in Belgium.

CONTENTS

1   INTRODUCTION

1. DEFINITION AND OBJECTIVES OF BIOLOGICAL MONITORING OF EXPOSURE TO INDUSTRIAL CHEMICALS

The ultimate objective of industrial toxicology is the prevention of health impairment that may result from exposure to chemicals at workplaces. This implies the definition of permissible levels of exposure, that is, levels that according to the present status of knowledge are estimated to cause no adverse health effects during the lifetime of the workers, and the regular assessment of the possible health risk associated with exposure by comparing the current or the integrated exposure with the permissible exposure limits.

Biological monitoring of exposure to industrial chemicals assesses the health risk through the evaluation of the internal exposure of the organism (i.e., the internal dose) by a biological method. Biological monitoring of exposure is complementary to the two other monitoring programs which are carried out to evaluate the health risk associated with exposure to occupational pollutants, namely ambient monitoring and health surveillance or biological monitoring of early effects (Lauwerys 1984a).

The basis of these monitoring programs is defined by following up the fate of a chemical exerting systemic biological effects from the environment to the target molecules in the organism (Figure 1).

Once absorbed and present in the circulation, the chemical is distributed to different compartments of the body. It may be eliminated unchanged in urine or in expired air. Organic chemicals usually undergo a biotransformation to more water-soluble compounds that are more easily excreted via urine or bile than the parent compound. The chemical or its metabolites may bind reversibly or irreversibly to sites on the target molecules. Binding to noncritical sites induces nonadverse effects that may or may not be reversible. Binding to critical sites may give rise to adverse health effects at least when the amount bound has reached a certain level and the repair mechanisms are inadequate or insufficient. This leads to the development of preclinical lesions, at an early stage, and to clinical lesions at a more advanced stage of intoxication.

1

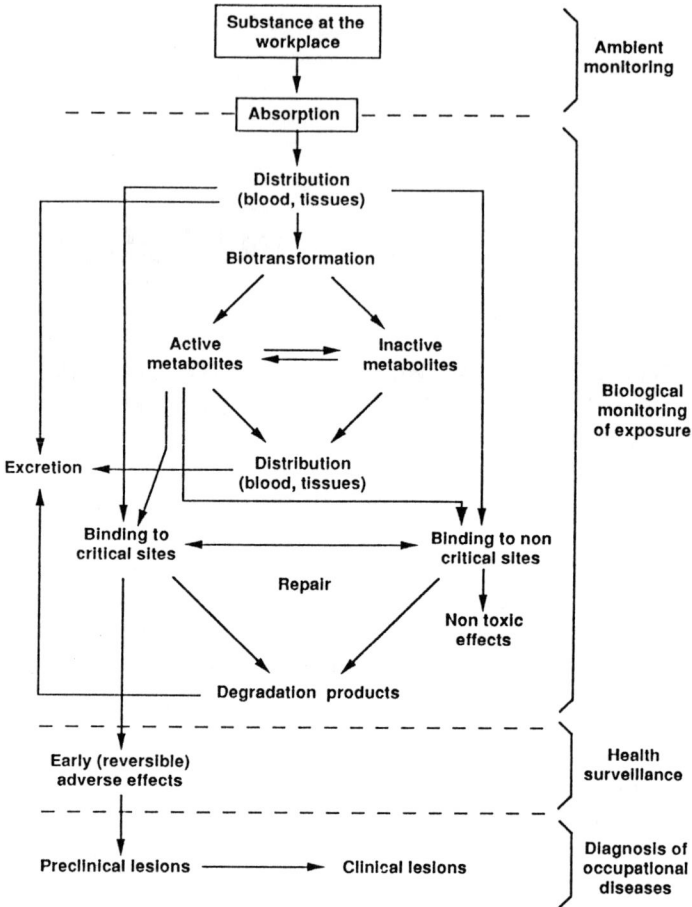

FIGURE 1 Fate of a chemical from the environment to the target molecules in the organism (From Lauwerys, 1991).

Assessing the health risk of populations exposed to occupational or environmental pollutants may be carried out by three types of monitoring — i.e., ambient monitoring, biological monitoring, and health surveillance. These correspond to the different levels shown on the pathway illustrated in Figure 1.

Ambient Monitoring. Ambient monitoring assesses the health risk by measuring the external exposure to the chemical, i.e., the concentration in air, food, water, and the like. In industry, ambient monitoring usually means monitoring the airborne concentration of the chemical. Depending on the type of sampling system selected — stationary or personal — the estimate of the risk may be carried out on a group or individual basis.

The presence of a health hazard is estimated by reference to environmental exposure limits, such as threshold limit value and time-weighted average.

Biological Monitoring of Exposure. Biological monitoring of exposure assesses the health risk through the evaluation of the internal dose. The main goal of biological monitoring of exposure is to ensure that the current or past exposure of the workers is "safe" (i.e., does not entail an unacceptable health risk). It is essentially a preventive medical activity. The presence of a risk is appreciated by reference to permissible levels in biological media, i.e., biological limit values.

Health Surveillance or Biological Monitoring of Effects. Health surveillance must be clearly distinguished from ambient and biological monitoring of exposure. Whereas the latter attempts to detect unhealthy exposure conditions (health risk), health surveillance evaluates the health status and aims at identifying individuals with early signs of adverse health effects, i.e., effects which are likely to be reversible or which do not progress to significant functional impairments when the exposure conditions are improved. Health surveillance must also be distinguished from the diagnosis of occupational diseases, which is the consequence of inadequate preventive programs.

In practice, these three types of monitoring (ambient monitoring, biological monitoring of exposure, health surveillance) are often applied simultaneously. Many chemicals may exert both local and systemic effects. The prevention of acute toxic effects on the respiratory tract or on the eye mucosa can only be prevented by keeping the airborne concentration of the irritant substance below a certain level. Likewise, biological monitoring is not usually indicated for detecting peak exposure to dangerous chemicals. On the other hand, an internal dose of a chemical considered safe by reference to the currently proposed biological limit values may still cause harmful effects in particularly susceptible individuals. Furthermore, the distinction between adverse and nonadverse biological effects is not always clear cut (Berlin et al 1979).

Biological monitoring of exposure attempts to estimate the internal dose. But depending on the chemical and the analyzed biological parameter, the term internal dose may cover different concepts (Lauwerys and Bernard 1985).

First, internal dose may mean the amount of chemical recently absorbed. Hence, a biological parameter may reflect the amount of chemical absorbed either shortly before sampling (e.g., the concentration of a solvent in the alveolar air or in blood during the work shift) or during the preceding day (e.g., the concentration of a solvent in alveolar air or in blood collected 16 h after the end of exposure) or during the last month when the chemical has a long biological half-life (e.g., the concentration of some metals in urine or in blood). Internal dose may also mean the amount of chemical stored in one or in several body compartments or in the whole body. This usually applies to cumulative toxic chemicals. For example, the concentration of polychlorinated biphenyl (PCB) in blood is a reflection of the amount accumulated in the main sites of deposition (i.e., fatty tissues). Finally, with ideal biological monitoring tests, the internal dose means the amount of chemical bound to the critical sites of action. Such tests can be developed, for instance, when the critical sites are easily accessible (e.g., hemoglobin in case of exposure to carbon monoxide) or when the chemical interacts with a blood constituent in a similar way as with the critical target molecule (e.g., hemoglobin adducts).

2. CLASSIFICATION OF BIOLOGICAL MONITORING METHODS

The biological tests currently used for the biological monitoring of exposure to industrial chemicals can thus be classified in three categories, but of different importance (Bernard and Lauwerys 1986): 1) determination of the chemical or its metabolites in biological fluid; 2) quantification of nonadverse biological effects related to the internal dose; and 3) direct measurement of the amount of active chemical interacting with the target molecules.

The great majority of the biological tests available for monitoring exposure belongs to *the first category*. According to their selectivity, these tests can be classified into two subgroups, the selective tests based on the direct measurement of the unchanged chemicals or their metabolites in biological media and the nonselective tests which are used as nonspecific indicators of exposure to a group of chemicals. As an example of nonselective exposure tests, one can cite the determination of diazo-positive metabolites in urine for monitoring exposure to aromatic amines, the analysis of thioethers in urine, and the determination of the mutagenic activity of urine.

Because of their lack of specificity and the existence of a large individual variability, these tests usually cannot be used to monitor exposure on an individual basis. It is, however, possible that when an adequate control group is used as reference, they may be useful as qualitative tests to identify groups at risk.

The *second category* of tests includes those based on the quantification of nonadverse effects which are related to the internal dose. The development of these tests usually requires some knowledge of the mechanism of action of the chemical; this explains that their number is still limited. As example of tests relying on the measurement of nonadverse effects one can cite the inhibition of pseudocholinesterase activity.

Contrary to the preceding exposure tests, those belonging to the *third category* directly estimate the amount of chemical interacting with the sites of action. When they are feasible, i.e., when the target site is easily accessible, these tests assess the health risk more accurately than any other monitoring procedure. A test of this kind has been used in occupational medicine for a long time, i.e., the determination of carboxyhemoglobin. Progress in this monitoring approach is to be expected, namely with the development of a new generation of tests based on immunological or gas chromatography-mass spectrometry techniques. The latter have the potential of detecting with a great specificity and sensitivity subtle alterations induced in the target molecules (e.g., DNA) by reactive chemicals (e.g., mutagens and carcinogens). The DNA adducts can be measured either in hydrolysates of DNA molecules (e.g., in white blood cells) or in degradation products of DNA released in body fluids (e.g., urine). However, these techniques are currently at an early stage of development and they have not yet been routinely applied to groups of workers exposed to industrial carcinogens and mutagens. Much research is still needed before these tests can be introduced in the routine biological monitoring of industrial workers.

3. PRINCIPAL ADVANTAGES OF BIOLOGICAL MONITORING

Biological monitoring of exposure may offer several advantages over environmental monitoring to evaluate the internal dose and hence to estimate the health risk (Lauwerys 1984b).

The greatest advantage of biological monitoring is the fact that the biological parameter of exposure is more directly related to the adverse health effects which one attempts to prevent than any environmental measurement. Therefore, it may offer a better estimate of the risk than ambient monitoring.

Biological monitoring takes into consideration absorption by routes other than the lungs. Many industrial chemicals can enter the organism by absorption through the skin or the gastrointestinal tract. For example, it has been demonstrated that in an acrylic fiber factory, skin absorption of the solvent, dimethylformamide, which is used for the dissolution of the polymer, is more likely to be absorbed by the cutaneous route than by inhalation (Lauwerys et al 1980). Adamsson et al (1979) studied the elimination of cadmium in feces in a group of male workers exposed to cadmium oxide dust in a nickel-cadmium battery factory. Since the cadmium concentration in air (total dust) measured with personal sampler did not exceed 16 $\mu g/m^3$, it was estimated that cadmium naturally occurring in food and cigarettes, cadmium excreted from the gastrointestinal tract, and cadmium transported from the lungs by mucociliary clearance to the gastrointestinal tract only could explain up to 100 μg of the cadmium in the feces. Since even among nonsmokers much higher values for fecal cadmium were recorded, this was interpreted as being the result of ingestion of cadmium from contaminated hands and other body surfaces. Among the smokers, direct oral contact with contaminated cigarettes or pipes is an additional factor. Our results (Roels et al 1982) on the intensity of hand contamination in workers from an electric condenser factory support Adamsson's findings. In a few workers exposed to cadmium, we have determined at different times of the day the amount of cadmium which could be collected after rinsing one hand with 500 ml of slightly acidified water. Before entering the canteen for lunch, up to 280 μg cadmium could be mobilized from one hand (versus less than 7 μg in the control subjects). These observations demonstrate clearly that in industry, the amount of cadmium and probably of any other element which enters the organism does not only depend on the amount inhaled, but also on the amount ingested.

Even if there exists a relationship between the airborne concentration, the overall dustiness of the workplace, and hence the amount of the industrial pollutant entering the organism by any route, one cannot expect that the determination of the airborne concentration may allow an estimate of the total amount of the chemical absorbed by the exposed workers (Lauwerys 1980). Firstly, personal hygiene habits (hand washing, smoking at the workplace, etc.) vary from one person to another. Secondly, it is well known that great individual variation exists in the absorption rate of a chemical through the lungs, the skin, or the gastrointestinal tract. For example, a study by Flanagan et al (1978) indicates, as it was demonstrated previously in animals, that subjects with low iron stores absorb considerably more cadmium (8.9%) through the gastrointestinal tract than persons with normal iron stores (2.4%). Oral absorption of cadmium is therefore higher in females than in males.

Even if strict personal hygiene measures can be implemented so that the pollutant can enter the organism only by inhalation (in addition to the amount transported from the lungs by mucociliary clearance to the gastrointestinal tract), there is no reason to postulate the existence of a relationship between the airborne concentration and the amount absorbed. This has been clearly demonstrated for lead by King et al (1979).

Many physicochemical and biological factors (particle size distribution, ventilatory parameters, etc.) preclude the existence of such a correlation. For example, a physical load of 100 W increases by a factor of 2 to 3 the respiratory uptake of trichloroethylene by comparison with the uptake at rest (Monster et al 1976/7). The daily uptake of xylene by volunteers exposed to the same time-weighted average concentration varies with the environmental conditions (constant or peak exposures) and the work load (Riihimaki et al 1979). A biological parameter may take all these various toxicokinetic factors into consideration.

Because of its capability to evaluate the overall exposure (whatever the route of entry), biological monitoring presents the advantage that it can be used to test the efficiency of various protective measures such as gloves, masks, and barrier creams (Lauwerys et al 1980).

Another advantage of biological monitoring is the fact that the nonoccupational background exposure (leisure activity, residency, dietary habits, smoking, etc.) may also be expressed in the biological level. The organism integrates this total external (environmental and industrial) exposure into one internal load (Zielhuis 1979). From all the above reasons, it is clear that for many industrial pollutants, the respect of any concentration in air may not necessarily prevent an undue intake by the exposed workers.

4. CONDITIONS AND LIMITATIONS OF BIOLOGICAL MONITORING

A rational biological monitoring is only possible when sufficient toxicological information has been gathered on the mechanism of action and/or the metabolism (absorption, biotransformation, distribution, excretion) of xenobiotics to which workers may be exposed (Lauwerys 1991). When a biological monitoring method is based on the determination of the chemical or its metabolite in biological media, it is essential to know how the substance is absorbed via the lung, the gastrointestinal tract, and the skin, and subsequently how it is distributed to the different compartments of the body, biotransformed, and finally eliminated. It is also important to know whether or not the chemical may accumulate in the body. These different kinetic aspects must be kept in mind when selecting the time of sampling.

Biological monitoring of exposure is of practical value only when certain relationships between external exposure, internal dose, and adverse effects are known. Normally, biological monitoring of exposure cannot be applied for assessing exposure to substances that exhibit their toxic effects at the sites of first contact (e.g., primary lung irritants) and are poorly absorbed. In this situation, the only useful quantitative relationship is that between external exposure and the intensity of the local effects. Usually, for such substances the internal dose cannot be quantitatively related to the risk of adverse effect.

For the other chemicals that are significantly absorbed and exert a systemic toxic action, a biological monitoring test may provide different information, depending on our current knowledge of the relationships among external exposure, internal exposure, and the risk of adverse effects as illustrated in Figure 2. If only the relationship between external exposure and the internal dose is known, the biological parameter can be used as an index of exposure, but it provides little information on the health risk (situation a, Figure 2). In other terms, biological monitoring performed under these conditions is much more an assessment of the exposure intensity than of the potential health risk. But if a quantitative relationship has been established between internal dose and adverse effects (c), i.e., if the internal dose effects and the internal dose-response relationships are known, biological monitoring allows for a direct health risk assessment and thus for an effective prevention of the adverse effects. It is indeed possible to derive a biological permissible value (biological limit value, BLV) from these dose effects and dose-response relationships, which is essential to make biological monitoring operational. Unfortunately, the majority of the published studies in this field have focused on the relationship between the internal dose and the external exposure rather than on that between the internal parameter reflecting the internal dose and the adverse effects. Consequently, for many chemicals, the latter relationship is insufficiently documented for a reliable estimation of the BLV. In other words, the biological permissible values are often derived indirectly from the exposure limits in air, through relationships (a) and (b) as shown in Figure 2, a method which is obviously much less reliable than that based on the knowledge of the internal dose effects and the internal dose-response relationships.

The relationships described above may be modified by various factors which influence the fate of an industrial chemical *in vivo*. Metabolic interactions can be predicted when workers are exposed simultaneously to chemicals which are biotransformed through identical pathways. Exposure to industrial chemicals which modify the activity of the biotransformation enzymes (e.g., microsomal enzyme inducers or inhibitors) may also influence the fate of another compound. Furthermore, metabolic interferences may occur between industrial agents and alcohol, food additives, pesticide residues, drugs, or even tobacco. Several biological conditions (sex, weight, fatty mass, pregnancy, diseases, etc.) may also modify the metabolism of an industrial chemical. In summary, the relationship between total uptake of an industrial chemical, its concentration or that of its metabolites in biological media, and the risk of adverse effects may be modified by various environmental and biological factors. They may have to be taken into consideration when interpreting the results of biological exposure tests.

Whatever the biologic parameter measured (the substance itself, its metabolite(s), or a biological effect), the test must be sufficiently sensitive (not too many false negatives) and specific to be of practical value (not too many false positives).

Other conditions should be fulfilled before attempting to implement biological monitoring methods in industry. The ethical aspects must not be neglected; in particular the collection of biological specimens cannot involve any health risk for the workers and the individual results should be considered confidential (Zielhuis 1978).

The selected parameter should be sufficiently stable to allow storage of the biological sample for a certain period of time, and it should be amenable to a nontime-consuming analysis by a not too sophisticated technique. The sensitivity, precision and accuracy of the analysis should be satisfactory. Several intercomparison programs for the analysis of industrial chemicals in biological material have indeed stressed the analytical difficulties sometimes associated with these measurements (Lauwerys et al 1975).

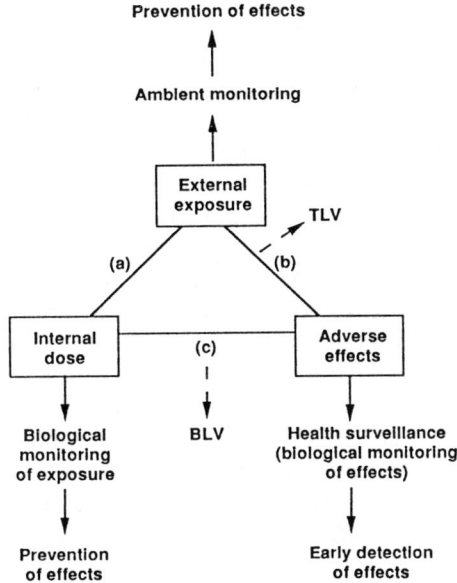

FIGURE 2 Types of monitoring in occupational or environmental health protection (From Lauwerys and Bernard, 1985).

For many industrial chemicals, one or all of the preceding conditions are lacking, which limits the possibilities of specific biological monitoring. It is therefore evident that environmental monitoring still has an important role to play for evaluating and, hence, for preventing excessive exposure to industrial pollutants. Moreover, as already stressed above, the prevention of acute toxic effects on the respiratory tract or the eye mucosa can only be prevented by keeping the airborne concentration of the irritant substances below a certain level. Local effects of industrial chemicals do not lend themselves to a biological surveillance program. Likewise, biological monitoring is usually not indicated for detecting peak exposure to dangerous chemicals (e.g., AsH_3, CO, HCN). Furthermore, identification of emission sources and the evaluation of the efficiency of engineering control measures are usually best performed by ambient air analysis.

In summary, both environmental and biological monitoring programs should not be regarded as opposite, but on the contrary, as truly complementary. They should be integrated as much as possible to assure low levels of contaminants for the continued health of workers (Carnow 1976).

5. INTERPRETATION OF THE RESULTS

The results of a biological monitoring program can be interpreted on an individual basis. However, this interpretation is possible only if the intraindividual variability of the parameter is not too great and its specificity is sufficiently high.

The results may also be interpreted on a group basis by considering their distribution as shown in Figure 3 (Bernard and Lauwerys 1987). If all the observed values are below the biological permissible value, the working conditions are satisfactory (curve 1). If all or the majority of the results are above the biological permissible value, the overall exposure conditions must certainly be corrected (curve 2). A third situation may also occur: the majority of the workers may have values below the biological permissible level, but a few of them have abnormally high values (the distribution is bi- or polymodal) (curve 3). Two interpretations can be put forward. Either the subjects exhibiting the high values perform activities exposing them to higher levels of the pollutant, in which case the biological monitoring program has identified job categories for which work conditions need to be improved or these workers do not perform different activities and, in this case, their higher internal dose must result from different hygiene habits or nonoccupational exposure.

6. PRACTICAL CONSIDERATIONS

Before specific compounds are dealt with a few additional practical remarks are relevant (Lauwerys 1975).

Some industrial chemicals have a long biological half-life in various body compartments, and the time of sampling (e.g., blood, urine) may not be critical. For other chemicals, the time of sampling is, on the contrary, critical because following exposure, the compound and/or its metabolites may be rapidly eliminated from the organism. In these cases, the biological sample is usually collected during exposure, at the end of the exposure

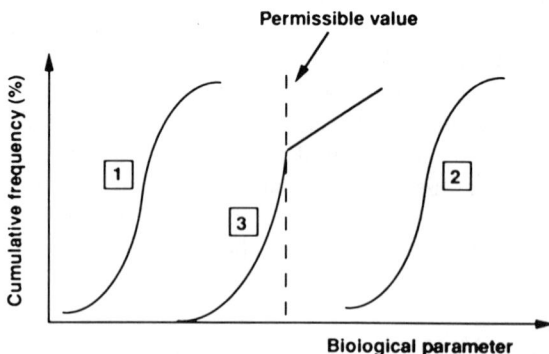

FIGURE 3   Interpretation of the results in biological monitoring of exposure (From Bernard and Lauwerys, 1987).

period, or sometimes just before the next working shift (16 h after the end of exposure) or even before resuming work after the weekend (i.e., 60 to 64 h after the last exposure).

For cumulative industrial chemicals (e.g., cadmium, lead, PCB), it is recommended to assess the baseline internal dose before exposure. This is also justified for some biological markers which exhibit a large interindividual variation in the control population and for which the relative change following exposure is more relevant than its absolute value (e.g., pseudocholinesterase in plasma).

When biological monitoring consists in sampling and analyzing urine, it is usually performed in "spot" specimens because routine collection of 24-h samples from workers is impractical (Elkins et al 1974). It is usually advisable to correct the results for the dilution of the urine. Two methods of correction have been routinely used: a) expression of the results per gram of creatinine or b) adjustment to a constant specific gravity.

Although there is no general superiority for creatinine adjustment over specific gravity, creatinine correction is better for concentrated and dilute samples (Elkins et al 1974). Furthermore, in the case of glucosuria and probably proteinuria, the specific gravity adjustment may give results that are too erroneous. Whatever the method of correction, analyses performed on too dilute urine samples (specific gravity less than 1010, creatinine concentration less than 0.3 g/l) are not reliable and should be repeated. In some circumstances, it may be feasible to report results in excretion rate (i.e., quantity/time unit) or one variant (standardization for a diuresis of 1 ml/min) (Araki et al 1986). Since creatinine excretion depends to a certain extent on urinary flow, Greenberg and Levine (1989) have also proposed to correct creatinine concentration in "spot" urine for the effects of varying hydration. One must, however, recognize that these standardization procedures, which suppose the collection of urine during a known period of time, are usually too elaborate for the routine control of workers. Furthermore, there are compounds for which the expression of the urinary results in excretion rate does not improve the accuracy of the exposure estimate. For example, the exposure to methanol, a solvent rapidly metabolized *in vivo,* is better appreciated by expressing urinary methanol in mg/l rather than in excretion rate (mg/h) (Sedivec et al 1981).

When the large interindividual variability and/or the high "background" level of the biological parameter selected makes the interpretation of a single measurement difficult, it is sometimes useful to analyze biological material collected before and after the exposure period. The change in the biological parameter due specifically to exposure can sometimes be better assessed.

When expired air is analyzed, it is preferable to collect end-exhaled samples, which represent alveolar air.

REFERENCES

E. Adamsson, M. Piscator and K. Nogawa. Pulmonary and gastrointestinal exposure to cadmium oxide dust in a battery factory. *Environ. Health Perspect.* 28:219, 1979.

S. Araki, H. Aono and K. Murata. Adjustment of urinary concentration to urinary volume in relation to erythrocyte and plasma concentrations: an evaluation of urinary heavy metals and organic substances. *Arch. Environ. Health* 41:171, 1986.

A. Berlin, A. Wolff and Y. Hasegawa. The use of biological specimens for the assessment of human exposure to environmental pollutants. *Proceedings of the International Workshop at Luxembourg,* 18-22 April 1977. Martinus Nijhoff, The Hague, 1979.

A. Bernard and R. Lauwerys. Present status and trends in biological monitoring of exposure to industrial chemicals. *J. Occup. Med.* 28:559, 1986.

A. Bernard and R. Lauwerys. General principles for biological monitoring of exposure to chemicals. In: *Biological Monitoring of Exposure to Chemicals: Organic Compounds.* Eds.: M. Ho and H. Dillon. John Wiley & Sons, New York, 1987.

B. Carnow. Discussion of TLV's for lead. Health Effects of Occupational Lead and Arsenic Exposure: A Symposium. USHEW, NIOSH Pub. No. 76-134, p. 186, 1976.

H. Elkins, L. Pagnotto and H. Smith. Concentration adjustments in urinalysis. *Am. Ind. Hyg. Assoc. J.* 35:559, 1974.

P. Flanagan, J. McLellan, J. Haist et al. Increased dietary cadmium absorption in mice and human subjects with iron deficiency. *Gastroenterology* 74:841, 1978.

G. Greenberg and R. Levine. Urinary creatinine excretion is not stable: a new method for assessing urinary toxic substance concentrations. *J. Occup. Med.* 31:832, 1989.

E. King, A. Conchie, D. Hiett and B. Milligan. Industrial lead absorption. *Ann. Occup. Hyg.* 22:213, 1979.

R. Lauwerys, J. Buchet, H. Roels et al. Intercomparison program of lead, mercury and cadmium analysis in blood, urine and aqueous solutions. *Clin. Chem.* 27:551, 1975.

R. Lauwerys. Biological criteria for selected industrial toxic chemicals: a review. *Scand. J. Work Environ. Health* 1:139, 1975.

R. Lauwerys. Current use of ambient and biological monitoring: reference workplace hazards, cadmium. International Seminar on the Assessment of Toxic Agents in the Workplace. Roles of Ambient and Biological Monitoring. OSHA, CCE, NIOSH, CCE Luxembourg, 1980.

R. Lauwerys. Basic concepts of human exposure monitoring. In: *Monitoring Human Exposure to Carcinogenic and Mutagenic Agents.* Eds.: A. Berlin et al. IARC Scientific Pub. No. 59. Lyon, p. 39, 1984a.

R. Lauwerys. Objectives of biological monitoring in occupational health practice. In: *Biological Monitoring and Surveillance of Workers Exposed to Chemicals.* Eds.: A. Aitio et al. Hemisphere, Washington, D.C., p. 3, 1984b.

R. Lauwerys and A. Bernard. La surveillance biologique de l'exposition aux toxiques industriels. Position actuelle et perspectives de développement. *Scand. J. Work Environ. Health* 11:155, 1985.

R. Lauwerys. Occupational Toxicology. In: *Casarett and Doull's Toxicology. The Basic Science of Poisons.* 4th ed. Eds.: M. Amdur, J. Doull, C. Klaassen. Pergamon, New York, 1991.

R. Lauwerys, A. Kivits, M. Lhoir et al. Biological surveillance of workers exposed to dimethylformamide and the influence of skin protection on its percutaneous absorption. *Int. Arch. Occup. Environ. Health* 45:189, 1980.

A. Monster, G. Boersma and W. Duba. Pharmacokinetics of trichloroethylene in volunteers, influence of workload and exposure concentration. *Int. Arch. Occup. Environ. Health* 38:87, 1976/7.

V. Riihimaki, P. Pfaffli and K. Savolainen. Kinetics of m-xylene in man. Influence of intermittent physical exercise and changing environmental concentrations on kinetics. *Scand. J. Work Environ. Health* 5:232, 1979.

H. Roels, J.-P. Buchet, J. Truc et al. The possible role of direct ingestion on the overall absorption of cadmium or arsenic in workers exposed to CdO or As_2O_3 dust. *Am. J. Ind. Med.* 3:53, 1982.

V. Sedivec, M. Mraz and J. Flek. Biological monitoring of persons exposed to methanol vapours. *Int. Arch. Occup. Environ. Health* 48:257, 1981.

R. Zielhuis. Biological monitoring: guest lecture given at the 26th Nordic Symposium on Industrial Hygiene, Helsinki, October 1977. *Scand. J. Work Environ. Health* 4:1, 1978.

R. Zielhuis. General aspects of biological monitoring. In: *The Use of Biological Specimens for the Assessment of Human Exposure to Environmental Pollutants.* Eds.: A. Berlin, A. Wolff, Y. Hasegawa. Martinus Nijhoff, The Hague, p. 341, 1979.

2 BIOLOGICAL MONITORING OF EXPOSURE TO INORGANIC AND ORGANOMETALLIC SUBSTANCES

1. ALUMINUM

Toxicokinetics

For the general population, food constitutes the main source of exposure to aluminum. The aluminum content of most foods is less than 1 mg/100 g (Ott 1985). The daily food intake usually ranges from less than 10 mg to 160 mg (Elinder and Sjögren 1986; Lauwerys 1990). The use of aluminum-lined vessels slightly increases the content of aluminum in food; this seems particularly true when acidic foodstuffs are stored in aluminum utensils. Orally administered aluminum compounds have been extensively used as drugs. It is generally accepted that in healthy persons, aluminum is poorly (less than 0.1%) absorbed by the oral route (Kaehny et al 1977; Ott 1985; Ihle and Becker 1985). The absorption is, however, substantially enhanced by concomitant ingestion of some acids such as citric acid (Weberg and Berstad 1986; Wills and Savory 1985). There is no evidence of dermal absorption.

In the occupational setting, some absorption may occur following very prolonged inhalation of high concentrations of aluminum dust (CEC 1991). However, there is no available information on the rate of aluminum absorption from the respiratory tract.

The major portion of aluminum in the blood compartment is bound to serum proteins (transferrin). The main sites of aluminum deposition are the bones and lungs, but small amounts can also be found in the muscles, kidneys, liver, brain, etc. (Skalsky and Carchman 1983; Charhon et al 1985; Lauwerys 1990; Eastwood et al 1990). The urine is the main route of excretion, explaining the risk of aluminum accumulation in humans with impaired renal function. The kinetics of aluminum excretion via the kidney suggests the existence of at least two compartments: one with a relatively rapid elimination rate, the other with a slower rate, probably after redistribution from a major site of deposition (Mussi et al 1984; Sjögren et al 1985, 1988). In volunteers exposed for 1 d to aluminum-containing welding fumes, the overall biological half-life of the metal in urine was about 8 h; among subjects exposed for less than 1 year, the overall half-time for urinary concentration was a few days, while welders exposed for more than 10 years had a calculated overall half-life of $\geq$6 months or even in the order of years (Sjögren et al 1983, 1985, 1988). In workers exposed to aluminum flake powders, Ljunggren et al (1991) observed urinary concentrations of aluminum 80-90 times higher than in controls, and calculated that the half-life of aluminum in urine was 5 to 6 weeks based on 4 to 5 weeks of nonexposure. Among retired workers the half-lives varied from less than 1 up to 8 years and were related to the number of years since retirement. This suggests that aluminum is retained and stored in several compartments of the body and excreted from these compartments at different rates over many years. Elinder et al (1991) have recently provided further evidence for accumulation of aluminum in the body and particularly the skeleton of healthy persons as a result of long-term occupational exposure from aluminum welding.

Biological monitoring

Because of the very low level of the element in biological fluids and its ubiquitousness, sample contamination is a major analytical problem and the normal concentrations reported in the literature vary greatly.

However, the nonoccupational exposure data are indicative of a normal mean plasma or serum concentration of $\leq$1 µg/100 ml (de Wolff and Van der Voet 1986; Guillard et al 1984; Versieck and Cornelis 1980, 1989; Suzuki et al 1989) and a urinary concentration below 30 µg/l (Schaller and Valentin 1984). According to Elinder et al (1991), the concentration found in urine from nonoccupationally exposed normal subjects is below 10 µg/l. In a population of healthy Italians, Minoia et al (1990) reported mean reference values of 0.6 µg/100 ml (range: 0.03-0.75) in serum (n: 916) and 10.9 µg/l (range: 2.3-19.5) in urine (n: 766).

The concentration of the metal in serum and urine is determined by the intensity of current exposure and the amount accumulated in the body (Greger and Baier 1983; Winney et al 1986), but the relationship between these parameters, which is greatly dependent on the integrity of renal function, has not yet been fully characterized.

Blood analysis

As expected, in subjects with severe renal insufficiency serum aluminum is mainly an indicator of body burden. Patients undergoing dialysis are at risk of bone toxicity or encephalopathy when their serum aluminum concentration persistently exceeds 19 µg/100 ml (Cundy and Kanis 1983; Winney et al 1986), or even below according to some authors (Verbeelen et al 1983; Rovelli et al 1988). Therefore, in order to protect dialysis patients the CEC (1986) has recommended that a level of 20 µg/100 ml plasma should never be exceeded, a level of >10 µg/100 ml should lead to an increased monitoring frequency and health surveillance, and >6 µg/100 ml should be considered as an excessive build-up of the aluminum body burden.

In case of long-term exposure, a correlation between aluminum concentration in serum and aluminum concentration in bones has been observed (Van der Meulen et al 1984).

Urine analysis

In subjects with normal renal function, urinary excretion of aluminum is a more sensitive indicator of aluminum exposure than its concentration in serum. Hence in case of occupational exposure, the concentration of aluminum in urine may be increased, whereas plasma or serum levels hardly differ from those found in control subjects (Mussi et al 1984; Valentin et al 1976; Alessio et al 1983; Savory and Wills 1988). Occupational exposure to fumes seems to produce higher urinary levels of aluminum than exposure to dust (Mussi et al 1984).

The urinary excretion of aluminum among welders was shown to be related to the number of years of exposure to aluminum (Sjögren et al 1988). In practice, in workers chronically exposed to aluminum, the concentration in urine collected 1 or 2 d after the end of the exposure (e.g., after a weekend of nonexposure) is probably a good indicator of the amount stored in the body. In workers newly exposed to aluminum compounds, concentration in urine is likely to be more influenced by exposure intensity than by body burden, at least when the urine sample is collected at the end of the shift. The relationship between external exposure and urinary concentration may vary depending on the bioavailability of various aluminum compounds, but so far this aspect has not been extensively investigated.

Nevertheless, the limited available data (Sjögren et al 1985) suggest that for some aluminum compounds to which workers may be exposed (e.g., aluminum-containing welding fumes, etc.), a mean urinary concentration of 200 µg/g creatinine at the end of the shift corresponds to a time-weighted average (TWA) exposure of approximately 5 mg/m^3. However, this only applies to workers exposed during less than a few weeks. The Deutsche Forschungsgemeinschaft (DFG, 1991) has adopted 200 µg/l urine (end-of-shift) as the biological tolerance value for aluminum.

In summary, the concentration of aluminum in urine collected at the end of the shift is likely to mainly reflect recent exposure. A tentative biological limit value of 150 µg/g creatinine is proposed.

REFERENCES

L. Alessio, I. Mussi, G. Calzaferri and M. Buratti. Comportamento dell'alluminio plasmatico e urinario in soggetti professionalmente esposti. VI. Convegno sulla patologia da tossici ambentali ed occupazionali, Cagliari, 26-27 maggio 1983. Atti edigraf Torino 285-292, 1983.

CEC. Resolution of the council and the representatives of the member states, meeting within the council of 16 June 1986, concerning the protection of dialysis patients by minimizing the exposure to aluminium. 86/C 184/09. Official Journal of the European Communities C 184 vol 29, 23 July 1986.

CEC. Alzheimer's disease: has aluminium any causal role? Toxicity of aluminium. Maximal admissible concentration of aluminium in drinking water. Scientific Advisory Committee to examine the toxicity and ecotoxicity of chemical compounds. CSTE/90/ 22/COM. Luxembourg, 1991.

A. Charhon, P. Chavassieux, P. Meunier and M. Accominotti. Serum aluminium concentration and aluminium deposits in bone in patients receiving haemodialysis. *Br. Med. J.* 290:1613, 1985.

T. Cundy and J. Kanis. Serum aluminium measurements in renal bone disease. *Lancet* 1168, 1983.

F. de Wolff and G. Van der Voet. Biological monitoring of aluminium in renal patients. *Clin. Chim. Acta* 160:183, 1986.

DFG. Deutsche Forschungsgemeinschaft. Maximum Concentrations at the Workplace and Biological Tolerance Values for Working Materials. Report No. XXVII. Commission for the Investigation of Health Hazards of Chemicals Compounds in the Work Area. VCH, Weinheim, 1991.

J. Eastwood, G. Levin, M. Paziannas et al. Aluminium deposition in bone after contamination of drinking water supply. *Lancet* 336:462, 1990.

C.-G. Elinder and B. Sjögren. Aluminium. In: *Handbook on Toxicology of Metals* — Vol. II. Eds.: L. Friberg, G. Nordberg, V. Vouk. 2nd edition. Elsevier, Amsterdam, 1986.

C.-G. Elinder, L. Ahrengart, V. Lidum et al. Evidence of aluminium accumulation in aluminium welders. *Br. J. Ind. Med.* 48:735, 1991.

J. Greger and M. Baier. Excretion and retention of low or moderate levels of aluminium by human subjects. *Fd. Chem. Toxicol.* 21:473, 1983.

O. Guillard, K. Tiphaneau, D. Retts and A. Piriou. Improved determination of aluminium in serum by electrothermal atomic absorption spectrometry and zeeman background correction. *Anal. Lett.* 17:1593, 1984.

B. Ihle and G. Becker. Gastrointestinal absorption of aluminium. *Am. J. Kidney Dis.* 6:302, 1985.

W. Kaehny, A. Hegg and A. Alfrey. Gastrointestinal absorption of aluminum from aluminum containing antacids. *New Engl. J. Med.* 296:1389, 1977.

R. Lauwerys. *Toxicologie Industrielle et Intoxications Professionnelles.* 3rd ed. Masson, Paris, 1990.

K. Ljunggren, V. Lidums and B. Sjögren. Blood and urine concentrations of aluminium among workers exposed to aluminium flake powders. *Br. J. Ind. Med.* 42:106, 1991.

C. Minoia, E. Sabbioni, P. Apostoli et al. Trace element reference values in tissues from inhabitants of the European Community. I. A study of 46 elements in urine, blood and serum of Italian subjects. *Sci. Total Environ.* 95:89, 1990.

I. Mussi, G. Calzaferri, M. Buratti and L. Alessio. Behaviour of plasma and urinary aluminium levels in occupationally exposed subjects. *Int. Arch. Occup. Environ. Health* 54:155, 1984.

S. Ott. Aluminium accumulation in individuals with normal renal function. *Am. J. Kidney Dis.* 6:297, 1985.

E. Rovelli, L. Luciani, C. Pagani et al. Correlation between serum aluminium concentrations and signs of encephalopathy in a large population of patients dialyzed with aluminium free fluids. *Clin. Nephrol.* 29:294, 1988.

J. Savory and M. Wills. Biological monitoring of aluminium. In: *Biological Monitoring of Toxic Metals. Rochester Series on Environmental Toxicity.* Eds.: Th. Clarkson, L. Friberg, G. Nordberg, P. Sager. Plenum Press, New York, 1988.

K. Schaller and H. Valentin. Biological indicators for the assessment of human exposure to industrial chemicals: Aluminium. Eds.: L. Alessio, A. Berlin, M. Boni, R. Roi. Industrial Health and Safety. Commission of the European Communities. Luxembourg. Eur 8903 En, 1984.

B. Sjögren, I. Lundberg and V. Lidums. Aluminium in the blood and urine of industrially exposed workers. *Br. J. Ind. Med.* 40:301, 1983.

B. Sjögren, V. Lidums, M. Hakansson and L. Hedstrom. Exposure and urinary excretion of aluminium during welding. *Scand. J. Work Environ. Health* 11:39, 1985.

B. Sjögren, G.-G. Elinder, V. Lidums and G. Chang. Uptake and urinary excretion of aluminium among welders. *Int. Arch. Occup. Environ. Health* 60:77, 1988.

H. Skalsky and R. Carchman. Aluminium homeostasis in man. *J. Am. Coll. Toxicol.* 2:405, 1983.

Y. Suzuki, S. Imai and T. Kamiki. Fluorimetric determination of aluminium in serum. *Analyst* 114:839, 1989.

H. Valentin, P. Preusser and K. Schaller. Die Analyse von Aluminium im Serum und Urin zur Überwachung exponierter Personen. *Int. Arch. Occup. Environ. Health* 38:1, 1976.

J. Van der Meulen, P. Bezemer, P. Lips and P. Oe. Individual differences in gastrointestinal absorption of aluminium. *New Engl. J. Med.* 311:1322, 1984.

D. Verbeelen, J. Smeyers-Verbeke, J. Sennesael and D. Massarf. Serum aluminium measurements in renal bone disease. *Lancet* 1168-1169 i, 1983.

J. Versieck and R. Cornelis. Normal levels of trace elements in human blood plasma or serum. *Anal. Chim. Acta* 116:217, 1980.

J. Versieck and R. Cornelis. *Trace Elements in Human Plasma or Serum.* CRC Press, Boca Raton, FL, 1989.

R. Weberg and A. Berstad. Gastrointestinal absorption of aluminium from single doses of aluminium containing antacids in man. *Eur. J. Clin. Invest.* 16:428, 1986.

M. Wills and J. Savory. Water content of aluminium, dialysis dementia, and osteomalacia. *Environ. Health Perspect.* 63:141, 1985.

R. Winney, J. Cowie and J. Robson. Role of plasma aluminium in the detection and prevention of aluminium toxicity. *Kidney International* 29 suppl. 18:S91, 1986.

2. ANTIMONY

Toxicokinetics

Inorganic antimony enters the organism by the oral and pulmonary routes and seems to be rapidly excreted mainly in urine and feces. Unlike inorganic arsenic, inorganic trivalent antimony is not methylated *in vivo*. It is excreted in the bile after conjugation with glutathione and also in urine. A significant proportion of the amount excreted in the bile undergoes an enterohepatic circulation (Buchet and Lauwerys 1985; Bailly et al 1991). Pentavalent compounds are mainly excreted in urine, while trivalent compounds in feces (Edel et al 1983; Gelhorn et al 1986). The initial excretion of antimony is rapid, but increased levels found in lungs of deceased smelter workers 20 years after retirement suggest a long retention of some antimony compounds. Elevated concentration of antimony have also been observed in kidneys, liver, and thyroid (Gerhardson et al 1982; Elinder and Friberg 1986; Norseth and Martinsen 1988).

The metabolism of various organic antimony derivatives used as antiparasitic drugs has been investigated, but the usefulness of this information for occupational health monitoring is limited.

Biological monitoring

In control subjects, the average concentration of antimony has been reported to be 0.3 (Mansour et al 1967) and 0.06 μg/100 ml (Lüdersdorf et al 1987) in blood and 0.08 μg/100 ml in serum (Wester 1973). According to Versiek and Cornelis (1989) the normal mean values for antimony in human serum or plasma circle around 0.05 to 0.1 μg/100 ml. Minoïa et al (1990) established a mean reference value of 0.216 μg/100 ml (range: 0.003-0.35) in blood (n: 27) and 0.05 μg/100 ml (range: 0.001-0.17) in serum (n: 22) and in another study (Blekastad et al 1984) a mean value ≤0.03 μg/100 ml was measured in 281 sera. Wester (1973) observed that the 24-h urinary excretion of antimony (n: 16) ranged from 0.5 to 2.6 μg. Mansour et al (1967) found a mean urinary concentration of 6.2 μg/l (n: 3), while, according to Lüdersdorf et al (1987), the urinary concentrations in 51 control subjects varied between 0.2 and 0.7 μg/l. These latter results are in agreement with the experience of Bailly et al (1991) (<1 μg/g creatinine) and the reference value of Minoïa et al (1990) (mean: 0.79 μg/l; range: 0.19-1.1; n: 360).

Urine analysis

Several authors have reported that workers exposed to antimony excrete significantly more antimony in urine than control workers (Brieger et al 1954; Cooper et al 1968; Klucik and Kemka 1960; Smith and Griffiths 1982; Lüdersdorf et al 1987; Bailly et al 1991).

A biological threshold limit value for antimony in urine is difficult to propose, but a preliminary study on workers exposed to pentavalent antimony suggests that determination of its concentration in urine may be used to assess the intensity of recent exposure and that, on average, an airborne concentration of antimony of about 0.5 mg/m^3 leads to an increase in urinary antimony concentration of 35 μg antimony/g creatinine during the shift (Bailly et al 1991). This should be confirmed by further studies.

REFERENCES

R. Bailly, R. Lauwerys, J.-P. Buchet et al. Experimental and human studies on antimony metabolism: their relevance for the biological monitoring of workers exposed to inorganic antimony. *Br. J. Ind. Med.* 48:93, 1991.

V. Blekastad, J. Jonsen, E. Steinnes and K. Helgeland. Concentrations of trace elements in human blood serum from different places in Norway determined by neutron activation analysis. *Acta Med. Scand.* 216:25, 1984.

H. Brieger, C. Semisch, J. Stasney and M. Piatnek. Industrial antimony poisoning. *Ind. Med. Surg.* 23:521, 1954.

J.-P. Buchet and R. Lauwerys. Study of inorganic arsenic methylation by rat liver in vitro: relevance for the interpretation of observations in man. *Arch. Toxicol.* 57:125, 1985.

D. Cooper, E. Pendergrass, A. Vorwald et al. Pneumoconiosis among workers in an antimony industry. *Am. J. Roentgenol. Radium Ther. Nucl. Med.* 103:495, 1968.

J. Edel, E. Marafante, E. Sabbioni and L. Manzo. CEP Consultants Ltd, Edinburgh U.K. International Conference on Heavy Metals in the Environment. Heidelberg, Sept. 1983.

C.-G. Elinder and L. Friberg. Antimony. In: *Handbook on the Toxicology of Metals. Vol II.* Eds.: L. Friberg, G. Nordberg, V. Vouk. 2nd edition. Elsevier, Amsterdam, 1986.

A. Gelhorn, N. Tupikova and H. Van Dyke. Tissue distribution and excretion of 4 organic antimonials after single or repeated administrations to normal hamsters. *J. Pharmacol.* 87:169, 1986.

L. Gerhardson, D. Brune, C. Nordberg and P. Webster. Antimony in lungs, liver and kidney tissue from deceased smelter worker. *Scand. J. Environ. Health* 8:201, 1982.

I. Klucik and R. Kemka. Vylucovanie antimonu u robotnikov antimonovej hute. *Pracovni Lekarstvi* 12:133, 1960.

E. Lüdersdorf, A. Fuchs, P. Mayer and G. Skulsuksai. Biological assessment of exposure to antimony and lead in the glass-producing industry. *Int. Arch. Occup. Environ. Health* 59: 469, 1987.

M. Mansour, A. Rassoul and A. Schulert. Anti-bilharzial antimony drugs. *Nature* 214:819, 1967.

C. Minoïa, E. Sabbioni, P. Apostoli et al. Trace element reference values in tissues from inhabitants of the European Community. I. A study of 46 elements in urine, blood and serum of Italian subjects. *Sci. Total Environ.* 95:89, 1990.

T. Norseth and I. Martinsen. Biological monitoring of antimony. In: *Biological Monitoring of Toxic Metals.* Eds.: T. Clarkson, L. Friberg, G. Nordberg, P. Sager. Plenum Press, New York, 1988.

B. Smith and M. Griffiths. Determination of lead and antimony in urine by atomic absorption spectroscopy with electrothermal atomization. *Analyst* 107:253, 1982.

J. Versieck and R. Cornelis. *Trace Elements in Human Plasma or Serum.* CRC Press, Boca Raton, FL, 1989.

P. Wester. Trace elements in serum and urine from hypertensive patients before and during treatment with chlorthalidone. *Acta Med. Scand.* 194:505, 1973.

3. ARSENIC

Toxicokinetics

In industry, workers are usually exposed to inorganic arsenic compounds. Some organoarsenicals are used as pesticides (mainly monomethylarsonic acid and cacodylic acid) and as feed additives for poultry and swine. Arsenic and its compounds may enter the human organism through ingestion or inhalation, the rate of absorption being highly dependent on the solubility of the compound and probably also on the valence state of arsenic (Georis et al 1990). Little is known about the cutaneous absorption in man, but skin might also be a possible route of

absorption of arsenic (Robinson 1975; Garb and Hine 1977). Once absorbed,
arsenic is fairly rapidly cleared from the blood and widely distributed to the
tissues. Whatever the arsenic compound to which the workers are exposed, the
absorbed arsenic is rapidly eliminated through the kidney either unchanged or
after biotransformation. For example, it has been demonstrated that in man
organoarsenicals present in marine organisms and cacodylic acid are excreted
unchanged, monomethylarsonic acid is slightly (around 10%) methylated into
cacodylic acid before excretion, while following exposure to inorganic tri- or
pentavalent arsenic the urinary excretion consists of 10 to 20% inorganic arsenic,
10 to 20% monomethylarsonic acid and 60 to 80% cacodylic acid (Buchet et al
1980, 1981a; Crecelius 1977). Following single oral exposure to a low dose of
arsenate, 25 and 45% of the administered dose is excreted in urine within 1 and
4 d, respectively (Buchet et al 1980). Observations made on a subject acutely
intoxicated by inorganic arsenic, however, indicate that a significant fraction of
the absorbed inorganic arsenic can also be excreted by the biliary route (Mahieu
et al 1987). The metabolic handling of inorganic trivalent arsenic is summarized
in Figure 4. One point has now been clearly demonstrated: when exposure has
been to inorganic arsenic, the only arsenic species excreted in urine are monomethylarsonic acid (MMA), cacodylic acid (DMA), and unchanged inorganic
arsenic. Monomethylarsonic acid and cacodylic acid are much less toxic than
inorganic arsenic and therefore the *in vivo* methylation process represents a true
detoxication mechanism. While the urinary excretion occurs in the form of
inorganic species during the first hours after the start of exposure, the methylation
process is rapidly triggered and this leads after about 8 h to a preponderant
excretion of the methylated species, mainly cacodylic acid (Buchet et al 1981a,b).
This is in agreement with a recent study on workers exposed to As_2O_3 which
showed that only for inorganic arsenic was the mean concentration in urine

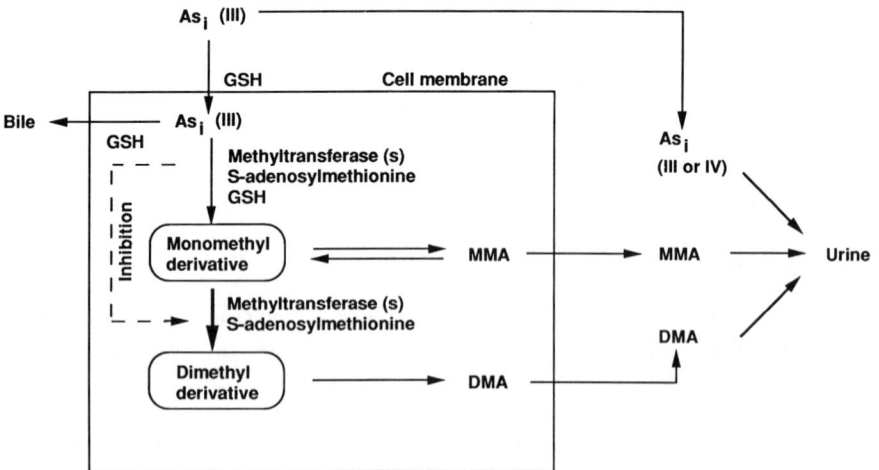

FIGURE 4 Main metabolic pathways of inorganic arsenic.

collected after the shift significantly higher than that found the next day morning: the proportion of inorganic arsenic in urine falling from 25% of the total after the shift to 14% before the next shift (Offergelt et al 1992). A similar observation has been reported by Yamauchi et al (1989).

A high uptake of trivalent inorganic arsenic can transiently inhibit the second methylation reaction, but so far this has only been observed after acute intoxication (Mahieu et al 1981; Lovell and Farmer 1985; Buchet and Lauwerys 1985; Farmer and Johnson 1990).

Biological monitoring

In healthy, nonoccupationally exposed subjects, mean serum and urine levels of total arsenic vary greatly depending on the level of the seafood content in the diet as well as on the arsenic content in drinking water (Valentine et al 1979; Versieck 1985; Foa et al 1987; Versieck and Cornelis 1989; Minoïa et al 1990; Yamato 1988).

For instance, Vahter (1988) has estimated the normal concentrations of arsenic in blood to be in the range 0.1-0.4 µg/100 ml, while, in subjects consuming arsenic-contaminated drinking water, blood arsenic (As) levels may reach values as high as 5-6 µg/100 ml or even higher. A single meal of certain fish species or crustaceans may give rise to urinary arsenic concentrations of more than 1000 µg/l (Norin and Vahter 1981). Some marine organisms may effectively contain very high concentrations of organoarsenicals of negligible toxicity which are also rapidly excreted (half-life about 18 h) by the renal route, apparently without transformation (Buchet et al 1980).

Hence, when the workers have not been instructed to refrain from eating fish or shellfish for 2 to 3 d before urine collection, misleading results can be obtained by measuring total arsenic concentration. This is illustrated in Table 1, which shows that, in persons nonoccupationally exposed to arsenic, but who have ingested a fish meal the day before urine collection, very high amounts of total arsenic can be found in urine, whereas the sum of inorganic arsenic and its methylated derivatives does not differ markedly from the value found in subjects who have not eaten fish.

Urine analysis

While in the past the biological monitoring of workers chronically exposed to inorganic arsenic in industry has frequently been carried out by measuring the total amount of arsenic present in urine collected at the end of the shift or at the beginning of the next shift (Pinto et al 1976), it is now well established that the determination of inorganic arsenic (iAs), monomethylarsonic acid, and cacodylic acid in urine is the method of choice for the biological monitoring of workers exposed to inorganic arsenic since these determinations are not influenced by the presence of organoarsenicals from marine origin.

Arsenic concentration in urine mainly reflects the recent exposure. In persons nonoccupationally exposed to inorganic arsenic, the sum of these three metabolites

Table 1
Influence of Fish Consumption on the Arsenic Content of Urine (µg/l)

Fish Meal	Inorganic Arsenic	Monomethyl-Arsonic Acid	Cacodylic Acid	Total Arsenic
None*	1.2	0.3	5.3	7.7
Fresh cod				
subject 1	3.0	0	0	185
subject 2	1.5	2.1	17.4	170
Cured herring	0.6	0	13.2	113
Trout	1.0	1.3	14.0	131
Plaice				
subject 1	0.4	0.2	11.3	18,800
subject 2	0.2	0	11.4	14,070
Sole	0.2	0	3.5	150

*Mean of 5 subjects.

From Buchet et al 1980.

(iAs, MMA, DMA) in urine is usually less than 10 µg/l in Europe, somewhat higher in the U.S. (Smith et al 1977) and around 50 µg/l or even higher (Yamato 1988) in Japan (Vahter et al 1986).

Controversial results have been reported with regard to the relationship between exposure to airborne inorganic arsenic compounds and the level of arsenic in urine. Pinto et al (1976), who studied copper smelter workers exposed to As_2O_3, obtained a satisfactory linear correlation between airborne and urinary arsenic concentrations averaged over the week and have reported that, on a group basis, exposure to an airborne arsenic level of 50 $\mu g/m^3$ corresponds approximately to urinary arsenic concentration of about 170 µg total arsenic/l urine. Landrigan et al (1982) also observed a good correlation between the mean total arsenic exposure (workers exposed to arsine gas and arsenic trioxide) and the average total arsenic excretion in urine (24-h collection) in a group of 48 lead acid battery workers; an exposure to 50 µg As/m^3 air leading to a mean arsenic excretion of about 133 µg/l urine. Vahter et al (1986) have estimated that this TWA exposure to soluble inorganic arsenic (e.g., in the form of As_2O_3) would lead to an average urinary excretion of 130 µg As/g creatinine or 195 µg As/l (sum of iAs, MMA, DMA) (urine collected 16 h after the end of exposure, linear relationship). These estimates are higher than those of Enterline et al (1987) and still much higher than those of Smith et al (1977) and Offergelt et al (1992). Smith et al (1977) measured the concentrations of iAs and its methylated metabolites (MMA + DMA) in morning urine specimens collected from 82 copper smelter workers. According to this last study, 75 µg/l would be the value corresponding to a TWA exposure to 50 µg As/m^3. Following the investigations of Offergelt et al (1992), which were undertaken to clarify the relationship between exposure to

As$_2$O$_3$ fume and dust and the urinary excretion of iAs, MMA and DMA, TWA exposures to 50 and 200 µg/m^3 lead to mean urinary concentrations of the sum of As metabolites (iAs + MMA + DMA) in the postshift urine sample of 54 and 88 µg As/g creatinine (Offergelt et al 1992).

The discrepancies between these different studies may partly be due to the different contribution of oral exposure to inorganic arsenic in various industries and underline the fact that the airborne concentration of arsenic is not a satisfactory indicator of the total amount of arsenic absorbed on an individual basis. It is also likely that in some of the above studies (Pinto et al 1976; Vahter et al 1986) the method used for air sampling was not appropriate for collecting As$_2$O$_3$ in the vapour phase (Costello et al 1983; Demange et al 1992).

When exposure is to poorly soluble inorganic arsenic compounds (e.g., gallium arsenide), the determination of arsenic in urine will reflect the amount absorbed, but not necessarily the total dose delivered to the lung or the gastrointestinal tract. Yamauchi et al (1989) measured only a very small increase in arsenic concentration in workers employed in a gallium arsenide plant compared to the smelter workers.

When exposure is to an organic arsenical other than monomethylarsonic acid and cacodylic acid, measurement of total arsenic acid in urine may be used to monitor exposure provided that the dietary habits have been recorded. As stated above, these organoarsenicals are usually excreted unchanged and represent a less important occupational health hazard than inorganic arsenic.

Hair and nail analyses

Because of their high content of keratin, whose sulfhydryl groups bind trivalent iAs, arsenic content of hairs or fingernails seem to be good indicators of the amount of inorganic arsenic absorbed during the growth period of the hair or the nail. Organic arsenic from marine origin does not appear to be taken up in hair (nail) to the same degree as inorganic arsenic. This implies that arsenic in hair (nail) reflects exposure to inorganic arsenic only. Moreover, determination of arsenic concentration along the length of the hair may provide information on the time of exposure and the length of the exposure period. In the industrial environment, it is difficult to distinguish between endogenous arsenic and arsenic externally deposited on the hair (Yamamura and Yamauchi 1980). As confirmed by a recent study (Yamauchi et al 1989), there is presently no reliable method available to remove exogenous arsenic. Agahian et al (1990) have claimed to have found an improved new washing technique able to remove 98% of exogenous arsenic on nail. Unless other investigations confirm this claim, it seems that the determination of arsenic in hair as well as in nail is more useful for evaluating the environmental exposure of the general population to inorganic arsenic than for estimating the exposure of workers. The arsenic levels in hair of nonoccupationally exposed adults are usually below 2 mg/kg (Valentine et al 1979) or even less than 1 mg/kg (Vahter et al 1986).

Blood analysis

As in urine, the arsenic concentration in blood reflects mainly recent exposure. The biological half-life of arsenic in blood is about 60 h (Mahieu et al 1981). There is no sufficient information in the literature to establish the relationship between the intensity of arsenic exposure and its concentration in blood nor to judge the value of measuring the different arsenic metabolites in blood.

In conclusion, the method of choice for monitoring industrial exposure to inorganic arsenic is the determination of iAs, MMA, and DMA in urine. As pointed out by Vahter et al (1986), the sampling should not be carried out on the first working day following a weekend, a holiday, or a period of considerably different exposure, since it takes a day or two before a steady state is achieved. Since inorganic arsenic is recognized as a human carcinogen, it is difficult to propose a biological exposure indice for long-term exposure. NIOSH (1978) has proposed to reduce the maximum allowable concentration of inorganic arsenic in air to 2 µg/m³. At this exposure level, no significant increase of urinary arsenic concentration above the background level can be expected (Lauwerys 1982).

According to the DFG (1991), exposures to 50, 100, and 200 µg/m³ lead to mean urinary concentrations of arsenic of 175, 330, and 640 µg/l, but the recent study of Offergelt et al (1992) tends to suggest that the latter are overestimated. According to these authors, 54, 69, and 88 µg/g creatinine (sum of iAs, MMA, DMA; end-of-shift) are the mean urinary concentrations corresponding to TWA of 50, 100, and 200 µg/m³, respectively. As data on early effects suggest that preclinical effects can occur at urinary concentrations of arsenic as low as 71 µg/l, ACGIH has proposed a biological exposure indice for arsenic and soluble compounds including arsine of 50 µg/g creatinine (iAs and methylated metabolites, end-of-workweek) or 75.5 µmol/mol creatinine (ACGIH 1991-1992). This biological exposure index does not apply for gallium arsenide. The official TWA in the U.S. (OSHA) is 10 µg/m³. At this exposure level and according to the study of Offergelt et al (1992) the urinary concentration should not exceed 30 µg/g creatinine (sum of iAs, MMA, DMA; end-of-shift).

REFERENCES

ACGIH. American Conference of Governmental Industrial Hygienists. *Threshold Limit Values for Chemical Substances and Physical Agents and Biological Exposure Indices.* Cincinnati, 1991-1992.

A. Abdelehani, A. Anderson, M. Jaghabir and F. Mather. Arsenic levels in blood, urine and hair of workers applying monosodium methanearsonate. *Arch. Environ. Health* 41:163, 1986.

B. Agahian, J. Lee, J. Nelson and R. Johns. Arsenic levels in fingernails as a biological indicator of exposure to arsenic. *Am. Ind. Hyg. Assoc. J.* 51:646, 1990.

J.-P. Buchet, R. Lauwerys and H. Roels. Comparison of several methods for the determination of arsenic compounds in water and in urine. *Int. Arch. Occup. Environ. Health* 46:11, 1980.

J.-P. Buchet and R. Lauwerys. Study of inorganic arsenic methylation by rat liver in vitro: Relevance for the interpretation of observations in man. *Arch. Toxicol.* 57:125, 1985.

J.-P. Buchet, R. Lauwerys and H. Roels. Comparison of urinary excretion of arsenic metabolites after a single oral dose of sodium arsenite, monomethylarsonate, or dimethylarsinate in man. *Int. Arch. Occup. Environ. Health* 48:71, 1981a.

J.-P. Buchet, R. Lauwerys and H. Roels. Urinary excretion of inorganic arsenic and its metabolites after repeated ingestion of sodium metaarsenite by volunteers. *Int. Arch. Occup. Environ. Health* 48:111, 1981b.

R. Costello, P. Eller and R. Hull. Measurement of multiple inorganic arsenic species. *Am. Ind. Hyg. Assoc. J.* 44:21, 1983.

E. Crecelius. Changes in the chemical speciation of arsenic following ingestion by man. *Environ. Health Perspect.* 19:147, 1977.

M. Demange, I. Vien, G. Hecht and M. Hery. Development of a method for sampling arsenic trioxide. *Cah. Notes Doc.* 146:63, 1992 (in French).

DFG. Deutsche Forschungsgemeinschaft. Maximum Concentrations at the Workplace and Biological Tolerance Values for Working Materials. Report No. XXVII. Commission for the Investigation of Health Hazards of Chemical Compounds in the Work Area. VCH, Weiheim, 1991.

P. Enterline, V. Henderson and G. Marsch. Exposure to arsenic and respiratory cancer — a reanalysis. *Am. J. Epidemiol.* 125:9, 1987.

J. Farmer and L. Johnson. Assessment of occupational exposure to inorganic As based on urinary concentration and speciation of arsenic. *Br. J. Ind. Med.* 47:342, 1990.

V. Foa, A. Colombi, M. Maroni and M. Buratti. Biological indicators for the assessment of human exposure to industrial chemicals: Arsenic. Eds.: L. Alessio, A. Berlin, M. Boni, R. Roi. Industrial health and safety. Commission of the European Communities. Luxembourg. Eur 1135 En, 1987.

L. Garb and C. Hine. Arsenical neuropathy residual effects following acute industrial exposure. *J. Occup. Med.* 19:567, 1977.

B. Georis, A. Cardenas, J.-P. Buchet, R. Lauwerys. Inorganic arsenic methylation by rat tissues slices. *Toxicology* 63:73, 1990.

P. Landrigan, R. Costello and W. Stringer. Occupational exposure to arsine. An epidemiologic reappraisal of current standards. *Scand. J. Work Environ. Health* 8:169, 1982.

R. Lauwerys. Biological monitoring of exposure to arsenic. In: *International Course on Biological Monitoring of Exposure to Industrial Chemicals.* Hemisphere, Washington, D.C., 1982.

M. Lovell and J. Farmer. Arsenic speciation in urine from humans intoxicated by inorganic arsenic compounds. *Hum. Toxicol.* 4:203, 1985.

P. Mahieu, J.-P. Buchet, H. Roels and R. Lauwerys. The metabolism of arsenic in humans acutely intoxicated by As_2O_3. Its significance for the duration of BAL therapy. *Clin. Toxicol.* 18:1067, 1981.

P. Mahieu, J.-P. Buchet and R. Lauwerys. Evolution clinique et biologique d'une intoxication orale aigue par l'anhydride arsenieux et considérations sur l'attitude thérapeutique. *J. Toxicol. Clin. Exp.* 7:273, 1987.

C. Minoïa, E. Sabbioni, P. Apostoli et al. Trace element reference values in tissues from inhabitants of the European Community. I. A study of 46 elements in urine, blood and serum of Italian subjects. *Sci. Total Environ.* 95:89, 1990.

NIOSH. Summary of NIOSH Recommendations for Occupational Health Standards, National Institute of Occupational Safety and Health, Cincinnati, 1978.

H. Norin and M. Vahter. A rapid method for the selective analysis of total urinary metabolites of inorganic arsenic. *Scand. J. Work Environ. Health* 7:38, 1981.

J. Offergelt, H. Roels, J.-P. Buchet et al. Relationship between airborne arsenic trioxide and urinary excretion of inorganic arsenic and its methylated metabolites. *Br. J. Ind. Med.* 49:387, 1992.

OSHA. Occupational Safety and Health Administration. Occupational Safety and Health Standards. Subpart 2. Toxic and Hazardous Substances. Code of Federal Regulations 29 (Part 1910.1000), p. 673, 1981.

S. Pinto, M. Varner, M. Nelson et al. Arsenic trioxide absorption and excretion in industry. *J. Occup. Med.* 18:677, 1976.

T. Robinson. Arsenic polyneuropathy due to a caustic arsenical paste. *Br. Med. J.* 3:139, 1975.

T. Smith, I. Crecelius and J. Reading. Airborne arsenic exposure and excretion of methylated arsenic compounds. *Environ. Health Perspect.* 19:89, 1977.

M. Vahter. Arsenic. In: *Biological Monitoring of Toxic Metals. Rochester Series on Environmental Toxicity.* Eds.: Th. Clarkson, L. Friberg, G. Nordberg, P. Sager. Plenum Press, New York, 1988.

M. Vahter, L. Friberg, B. Rahnster et al. Airborne arsenic and urinary excretion of metabolites of inorganic arsenic among smelter workers. *Int. Arch. Occup. Environ. Health* 57:79, 1986.

J. Valentine, H. Kang and G. Spivey. Arsenic levels in human blood, urine and hair in response to exposure via drinking water. *Environ. Res.* 20:24, 1979.

J. Versieck. Trace elements in body fluids and tissues. *CRC Crit. Rev. Clin. Lab. Sci.* 22:97, 1985.

J. Versieck and R. Cornelis. *Trace Elements in Human Plasma or Serum.* CRC Press, Boca Raton, FL, 1989.

Y. Yamamura and H. Yamauchi. Arsenic metabolites in hair, blood and urine in workers exposed to arsenic trioxide. *Ind. Health* 18:203, 1980.

N. Yamato. Concentrations and chemical species of arsenic in human urine and air. *Bull. Environ. Contam. Toxicol.* 40:633, 1988.

H. Yamauchi, K. Takahashi, M. Mashiko and Y. Yamamura. Biological monitoring of arsenic exposure of gallium arsenide and inorganic arsenic exposed workers by determination of inorganic arsenic and its metabolites in urine and air. *Am. Ind. Hyg. Assoc. J.* 50:606, 1989.

4. BARIUM

Toxicokinetics

Soluble compounds of barium are readily absorbed from the respiratory and gastrointestinal tracts. Excretion seems to occur through feces and urine. Insoluble barium compounds are practically unabsorbed (Reeves 1986).

Biological monitoring

The mean concentrations of barium in urine and whole blood measured by Minoïa et al (1990) in unexposed subjects were 2.7 µg/l (n: 35; range: 0.25-5.7) and 0.12

µg/100 ml (n: 25; range: 0.047-0.24). The levels observed by Dare et al (1984) in three control subjects were <5 µg/l in urine.

According to Zschiesche et al (1992), in nonoccupationally exposed subjects, the concentration of barium in urine and plasma are usually less than 20 µg/l and 0.8 µg/100 ml.

Blood and urine analyses

Exposure of welders to barium-containing fumes resulted in elevated barium urinary levels. After 3 h welding, barium in urine ranged between 31 and 234 µg/l (n: 5); the next morning, the concentrations were still elevated (29-110 µg/l) (Dare et al 1984). Since the airborne concentration of barium was not measured, the relationship between exposure and urinary level cannot be assessed.

In 18 arc welders, Zschiesche et al (1992) found a correlation between airborne concentration to barium (more than 90% soluble in acids) and postshift barium concentration in plasma (r: 0.63) and urine (r: 0.47), respectively.

REFERENCES

P. Dare, P. Hewitt, R. Hicks et al. Barium in welding fumes. *Ann. Occup. Hyg.* 28:445, 1984.
C. Minoïa, E. Sabbioni, P. Apostoli et al. Trace element reference values in tissues from inhabitants of the European Community. I. A study of 46 elements in urine, blood and serum of Italian subjects. *Sci. Total Environ.* 95:89, 1990.
A. Reeves. Barium. In: *Handbook on the Toxicology of Metals. Vol II*. 2nd ed. Eds.: L. Friberg, G. Nordberg, V. Vouk. Elsevier Science Publishers, Amsterdam, 1986.
W. Zschiesche, K.-H. Schaller and D. Weltle. Exposure to soluble barium compounds: an interventional study in arc welders. *Int. Arch. Occup. Environ. Health* 64:13, 1992.

5. BERYLLIUM

Toxicokinetics

Acute and chronic exposure to beryllium and its compounds by inhalation may lead to severe lung disturbances. There are no data on the deposition or absorption of inhaled beryllium in human beings, but it can be expected that, as with other inhaled particles, dose, size, and solubility of beryllium compounds are the important factors governing deposition and lung clearance (WHO 1990).

Experimental studies indicate a pulmonary half-life of 20 d for soluble compounds ($BeCl_2$), at least a third being transferred to the systemic circulation. Insoluble compounds (e.g., BeO) have a longer retention time in the lung; their pulmonary half-time has been estimated at about 1 year (Kjellström and Kennedy 1984).

The uptake of beryllium through the gastrointestinal tract and skin absorption contribute probably only very little to the total body burden of beryllium exposed persons. However, because of the skin effects elicited by beryllium compounds, this route might in some circumstances be of some significance (WHO 1990). In general, inhalation exposure to beryllium compounds results in long-term storage of appreciable amounts of beryllium in lung tissue (particularly in pulmonary lymph nodes) and in the skeleton, which is the ultimate site of storage. More soluble compounds are also translocated to liver, abdominal lymph nodes, spleen, heart, muscle, skin, and kidney (WHO 1990). Elimination of absorbed beryllium mainly occurs via the urine and to a minor degree in the feces.

Biological monitoring

Beryllium can be determined in blood and urine, but presently these analyses can only be used as qualitative tests to confirm exposure to the metal. It is not known to what extent the concentrations of beryllium in blood and urine may be influenced by recent exposure and by the amount already stored in the body. Furthermore, no relationship between internal dose and toxic effects has been established (Reeves 1986). It is difficult to interpret the limited published data on the excretion of beryllium in exposed workers, because usually the external exposure has not been adequately characterized and the analytical methods have different sensitivity and precision (De Nardi et al 1953; Dutra et al 1949; Klemperer et al 1951); for a more recent review concerning the analytical methods and problems of analysis see Apostolli et al (1989).

In ten persons without any occupational contact with beryllium, Stiefel et al (1980) reported a mean urinary concentration of 0.9 µg/l (SD: 0.5). Higher values (2 µg/l) were found in smokers inhaling nonfiltered smoke. In blood, the same authors found a mean value of 0.1 µg/100 g (SD: 0.04; n: 10). In control subjects Minoïa et al (1990) recently reported mean beryllium urinary and serum concentrations of 0.4 (0.04-0.76) µg/l (n: 579) and 0.015 (0.003-0.027) µg/100 ml, respectively. According to the Dutch Committee for Occupational Standards (1988), the average level in blood of nonexposed subjects is up to about 0.1 µg/100 ml, the 24-h loss by urine is about 1 µg. No statistical difference has been detected between alcohol drinkers and nondrinkers (Apostoli et al 1989).

Urine analysis

Observations made by Stiefel et al (1980) on eight newly exposed workers followed for 30 d suggest that beryllium in urine may be a reflection of the current exposure. A peak airborne concentration of approximately 8 µg/m^3 (duration of sampling not indicated) would lead to a urinary excretion of 4 µg/l (time of urine sampling unknown). Cammarano et al (1985) have reported higher beryllium levels in urine at the end of the shift (1.8 µg/l) than at the beginning of the shift (1.1 µg/l). It seems, however, that an increased urinary excretion of beryllium can still be detected for several years after the end of exposure (De Nardi et al 1953; Klemperer et al 1951) and, in workers removed from exposure, the variation in the amount of beryllium excreted daily is rather low (Klemperer et al 1951). This would suggest that, in the absence of recent exposure, beryllium in a single 24-h urine collection gives some evidence of the rate of which the metal is mobilized from the tissues (Klemperer et al 1951). However, this interpretation needs confirmation. Furthermore, a concentration of beryllium in urine below 0.05 μg/l is not an argument to exclude past exposure to beryllium (Klemperer et al 1951). Indeed, it has been reported that beryllium is not always detectable in workers exposed to beryllium in the past and who had developed pulmonary granulomatosis.

At present it is not possible to establish a clear relationship between exposure and body burden.

REFERENCES

P. Apostoli, S. Porru, C. Minoïa and L. Alessio. Biological indicators for the assessment of human exposure to industrial chemicals: Beryllium. Eds.: L. Alessio, A. Berlin, M. Boni, R. Roi. Commission of the European Communities. Industrial Health and Safety. Luxembourg. Eur 12174 EN, 1989.

G. Cammarano, G. Catenacci and C. Minoïa. Esposizione a metalli in addetti alla pulitura di caldaie ad olio combustibile in una centrale thermoelettrica. In: *48° Congresso Nazionale della Società Italiana di Medicina del Lavoro e Igiene Industriale,* pp. 179-183, Monduzzi Ed., 1985.

J. De Nardi, H. Van Ordstrand, G. Curtis and J. Zielinski. Berylliosis, summary and survey of all clinical types observed in a 12 year period. *Arch. Ind. Hyg. Occup. Med.* 8:1, 1953.

Dutch Expert Committee for Occupational Standard. Health-based recommended occupational exposure limits for beryllium and beryllium compounds. Directorate-General of Labour. RA 4/88, 1988.

F. Dutra, J. Cholak and D. Hubbard. The value of beryllium determinations in the diagnosis of berylliosis. *Am. J. Clin. Pathol.* 19:229, 1949.

T. Kjellstrom and P. Kennedy. Criteria document for Swedish Occupational Standards: Beryllium. Arbete Och Hälsa 2, 1984.

F. Klemperer, A. Martin and J. Van Riper. Beryllium excretion in humans. *Arch. Ind. Hyg. Occup. Med.* 4:251, 1951.

C. Minoïa, E. Sabbioni, P. Apostoli et al. Trace element reference values in tissues from inhabitants of the European Community. I. A study of 46 elements in urine, blood and serum of Italian subjects. *Sci. Tot. Environ.* 95:89, 1990.

A. Reeves. Beryllium. In: *Handbook on the Toxicology of Metals. Vol. II.* 2nd ed. Eds.: L. Friberg, G. Nordberg, V. Vouk. Elsevier, Amsterdam, 1986.

T. Stiefel, K. Schulze, H. Zorn and G. Tolg. Toxicokinetic and toxicodynamic studies of beryllium. *Arch. Toxicol.* 45:81, 1980.

WHO. IPCS. International Programme on Chemical Safety. Environmental Health Criteria 106, Beryllium. World Health Organization, Geneva, 1990.

6. CADMIUM

Toxicokinetics

In the occupational setting, inhalation is the major route of absorption of cadmium. The degree of absorption depends upon particle size and solubility of the cadmium compounds. An absorption rate of 25 to 50% has been estimated for cadmium fumes. The gastrointestinal tract can contribute significantly to exposure via contamination of hands and food. Moreover, a fraction of the dust particles deposited in the respiratory tract is cleared to the gastrointestinal tract. According to animal data about 2% of ingested cadmium is absorbed. This rate can be increased in case of low intake of iron, calcium, or protein. In humans, the oral absorption rate is in the range of 2 to 7% with values of up to 20% for people with very low iron stores. The absorption of cadmium through skin is negligible (Friberg et al 1985, 1986; Nordberg and Nordberg 1988).

Cadmium is a very cumulative toxin with a biological half-life of over 10 years in man (a half-life of more than 30 years has been shown in the muscles). In blood, over 70% of the cadmium is bound to red blood cells. Cadmium accumulates mainly in the kidney and in the liver, with about 50% of the body burden found in these two organs. Estimates of the average body burden of nonoccupationally exposed adults range from 5 to 40 mg. The body burden of smokers is about twice that of nonsmokers. In all tissues, cadmium is bound mainly to metallothionein, whose production is stimulated by cadmium exposure, but also other bivalent metals such as zinc, copper, and mercury. Cadmium is mainly excreted via the urine and the extent of urinary excretion increases with age at least until age 50-60 and then declines slowly. Cadmium can also be excreted by other routes (bile, gastrointestinal tract, saliva, hair, nails), but to a lesser extent than in urine (Lauwerys 1978). In man, the three main target organs after long-term exposure to cadmium are the lung, the bone and the kidney, though it is generally accepted that the kidney is the critical organ (i.e., the organ that exhibits the first adverse effect). The early adverse effects of cadmium on the kidney are usually an increased excretion of specific proteins in urine (Lauwerys 1979; Bernard and Lauwerys 1989, 1990; Bernard et al 1990; Chia et al 1989; Thun et al 1989).

Biological monitoring

Several approaches have been used to evaluate the critical internal dose of cadmium, i.e., the lowest dose that under chronic exposure conditions may lead to the occurrence of adverse functional changes in the kidney. It is now possible to measure directly by neutron activation the amount of the metal that has accumulated in the two main sites of deposition, i.e., the liver and the kidney (Al-Haddad et al 1981; McLellan et al 1975; Vartsky et al 1977; Scott and Chettle 1986; Smith et al 1986; Ellis et al 1985). However, the evaluation of tissue cadmium *in vivo* by neutron activation is not a routine procedure. Therefore, it is important to evaluate to what extent indirect biological indicators, such as cadmium levels in blood and urine or metallothionein levels in plasma and urine, can be used to estimate the internal dose. In nonoccupationally exposed adults who are nonsmokers, the cadmium level in blood is generally below 0.1 µg/100 ml (Nordberg and Nordberg 1988); it increases with age, but remains usually below 0.5 µg/100 ml whole blood. Minoïa et al (1990) found, in a population of 900 healthy Italians, a mean value of 0.06 µg/100 ml (range: 0.01-0.17), whereas considerably higher values (up to 0.8 µg/100 ml) have been found in smokers. Staessen et al (1990) measured, in London civil servants, mean whole blood concentrations of 0.07 µg/100 ml in nonsmokers and 0.15 µg/100 ml in smokers. In adults not occupationally exposed to the metal, the level in urine is usually below 2 µg cadmium/g creatinine. It is higher in smokers than in nonsmokers. Blood and urine cadmium concentrations are both influenced by current exposure and body burden, but, in some circumstances, the influence of one factor is predominant.

Urine analysis

The significance of cadmium in urine can be summarized as follows (Lauwerys et al 1976, 1979a) (Figure 5). At low exposure conditions (i.e., general environ-mental exposure or moderate occupational exposure), when the total amount of cadmium absorbed has not yet saturated all the available cadmium-binding sites in the body (in particular the kidney), the cadmium concentration in urine reflects mainly the cadmium level in the body and hence in the kidney (Figure 5, a and b). When integrated exposure has been so high as to cause a saturation of the binding sites, cadmium in urine may then be related partly to the body burden and partly to the recent exposure (Figure 5c). The relative importance of each factor depends on the intensity of exposure. Thus, in workers newly exposed to cad-mium, a time lag is observed before cadmium in urine correlates with exposure. This time interval will depend on the intensity of the integrated exposure to cadmium.

So, in low-level exposure, cadmium in urine is considered to mainly reflect the body burden, while in high-exposure conditions (saturation of binding sites) and without kidney damage, it is also significantly influenced by current exposure. When renal damage develops a considerable increase of urinary excretion occurs.

The World Health Organization (WHO 1981) has recommended that the urinary concentration of cadmium should not exceed 10 µg/g creatinine and that control measures should be applied if a value exceeds 5 µg/g creatinine. Accord-ing to OSHA (1990), exposure and work practice should be reassessed when cadmium concentrations exceed 5 µg/g creatinine.

The current ACGIH biological exposure indice is 10 µg/g creatinine. However, this substance is on the 1991 Notice of Intent to Establish, the proposed change for cadmium in urine being from 10 µg/g creatinine to 5 µg/g creatinine (ACGIH 1991-1992). The German biological tolerance value is 15 µg/g creatinine (DFG 1991).

In urine, cadmium is present mainly bound to metallothionein. Hence, the determination of the level of this protein in urine may provide the same information

A. No saturation of kidney binding sites

B. No saturation of kidney binding sites

C. Saturation of kidney binding sites
   no renal dysfunction

D. Saturation of kidney binding sites
   renal dysfunction

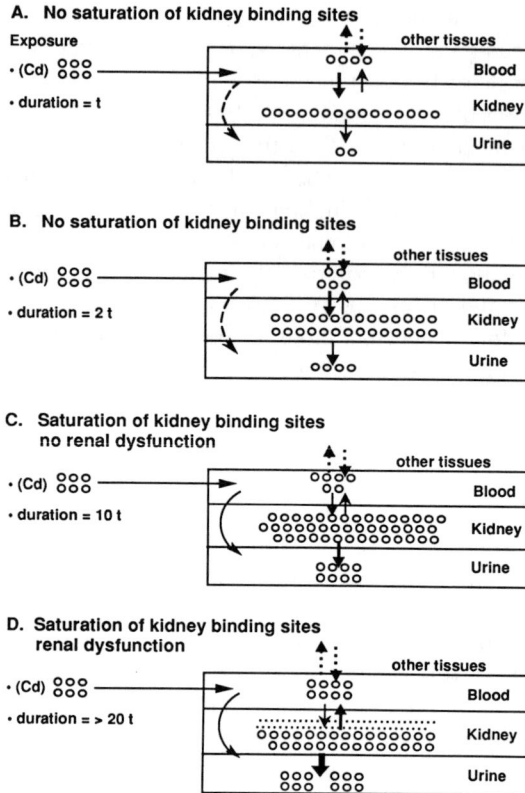

FIGURE 5 Evolution of urinary cadmium.

as the determination of cadmium, but metallothionein analysis presents an advantage over cadmium analysis in that it is not subject to external contamination (Chang et al 1980a). A radioimmunoassay has been developed for the determination of metallothionein in urine (Chang et al 1980b). Shaikh et al (1987, 1990) have confirmed that urinary metallothionein is a specific and sensitive indicator of increased cadmium body burden in occupationally exposed subjects.

Blood analysis

With regard to the significance of the cadmium level in blood, several arguments based on human observations (Lauwerys et al 1979a) and animal experiments (Bernard et al 1980) suggest that in workers moderately exposed to cadmium it mainly reflects the last few months of exposure. However, after a long period of exposure, the body burden plays more than a negligible role in determining the level of cadmium in blood (Ghezzi et al 1985; Shaikh et al 1990). In workers who have been removed from exposure at least for several months, cadmium in blood may prevalently reflect the body burden (Hassler et al 1983).

In conclusion, if under occupational exposure conditions cadmium in blood may be considered as mainly an indicator of recent exposure, the relative influence of the cadmium body burden may be more important in persons with previous exposure and persons who have accumulated large amounts of cadmium. Some studies suggest that the cumulative level of cadmium in blood, estimated by repeated measurements of the metal in blood, more correctly reflects the total individual exposure and may be used to identify individuals at risk of developing renal damage (Rogenfelt et al 1984; Järup et al 1988).

The World Health Organization (1981) has recommended that cadmium concentration in blood should not exceed 10 µg/l, and that control measures should be applied if the concentration exceeds 5 µg/l. The German biological tolerance value is 15 µg/l (DFG 1991). The newly proposed ACGIH biological exposure indice is 5 µg/l.

Hair analysis

The determination of cadmium in hair has also been proposed to evaluate exposure. This analysis is of limited practical value in industry because it is very difficult to distinguish between endogenous cadmium and cadmium externally deposited on the hair (Nishiyama and Nordberg 1972).

Feces analysis

Fecal cadmium is a good indicator of the daily intake of cadmium via food, but is without interest for the monitoring of occupationally exposed subjects.

We have performed several epidemiologic studies on cadmium-exposed male workers to estimate the critical levels of cadmium in blood, urine, and renal cortex (Lauwerys et al 1979b, 1980; Roels et al 1979, 1981a,b; Lauwerys and De Wals 1981; Bernard et al 1979; Buchet et al 1980). The main results can be summarized as follows: the prevalence of several biological signs of renal dysfunction (such as low- and high-molecular weight proteinuria) is systematically increased in groups of workers whose cadmium in urine exceeds 10 µg/g creatinine, corresponding to a cadmium concentration in renal cortex of approximately 200 ppm. However, a logistic regression analysis of the data suggests that some renal markers may become abnormal from thresholds of urinary cadmium between 3 and 5 µg/g of creatinine (Roels et al 1992).

The health significance of the presence of an isolated renal change (such as an increased urinary excretion of tubular enzymes) has not yet been assessed. It has been demonstrated, however, that the persistent finding of microproteinuria is predictive of an exacerbated age-related decline in renal function (Roels et al 1989). Therefore, an accumulation of cadmium in the body which would lead to a urinary excretion exceeding 10 µg/g creatinine should be prevented.

However, in view of the long biological half-life of cadmium, which precludes its rapid disappearance from the body after cessation of exposure and the possible occurrence of slight renal changes in some individuals excreting less than 10 µg cadmium/g creatinine, it is advisable to propose a lower biological exposure index for cadmium in urine. A value of 5 μg cadmium/g creatinine seems reasonable, provided it is applied together with a health-surveillance program, including the detection of early renal changes in order to identify the most susceptible workers. For blood, a value of 0.5 μg cadmium/100 ml whole blood is proposed as a tentative no-effect level for long-term exposure. It should be realized, however, that the above estimates of the critical concentrations of cadmium in urine or in renal cortex are derived from data collected in adult male workers.

There is indication that these thresholds may be lower for the general population mainly exposed to cadmium via food and tobacco smoking (Buchet et al 1990).

REFERENCES

ACGIH. American Conference of Governmental Industrial Hygienists. *Threshold Limit Values and Biological Exposure Indices.* Cincinnati, 1991-1992.

I. Al-Haddad, D. Chettle, J. Fletcher and J. Fremlin. A transportable system for measurement of kidney cadmium in vivo. *Int. J. Appl. Rad. Isotopes* 32:109, 1981.

A. Bernard, J.-P. Buchet, H. Roels et al. Renal excretion of proteins and enzymes in workers exposed to cadmium. *Eur. J. Clin. Invest.* 9:11, 1979.

A. Bernard, A. Goret, J.-P. Buchet et al. Significance of cadmium levels in blood and urine during long-term exposure of rats to cadmium. *J. Toxicol. Environ. Health* 6:175, 1980.

A. Bernard and R. Lauwerys. Cadmium Nag activity and β_2-microglobulin in the urine of cadmium pigment workers. *Br. J. Ind. Med.* 46:679, 1989.

A. Bernard and R. Lauwerys. Early markers of cadmium nephrotoxicity: biological significance and predictive value. *Toxicol. Environ. Chem.* 27:65, 1990.

A. Bernard, H. Roels, A. Cardenas and R. Lauwerys. Assessment of urinary protein-1 and transferin as early markers of cadmium nephrotoxicity. *Br. J. Ind. Med.* 47:559, 1990.

J.-P. Buchet, H. Roels, A. Bernard and R. Lauwerys. Assessment of renal function of workers exposed to inorganic lead, cadmium or mercury vapour. *J. Occup. Med.* 22:741, 1980.

J.-P. Buchet, R. Lauwerys, H. Roels et al. Renal effects of cadmium body burden of the general population. *Lancet* 336:699, 1990.

C. Chang, R. Lauwerys, A. Bernard et al. Metallothionein in cadmium exposed workers. *Environ. Res.* 23:422, 1980a.

C. Chang, R. Vander Mallie and J. Garvey. A radioimmunoassay for human metallothionein. *Toxicol. Appl. Pharmacol.* 55:94, 1980b.

K. Chia, C. Ong, H. Ong and G. Endo. Renal tubular function of workers exposed to low levels of cadmium. *Br. J. Ind. Med.* 46:165, 1989.

DFG. Deutsche Forschungsgemeinschaft. Maximum Concentrations at the Workplace and Biological Tolerance Values for Working Materials. Report No. XXVII. Commission for the Investigation of Health Hazards of Chemical Compounds in the Work Area. VCH, Weinheim, 1991.

K. Ellis, S. Cohn and T. Smith. Cadmium inhalation exposure estimates: their significance with respect to kidney and liver cadmium burden. *J. Toxicol. Environ. Health* 15:173, 1985.

L. Friberg, C.-G. Elinder, T. Kjellström and G. Nordberg. In: *Cadmium and Health: A Toxicological and Epidemiological Appraisal. Vol. I. Exposure, Dose and Metabolism.* CRC Press, Boca Raton, FL, 1985.

L. Friberg, T. Kjellström and G. Nordberg. Cadmium. In: *Handbook on the Toxicology of Metals. Vol. II.* 2nd ed. Eds.: L. Friberg, G. Nordberg, V. Vouk. Elsevier, Amsterdam, 1986.

I. Ghezzi, F. Toffoletto, G. Sesana et al. Behaviour of biological indicators of cadmium in relation to occupational exposure. *Int. Arch. Occup. Environ. Health* 55:133, 1985.

E. Hassler, B. Lind and M. Piscator. Cadmium and blood and urine related to present and past exposure. A study of workers in an alkaline battery factory. *Br. J. Ind. Med.* 420:420, 1983.

L. Järup, C.-G. Elinder and G. Spang. Cumulative blood. Cadmium and tubular proteinuria: dose-response relationship. *Int. Arch. Occup. Environ. Health* 60:223, 1988.

R. Lauwerys, J.-P. Buchet and H. Roels. The relationship between cadmium exposure or body burden and the concentration of cadmium in blood and urine in man. *Int. Arch. Occup. Environ. Health* 36:275, 1976.

R. Lauwerys. *Criteria (Dose-Effect Relationships) for Cadmium, Commission of the European Communities,* Pergamon Press, Oxford, 1978.

R. Lauwerys. Cadmium in man. In: *Chemistry, Biochemistry and Biology of Cadmium.* Ed.: M. Webb. Elsevier/North-Holland, Amsterdam, 1979.

R. Lauwerys, H. Roels, M. Regniers et al. Significance of cadmium concentration in blood and in urine in workers exposed to cadmium. *Environ. Res.* 20:375, 1979a.

R. Lauwerys, A. Bernard, J.-P. Buchet and H. Roels. Dose-response relationship for the nephrotoxic action of cadmium in man. In: *Proceedings of the International Conference on Management and Control of Heavy Metals in the Environment,* CEP Ltd, Edinburgh, 1979b.

R. Lauwerys, H. Roels, A. Bernard and J.-P. Buchet. Renal response to cadmium in a population living in a non-ferrous smelter area in Belgium. *Int. Arch. Occup. Environ. Health* 45:271, 1980.

R. Lauwerys and P. De Wals. Environmental pollution by cadmium and mortality from renal diseases. *Lancet* 1:383, 1981.

J. McLellan, B. Thomas, J. Fremlin and M. Harvey. Cadmium. Its in vivo detection in man. *Phys. Med. Biol.* 20:88, 1975.

C. Minoïa, E. Sabbioni, P. Apostoli et al. Trace element reference values in tissues from inhabitants of the European Community. I. A study of 46 elements in urine, blood and serum of Italian subjects. *Sci. Tot. Environ.* 95:89, 1990.

K. Nishiyama and G. Nordberg. Adsorption and elution of cadmium on hair. *Arch. Environ. Health* 25:92, 1972.

G. Nordberg and M. Nordberg. Biological monitoring of cadmium. In: *Biological Monitoring of Toxic Metals.* Eds.: Th. Clarkson, L. Friberg, G. Nordberg, P. Sager. Rochester Series on Environmental Toxicity. Plenum Press, New York, 1988.

OSHA. Occupational Safety and Health Administration. U.S. Department of Labor. OSHA proposed Rule and Notice of Hearing on Occupational Exposure to Cadmium. 55 FR 4052. USDOL, Washington, D.C., February 1990.

H. Roels, A. Bernard, J.-P. Buchet et al. The critical concentration of cadmium in renal cortex and in urine in man. *Lancet* 1:221, 1979.

H. Roels, R. Lauwerys, J.-P. Buchet et al. In vivo measurement of liver and kidney cadmium in workers exposed to this metal. Its significance with respect to cadmium in blood and urine. *Environ. Res.* 26:217, 1981a.

H. Roels, R. Lauwerys, J.-P. Buchet and A. Bernard. Environmental exposure to cadmium
    and renal function of aged women in three areas of Belgium. *Environ. Res.* 24:117,
    1981b.

H. Roels, R. Lauwerys, J.-P. Buchet et al. Health significance of cadmium induced renal
    dysfunction: a five-year follow-up. *Br. J. Ind. Med.* 46:755, 1989.

H. Roels, A. Bernard, A. Cardenas et al. Markers of early renal changes induced by
    industrial pollutants. III. Application to workers exposed to cadmium. *Br. J. Ind. Med.*
    50:37, 1993.

A. Rogenfelt, C.-G. Elinder and L. Jarup. A suggestion on how to use measurements of
    cadmium in blood as a cumulative dose estimate. *Int. Arch. Occup. Environ. Health*
    55:43, 1984.

M. Scott and D. Chettle. In vivo elemental analysis in occupational medicine. *Scand. J.
    Work Environ. Health* 12:81, 1986.

Z. Shaikh, C. Tohyama and C. Nolan. Occupational exposure to cadmium: effect on
    metallothionein and other biological indices of exposure and renal function. *Arch.
    Toxicol.* 59:360, 1987.

Z. Shaikh, K. Ellis, K. Subramanian and A. Greenberg. Biological monitoring for occupa-
    tional cadmium exposure the urinary metallothionein. *Toxicology* 63:53, 1990.

N. Smith, M. Topping, J. Steward and J. Fletcher. Occupational cadmium exposure in jig
    solderers. *Br. J. Ind. Med.* 43:663, 1986.

J. Staessen, N. Yeoman, A. Fletcher et al. Blood cadmium in London civil servants. *Int. J.
    Epidemiol.* 19:362, 1990.

M. Thun, A. Osorio, S. Shuber et al. Nephropathy in cadmium workers: assessment of risk
    from airborne occupational exposure to cadmium. *Br. J. Ind. Med.* 46:689, 1989.

D. Vartsky, K. Ellis, N. Chen and S. Cohn. A facility for in vivo measurement of kidney
    and liver cadmium by neutron capture prompt gamma ray analysis. *Phys. Med. Biol.*
    22:1085, 1977.

WHO. World Health Organization. Recommended Health-based Limits in Occupational
    Exposure to Heavy Metals. Report of a Study Group. Technical Report Series 647.
    Geneva, 1981.

7. CARBON DISULFIDE

Toxicokinetics

Carbon disulfide (CS_2) can be absorbed by all routes, but inhalation represents the
main route of absorption in occupational exposure (WHO 1979). In man, 70 to
90% of the amount absorbed is metabolized and mainly excreted in the urine in
the form of various sulfur metabolites. The remainder is excreted unchanged,
mainly via the lungs. Less than 1% of absorbed carbon disulfide is excreted
unchanged in the urine. Carbon disulfide binds to amino acids and proteins in
blood and tissues. It also reacts with glutathione (GSH) and is oxidized by
microsomal enzymes yielding reactive sulfur.

It has been shown by Vasak et al (1963) that some metabolites excreted in the
urine give a positive iodine-azide reaction. Some of these metabolites have been
isolated from human urine: thiourea, mercaptothiazolinone, 2-oxothiazolidine-4-
carboxylic acid, and 2-thiothiazolidine-4-carboxylic acid (TTCA) (Pergal et al
1972a,b; Van Doorn et al 1981) (Figure 6).

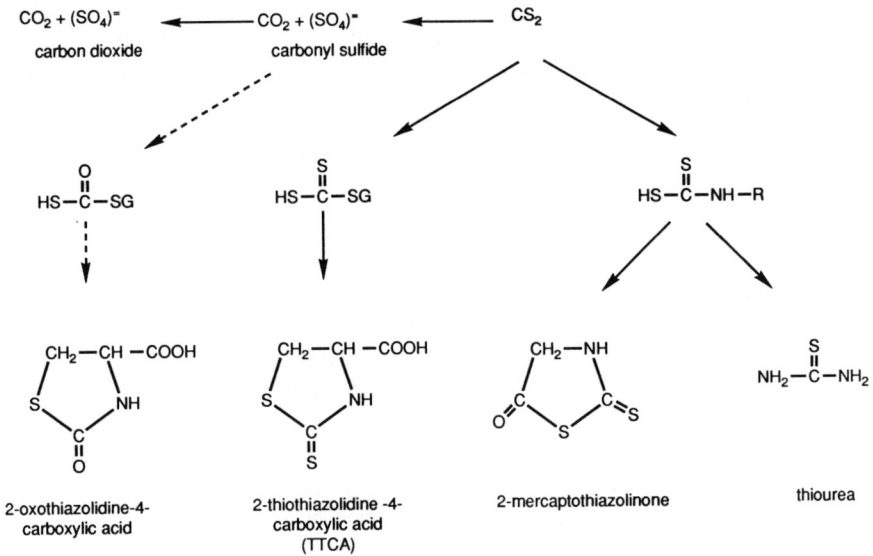

FIGURE 6 Main metabolic pathways of carbon disulfide.

Biological monitoring

Brugnone et al (1991) have measured mean values of 26.1 ng free CS_2/100 ml blood and 62.5 ng bound CS_2/100 ml blood in 42 subjects not occupationally exposed to carbon disulfide.

Several tests have been proposed for evaluating exposure to carbon disulfide: in blood, the determination of carbon disulfide, and in urine, the iodine-azide test, the determination of thiocompounds, the determination of specific metabolites, and the Antabuse test.

Blood analysis

The determination of carbon disulfide in blood does not seem to provide reliable information on the intensity of exposure.

Urine analysis

The iodine azide test. This test is based on the elapsed time between adding the iodine-azide reagent to urine until discoloration of the iodine. Vasak et al (1963) proposed an exposure index, E

$$E = C \log t$$

where C = creatinine concentration in mg/ml urine, t = discoloration time in seconds.

The test has a rather high detection limit, which restricts its use to exposure levels in excess of 50 mg/m^3 (WHO 1979) or even 100 mg/m^3 according to Rosier et al (1984).

Determination of thio compounds. An enhancement of urinary excretion of thio-compounds can be demonstrated in viscose workers, but this test lacks specificity (Rosier et al 1984; Van Poucke et al 1986, 1990).

Determination of specific metabolites. Van Doorn et al (1981) have demonstrated that the determination of TTCA in urine can be used as a biological monitoring test for exposure to carbon disulfide. This metabolite has a short biological half-life. The urine should therefore be sampled at the end of the shift. However, Van Poucke et al (1990) have demonstrated an accumulation of CS$_2$ metabolites following repeated exposure; the TTCA test was positive in preshift urine even after 32 to 63.5 h without exposure, and in highly exposed workers values tended to increase during consecutive days of exposure. This is not unexpected since CS$_2$ is highly soluble in fat and probably accumulates in the body during the work-week.

Although less than 6% of the absorbed CS$_2$ seems to be transformed into TTCA (Rosier et al 1987b), the latter has been shown to be a specific, sensitive, and reliable indicator of exposure to CS$_2$. TWA exposure as low as 1 ppm CS$_2$ is still associated with detectable amounts of TTCA in postshift urine samples (Rosier et al 1982, 1984, 1987b; Campbell et al 1985; Catenacci et al 1988; Meuling et al 1988; Van Poucke et al 1986). A concentration of TTCA in the end-of-shift urine of 5 mg/g creatinine corresponds to an exposure level of 10 ppm CS$_2$ (31 mg/m^3). This value has been adopted as BEI (Biological Exposure Indice) by the ACGIH (1991-1992). The German biological tolerance value is 8 mg/l (DFG 1991).

The Antabuse test. A method for the detection of subjects potentially more susceptible to the toxic action of carbon disulfide has been proposed by Djuric et al (1973) and Djuric (1984). The amount of diethyldithiocarbamate (DDC) ex-creted in urine within 4 to 5 h after the oral administration of 0.5 g of disulfuram (Antabuse) is determined. The susceptibility to carbon disulfide may be inversely related to the amount of DDC excreted. Besarabic (1978) has recommended that only persons with a DDC excretion of over 150 mg/g creatinine (within 4 h after the oral administration of 0.5 g Antabuse) be accepted for a job that requires exposure to carbon disulfide. The practical application of such a "selection" test is questionable.

Breath analysis

The determination of CS$_2$ in breath has also been proposed for the biological monitoring of workers exposed to CS$_2$. However, as pointed out by Rosier et al (1987a), the elimination of CS$_2$ in breath is rapid (initial fast decrease with a half-life of 1.1 min, then a slower decrease with a half-life of 110 min). Moreover, on the contrary to TTCA excretion, the amount of CS$_2$ eliminated by exhalation showed a high interindividual dispersion. For these reasons, CS_2 in exhaled air seems unsuitable as a biological exposure monitoring method.

In conclusion, TTCA determination in urine presently constitutes the method of choice for assessing exposure to CS_2.

REFERENCES

ACGIH. American Conference of Governmental Industrial Hygienists. *Threshold Limit Values and Biological Exposure Indices.* Cincinnati, 1991-1992.

M. Besarabic. Antabuse test and absenteeism in workers exposed to carbon disulfide. *Arh. Hig. Rada Toksikol.* 29:323, 1978.

F. Brugnone, L. Perbellini, G. Maranelli et al. Monitoring of carbon disulfide environmental pollution by its assay in human blood. The 1991 Eurotox Congress. Book of abstracts. Maastricht, The Netherlands. Sept. 1991.

L. Campbell, A. Jones and H. Wilson. Evaluation of occupational exposure to carbon disulfide by blood, exhaled air and urine analysis. *Am. J. Ind. Med.* 8:143, 1985.

G. Catenacci, F. Pugliese, S. Ghittori et al. Determinants of CS_2 in blood and urine and of 2 thio-thiazolidine-4-carboxylic acid in urine in exposed to low levels of CS_2. In: *Occupational Health in the Production of Artificial Organic Fibres.* Ed.: C. Braun. Arnhem, ENKA, p. 65-72, 1988.

DFG. Deutsche Forschungsgemeinschaft. Maximum Concentrations at the Workplace and Biological Tolerance Values for Working Materials. Report No. XXVII. Commission for the Investigation of Health Hazards of Chemical Compounds in the Work Area. VCH, Weinheim, 1991.

D. Djuric, A. Postic-Grujin, L. Graova-Leoposavic and V. Delic. Disulfiram as an indicator of human susceptibility to carbon disulfide. Excretion of diethyldithiocarbamate sodium in the urine of workers exposed to CS_2 after oral administration of disulfiram. *Arch. Environ. Health* 26:287, 1973.

D. Djuric. Proposal for the "Antabuse test". *G. Ital. Med. Lav.* 6:115, 1984.

W. Meuling, P. Bragt and C. Braun. Biological monitoring of CS_2: relation between exposure and urinary 2-thiothiazolidine-4-carboxylic acid (TTCA). In: *Occupational Health in the Production of Artificial Organic Fibres.* Ed.: C. Braun. Arnhem, Enka, p. 9, 1988.

M. Pergal, N. Vukojevic, N. Girin-Popov et al. Carbon disulfide metabolites excreted in the urine of exposed workers. I. Isolation and identification of 2-mercapto-2-thiazolinone-5. *Arch. Environ. Health* 25:38, 1972a.

M. Pergal, N. Vukojevic and D. Djuric. II. Isolation and identification of thiocarbamide. *Arch. Environ. Health* 25:42, 1972b.

J. Rosier, M. Van Hoorne, R. Grosjean et al. Preliminary evaluation of urinary 2-thio-thiazolidine-4-carboxylic acid (TTCA) levels as a test for exposure to CS_2. *Int. Arch. Occup. Environ. Health* 51:159, 1982.

J. Rosier, G. Billemont, C. Van Peteghem et al. Relationship between the iodine-azide test and the TTCA test for exposure to carbon disulfide. *Br. J. Ind. Med.* 41:412, 1984.

J. Rosier, H. Veulemans, R. Maaschelein et al. Experimental human exposure to carbon disulfide. I. Respiratory uptake and elimination of CS_2 under rest and physical exercise. *Int. Arch. Occup. Environ. Health* 59:233, 1987a.

J. Rosier, H. Veulemans, R. Maaschelein et al. Experimental human exposure to carbon disulfide. II. Urinary excretion of 2 thiothiazolidine-4-carboxylic acic 2 TTCA during and after exposure. *Int. Arch. Occup. Environ. Health* 59:243, 1987b.

R. Van Doorn, L. Delbressine, Ch. Leidekkers et al. Identification and determination of 2-thiothiazolidine-4-carboxylic acid in urine of workers exposed to carbon disulfide. *Arch. Toxicol.* 47:51, 1981.

L. Van Poucke, E. Tytgat, J. Rosier et al. Evaluatie van de persoonlijke bloodstelling aan CS$_2$ en H$_2$S in de viscose nyverheid. *Cah. Méd. Trav.* 23:239, 1986.

L. Van Poucke, C. Van Peteghem and M. Van Hoorne. Accumulation of carbon disulphide metabolites. *Int. Arch. Occup. Environ. Health* 62:479, 1990.

V. Vasak, M. Venecek and B. Kimmelova. The assessment of exposure of workers to carbon disulfide vapours. *Pracov. Lek.* 15:145, 1963 (in Czechoslovakian).

WHO. World Health Organization. Task group on environmental health criteria for carbon disulfide. Environmental Health Criteria 10, Carbon disulfide. Geneva, 1979.

8. CHROMIUM

Toxicokinetics

In industry, workers may be exposed to trivalent and hexavalent chromium compounds whose metabolic handling and toxicity are strikingly different.

Hexavalent Chromium Compounds

The toxicity of chromium is mainly attributed to hexavalent compounds which can be absorbed by the lung and gastrointestinal tract and even to a certain extent by the intact skin (Baranowska-Dutkiewicz 1981). It seems that the carcinogenic risk for the respiratory tract is principally related to the more insoluble hexavalent chromium compounds. Once absorbed, chromium is mainly excreted via urine and its excretion is rather rapid (Franchini et al 1975). The half-time of chromium in urine has been estimated to range from 15 to 41 h (Tossavainen et al 1980). Detailed kinetic studies, however, suggest the existence of three compartments showing half-lives of approximately 7 h, 15-30 d, and 3-5 years (Welinder et al 1983; Rahkonen et al 1983; Kalliomaki et al 1985; Aitio et al 1988).

With the exception of the lungs (in which chromium concentration increases with age), tissue levels of chromium decline with age (WHO 1988). Certain factors might be associated with increased urinary chromium levels (i.e., exercise, past employment in a chromium-exposed occupation, drinking beer, diabetic status), while red blood cell chromium level does not appear to be affected (Bukowski et al 1991).

Trivalent Chromium Compounds

In the trivalent state, chromium is poorly absorbed. Studying the distribution and kinetics of intravenous trivalent radioactive chromium in six human subjects, Lim et al (1983) found some accumulation in the liver, spleen, soft tissues, and bones. The data fitted to a model consisting of a plasma pool in equilibrium with fast ($T_{1/2} = 0.5$-12 h), medium (1-14 d) and slow (3-12 months) compartments (Lim et al 1983).

Biological monitoring

In persons nonoccupationally exposed to chromium, the concentration of chromium in serum or plasma and in urine usually does not exceed 0.05 µg/100 ml and 5 µg/g creatinine, respectively (Franchini et al 1975; Versieck 1985; Schermaier et al 1985; Lewis et al 1985; Kiilunen et al 1987; Versieck and Cornelis 1989; Minoïa et al 1990).

The values reported by WHO (1988) and based on the data of the U.S. EPA (1978) range from 0.02 to 7 µg/100 ml in serum and plasma and 0.5 to 5.4 µg/100 ml in red blood cells.

Hexavalent Chromium Compounds

Urine analysis

At present the determination of the concentration of chromium in urine seems to be the most practical biological monitoring method for assessing exposure to hexavalent chromium compounds.

Determination of chromium in the urine at the end of the workday seems to be a good indicator of recent exposure to soluble hexavalent chromium compounds (Mutti et al 1985; Cavalleri and Minoïa 1985). However, Franchini et al (1984) underlined the fact that urinary chromium is influenced by both recent and past exposure, whereas the difference between chromium values in spot urine samples collected at the end and before the workshift would reflect current exposure.

Bonde and Christensen (1991) did not detect increase in urine chromium concentrations across workshifts in workers exposed to low levels of hexavalent chromium (0.003 mg/m³; SD: 0.002 mg/m³). However, these workers had significantly increased levels of chromium (mean: 2.1 nmol/mmol creatinine or 0.966 µg/g creatinine) compared with referents (mean: 0.7 nmol/mmol creatinine or 0.322 µg/g creatinine).

Studies carried out by several authors (Berode and Guillemin 1977; Gylseth et al 1977; Sjögren et al 1983; Tola et al 1977) suggest the following relation: a TWA exposure of 0.025 or 0.05 mg/m³ hexavalent chromium is associated with an average urinary concentration at the end of the exposure period of 15 or 30 µg/g creatinine, respectively. This relation is only valid on a group basis. Following exposure to 0.025 mg/m³ hexavalent chromium, the lower 95% confidence limit value is approximately 5 µg/g creatinine. This estimate is in good agreement with the relationship observed by Mutti et al (1985): in subjects chronically exposed to hexavalent chromium a mean urinary chromium increase (pre-exposure value substracted from end-of-shift value) of 12.2 µg/g creatinine or a mean total concentration of 30 µg/g creatinine correspond to an air concentration of 50 µg/m³ chromium VI from welding fumes. Lindberg and Vesterberg (1983) also found a correlation between air and urinary chromium concentrations and determined that a postshift urinary chromium level of ≤100 nmol/l (5.2 µg/l) would reflect a TWA exposure to ≤2 µg/m³ chromium hexavalent from chromic acid. According to the results of Angerer et al (1987), urinary chromium in the order of 40 µg/l would correspond to an exposure of 100 µg/m³ chromium trioxide in stainless steel welders.

The ACGIH (1991-1992) has adopted as BEI for chromium VI (water-soluble fume) 10 µg total chromium/g creatinine as increase during shift and 30 µg/g creatinine at the end-of-shift (end-of-workweek) corresponding to a TWA of 0.05 mg/m³.

Blood analysis

In the blood, hexavalent chromium enters the red blood cells and is reduced to trivalent chromium. Chromium III does not succeed in passing cell membranes.

Some authors have studied the possibility of determining the erythrocyte chromium concentration (Lewalter et al 1985; Von Wiegand et al 1985; Korallus 1986). Too few data exist to assess the interest of blood chromium measurements, but the concentration of chromium in erythrocytes might be an indicator of the exposure intensity to hexavalent chromium during the lifetime of the red blood cells.

Angerer et al (1987) have suggested use of a combination of chromium concentration in plasma for its great diagnostical sensitivity to assess recent exposure and chromium concentration in erythrocyte which might specifically reflect the internal dose of carcinogenic hexavalent compounds. They observed that an average exposure to 100 µg/m³ hexavalent compounds resulted in mean plasma and erythrocyte levels of 1 µg/100 ml and about 0.06 µg/100 ml, respectively.

Trivalent Chromium Compounds

Urine analysis

As mentioned above, these compounds are poorly absorbed, but nevertheless increased chromium urinary concentrations have been observed in workers exposed to trivalent chromium, indicating that chromium III is absorbed at least to some extent (Kiilunen et al 1983; Aitio et al 1984; Saner et al 1984). According to Randall and Gibson (1987), urinary chromium may serve as an indice of chromium exposure and status in tannery workers. However, examining end-of-shift urinary levels, Mutti et al (1985) concluded that they did not correlate with exposure to trivalent chromium compounds. The results obtained by Cavalleri and Minoïa (1985) confirmed this observation.

Blood analysis

Randall and Gibson (1987) have also suggested that serum chromium might serve as an indice of exposure to chromium III in tannery workers, but this requires confirmation.

Hair analysis

Randall and Gibson (1989) have reported that trivalent chromium absorbed from leather-tanning compounds resulted in raised concentrations of chromium in hair, which might be used as an index of exposure.

In conclusion, plasma and urine levels of chromium seem to mainly reflect hexavalent chromium uptake; however, trivalent chromium ions probably contribute, at least to some extent, to these levels. The end-of-shift urinary chromium level, which mainly reflects recent exposure, but is also influenced by the body burden, should not exceed, after several weeks of exposure, 30 µg/g creatinine. The increase during the shift should not exceed 10 µg/g creatinine.

REFERENCES

ACGIH. American Conference of Governmental Industrial Hygienists. *Threshold Limit Values and Biological Exposure Indices.* Cincinnati, 1991-1992.

A. Aitio, J. Jarvisalo, M. Kiilunen et al. Urinary excretion of chromium as an indicator of exposure to trivalent chromium sulphate in leather tanning. *Int. Arch. Occup. Environ. Health* 54:241, 1984.

A. Aitio, J. Jarvisalo, M. Kiilunen et al. Chromium. In: *Biological Monitoring of Toxic Metals. Rochester Series on Environmental Toxicity.* Eds.: Th. Clarkson, L. Friberg, G. Nordberg, P. Sager. Plenum Press. New York, 1988.

J. Angerer, W. Amin, R. Heinrich et al. Occupational exposure to metals. I. Chromium exposure of stainless steel welders — Biological monitoring. *Int. Arch. Occup. Environ. Health* 59: 503, 1987.

B. Baranowska-Dutkiewicz. Absorption of hexavalent chromium by skin in man. *Arch. Toxicol.* 47:47, 1981.

M. Berode and M. Guillemin. Evaluation d'une exposition professionnelle au chrome par le dosage du chrome urinaire. *Méd. Soc. Prév.* 22:201, 1977.

J. Bonde and J. Christensen. Chromium in biological samples from low-Level exposed stainless steel and mild steel welders. *Arch. Environ. Health* 46:225, 1991.

J. Bukowski, M. Goldstein, L. Korn and B. Johnson. Biological markers in chromium exposure assessment: confounding variables. *Arch. Environ. Health* 46:230, 1991.

A. Cavalleri and C. Minoïa. Serum and erythrocyte chromium distribution and urinary elimination in persons occupationally exposed to chromium VI and chromium III. *G. Ital. Med. Lav.* 7:35, 1985.

I. Franchini, A. Mutti, F. Gardini and A. Borghetti. Excrétion et clearance rénale du chrome par rapport au degré et à la durée de l'exposition professionnelle. In: *Rein et Toxique.* Masson, Paris, p. 271, 1975.

R. Franchini, A. Mutti, E. Cavatorta et al. Biological indicators for the assessment of human exposure to industrial chemicals: chromium. Eds.: L. Alessio, A. Berlin, M. Boni, R. Roi. Commission of the European Communities. Industrial Health and Safety. Luxembourg. Eur 8903 EN, 1984.

B. Gylseth, N. Gundersen and S. Langard. Evaluation of chromium exposure based on a simplified method for urinary chromium determination. *Scand. J. Work Environ. Health* 3:28, 1977.

P.-L. Kalliomaki, K. Kalliomaki, E. Rahkonen and M.-L. Juntilla. Magnetopneumography — lung retention and clearance of manual metal arc welding fumes based on experimental and human data. In: *Biomagnetism: Application and Theory*. Eds.: H. Weinberg, G. Stroink, K. Katila. Pergamon Press, Oxford, p. 416, 1985.

M. Kiilunen, H. Kivisto, P. Ala-Laurila et al. Exceptional pharmacokinetics of trivalent chromium during occupational exposure to chromium lignosulfonate dust. *Scand. J. Work. Environ. Health* 9:265, 1983.

M. Kiilunen, J. Jarvisalo, O. Makitie and A. Aito. Analysis, storage, stability and reference values for urinary chromium and nickel. *Int. Arch. Occup. Environ. Health* 59:43, 1987.

V. Korallus. Biological activity of chromium (VI) — against chromium (III) compounds: New aspects of biological monitoring. In: *Chromium Symposium 1986: An Update*. Ed.: D. Serrone. Pittsburgh, Industrial Health Foundation Inc., p. 210, 1986.

J. Lewalter, V. Korallus, C. Harzdorf and H. Weidemann. Chromium bond detection in isolated erythrocytes: a new principle of biological monitoring of exposure to hexavalent chromium. *Int. Arch. Occup. Environ. Health* 55:305, 1985.

S. Lewis, T. O'Haver and J. Harnly. Determination of metals at the μg/l levels in blood serum by simultaneous multielement atomic absorption spectrometry with graphite furnale atomization. *Anal. Chem.* 57:2, 1985.

T. Lim, Th. Sargent III and N. Kusubov. Kinetics of trace element chromium III in the human body. *Am. J. Physiol.* 244:R 445, 1983.

E. Lindberg and O. Vesterberg. Monitoring exposure to chromic acid in chrome plating by measuring chromium in urine. *Scand. J. Work Environ. Health* 9:333, 1983.

C. Minoïa, E. Sabbioni, P. Apostoli et al. Trace element reference values in tissues from inhabitants of the European community. I. A study of 46 elements in urine, blood and serum of Italian subjects. *Sci. Total Environ.* 95:89, 1990.

A. Mutti, C. Pedroni, G. Arfini et al. Biological monitoring of occupational exposure to different chromium compounds at various valency states. In: *Carcinogenic and Metal Compounds Environmental and Analytical Chemistry and Biological Effects*. Eds.: E. Merian, R. Frei, W. Hardi, C. Schlatter. Gordon and Breach Science Publishers, London, p. 119, 1985.

E. Rahkonen, M.-L. Junttila, P. Kalliomaki et al. Evaluation of monitoring among stainless steel welders. *Int. Arch. Occup. Environ. Health* 52:243, 1983.

J. Randall and R. Gibson. Serum and urine chromium as indices of chromium status in tannery workers. *Proc. Soc. Exp. Biol. Med.* 185:16, 1987.

J. Randall and R. Gibson. Hair chromium as an index of chromium exposure of tannery workers. *Br. J. Ind. Med.* 46:171, 1989.

G. Saner, V. Yuzbasiyan and S. Cigdem. Hair chromium concentration and chromium excretion in tannery workers. *Br. J. Ind. Med.* 41:263, 1984.

A. Schermaier, L. O'Connor and K. Pearson. Semi-automated determination of chromium in whole blood and serum by zeeman electrothermal atomic absorption spectrophotometry. *Clin. Chim. Acta* 152:123, 1985.

B. Sjögren, L. Hedstrom and U. Ulfvarson. Urine chromium as an estimator of air exposure to stainless steel welding fumes. *Int. Arch. Occup. Environ. Health* 51:347, 1983.

S. Tola, J. Kilpio, M. Virtamo and K. Haapa. Urinary chromium as an indicator of the exposure of welders to chromium. *Scand. J. Work Environ. Health* 3:192, 1977.

A. Tossavainen, M. Norminen, P. Mutanen and S. Tola. Application of mathematical modeling for assessing the biological half-times of chromium and nickel in field studies. *Br. J. Ind. Med.* 37:285, 1980.

U.S. EPA. Reviews of the environmental effects of pollutants. III. Chromium. US Environmental Protection Agency, Washington D.C., 285 pp. ORNL/EIS-80; EPA 600/1-78-023. 1978.

J. Versieck. Trace elements in human body fluids and tissues. *CRC Crit. Rev. Clin. Lab. Sci.* 22:97, 1985.

J. Versieck and R. Cornelis. *Trace Elements in Human Plasma or Serum.* CRC Press, Boca Raton, FL, 1989.

H. Von Wiegand, H. Ottenwalder and H. Bolt. Die Chrombestimmung in Humanerythrozyten. Neue Grunlagen für eine biologisch Uberwachung Chromatexponieter. Arbeitmedizin, socialmedizin, Präventivmedizin 20:1, 1985.

H. Welinder, M. Littorin, B. Gullberg and S. Skerkving. Elimination of chromium in urine after stainless steel welding. *Scand. J. Work Environ. Health* 9:397, 1983.

WHO. IPCS. International Programme on Chemical Safety. Environmental Health Criteria 61. Chromium. World Health Organization, Geneva, 1988.

9. COBALT

Toxicokinetics

In the occupational setting, workers are mainly exposed to pure cobalt metal, oxides, or salts (e.g., in cobalt refinery) or to dust containing various proportions of cobalt in association with other substances, such as carbides (e.g., hard metals). Exposure mainly occurs by inhalation and to a certain extent also by the oral route. Slight dermal absorption of cobalt has also been reported following prolonged skin contact with cobalt dust.

Very limited data are available on the absorption of cobalt by the lung and the gastrointestinal tract in man. The absorption rate is certainly highly dependent on the solubility of cobalt compounds in biological media (Christensen and Mikkelsen 1985; Alessio and Dell'Orto 1988) which may be influenced by the concomitant presence of other substances. For example, it has been shown in animal experiments that the pulmonary absorption of cobalt is greatly increased in the presence of tungsten carbide (Lasfargue et al 1992). It has been estimated that the oral absorption rate can vary from 5 to 45% and that about 30% of cobalt inhaled as cobalt oxide can be absorbed (Elinder and Friberg 1986). Whatever the exposure route, cobalt is mainly excreted in urine and to a lesser extent via feces. The urinary elimination is characterized by a rapid phase of a few days duration followed by a second phase which may last a couple of years.

Biological monitoring

In nonoccupationally exposed subjects a wide range of cobalt concentrations have been reported in urine and plasma (or serum). The mean control values observed in a population of healthy Italians were 0.57 µg/l urine (n: 468; range: 0.18-0.96); 0.039 µg/100 ml blood (n: 441; range: 0.001-0.091) and 0.021 µg/100 ml serum (n: 405; range: 0.008-0.04) (Minoïa et al 1990). This is in agreement with the mean values estimated by IARC (1991): 0.01 to 0.03 µg/100 ml serum and 0.1 to

2 µg/l urine. Normal values are probably below 2 µg/g creatinine for urine and
0.05 µg/100 ml for serum or plasma (Cornelis et al 1975; Schumacher-Wittkopf
and Angerer 1981; Elinder and Friberg 1986; Versieck 1985; Versieck and
Cornelis 1989; Ferioli et al 1987; Angerer et al 1989). Increased urinary levels of
cobalt have been measured in persons taking multivitamin pills (Reynolds 1989,
cited in IARC 1991). Concerning patients with protheses made from cobalt and
chromium, several studies have reported elevated urinary and serum cobalt levels
while others have not found significant changes (for review, see IARC 1991).
Alexandersson (1988) found that smokers had significantly higher mean urinary
cobalt concentrations (0.6 µg/l; SD: 0.6) than nonsmokers (0.3 µg/l; SD: 0.1), but
there was no difference in the blood levels.

As shown by recent studies, the determinations of cobalt in blood and urine can
be used as indicators of exposure (Yamada et al 1987; Ichikawa 1987; Ichikawa
et al 1985, 1988; Perdrix et al 1983, 1987; Pellet et al 1984; Dupas et al 1989;
Catilina et al 1989; Posma and Dijstelberger 1985; Nicolaou et al 1987;
Alexandersson 1988; Raffn et al 1988; Stebbins et al 1992).

Urine analysis

In workers undergoing cobalt exposure, the urinary concentration increases proportionally more than that in blood. In practice the concentration of cobalt in urine
is mainly influenced by recent exposure, but in view of the biological half-life of
cobalt excretion (see above), under stable exposure conditions, its urinary concentration increases during the workweek.

From observations made on workers exposed to hard metal dust, Pellet et al
(1984) concluded that the difference between end- and beginning-of-shift urinary
cobalt concentrations reflects the day exposure, the concentration in the Friday
evening urine is an indicator of the cumulative exposure during the week, and the
level of cobalt in urine collected on Monday morning mainly reflects long-term
exposure.

Posma and Dijstelberger (1985) have estimated that a mean level of 10 µg/g
creatinine in the morning urine sample corresponds to a TWA of 0.1 mg/m³. In
1985, Molin Christensen and Mikkelsen suggested that exposure should be kept
at a level such that the urine concentrations of cobalt at the end of the workday
generally do not exceed 0.4 µg/mmol creatinine or 3.5 µg/g creatinine which falls
within their control values (0.004 to 1.21 µg/mmole creatinine). According to
Scansetti et al (1985), a TWA exposure of 0.1 mg/m³ would lead to a mean cobalt
concentration of about 30 µg/l urine at the end of the first workshift of the week
and about 60 µg/l at the end of the last shift.

Similar results were reported by Ichikawa et al (1985), who observed that for
a TWA exposure of 0.1 mg/m³ the cobalt concentration in urine collected toward
the end of a workshift on Wednesday or Thursday ranged from 59 to 78 µg/l (95%
confidence limits); a TWA of 50 µg/m³ lead to a mean concentration of 34 µg/l.
According to Angerer et al (1985), 50 and 200 µg cobalt/l of urine (no time of
sampling was reported but probably at end-of-shift) corresponded to exposure

levels of 0.1 and 0.5 mg/m^3. In a further study, Angerer (1989) suggested that an exposure to 50 µg/m^3 entails a mean urinary level of 30 µg/l.

Postshift urine samples were collected from 13 workers in the hammering, brazing, and tip-grinding areas in a small company producing carbide tip saw blades. Wet tip-grinding operators had the highest urine cobalt concentration (33.6 ± 6.5 µg/l) (mean ± SD) followed by the dry tip-grinding operators (12.2 ± 6.2 µg/l). The other workers had an even lower mean level (6.4 ± 3 µg/l). Urine cobalt and air exposure data showed a high degree of correlation; exposure to mean cobalt air levels of 0.01, 0.05, and 0.1 mg/m^3, leading to mean cobalt urinary concentrations of 7.27, 37.9, and 76.2 µg/l, respectively (Stebbins et al 1992).

In workers exposed to cobalt oxides, cobalt salts, or cobalt metal powder in a refinery, we have found that a TWA of 50 µg cobalt/m^3 leads to an average cobalt concentration of 33 and 46 µg/g creatinine in the urine collected at the end of the shift on Monday and Friday, respectively.

Blood analysis

For Ichikawa et al (1985), a TWA exposure to 0.1 mg/m^3 would result in blood levels ranging from 0.57 to 0.79 µg/100 ml. For the same exposure level, Posma and Dijstelberger (1985) reported 0.2-0.3 µg/100 ml serum. Angerer et al (1985) estimated that 1 and 2.5 µg/100 ml blood are the mean concentrations to be expected when the average exposure levels were 0.1 and 0.5 mg/m^3. In a further subsequent study, Angerer (1989) suggested that an exposure to 50 µg/m^3 would result in a mean blood level of 2.5 µg/l.

Other analyses

Increased toenail and pubic hair cobalt concentrations have also been measured in workers exposed to hard metal dusts (Nicolaou et al 1987).

The concentration of cobalt in blood and in urine below which the risk of adverse effects is negligible cannot yet be identified. It is even possible that the biological exposure limits must take into account the physicochemical properties of the compounds to which the workers are exposed (e.g., pure cobalt particles or cobalt in association with other substances, such as carbides).

REFERENCES

L. Alessio and A. Dell'Orto. Biological monitoring of cobalt. In: *Biological Monitoring of Toxic Metals. Rochester Series on Environmental Toxicity.* Eds.: Th. Clarkson, L. Friberg, G. Nordberg, P. Sager. Plenum Press, New York, 1988.

R. Alexandersson. Blood and urinary concentrations as estimators of cobalt exposure. *Arch. Environ. Health* 43:299, 1988.

J. Angerer, R. Heinrich, D. Szadkowski and G. Lehnert. Occupational exposure to cobalt powder and salts — biological monitoring and health effects. In: *International Conference "Heavy Metals in the Environment".* Ed.: S. Lekkas. Vol. 2, p. 11. Athens, Sept. 1985.

J. Angerer, R. Heinrich-Ramm and G. Lehnert. Occupational exposure to cobalt and nickel: biological monitoring. *Int. J. Environ. Anal. Chem.* 35:81, 1989.

J. Angerer. Cobalt. In: *Biologische Arbeitsstoff Toleranzwerte (BAT Werte), Arbeitsmedizinisch, Toxikologische Bergründungen.* Eds.: D. Henschler, G. Lehnert. VCH Verlag, Weinheim, p. 1, 1989.

D. Catilina, M. Catilina, D. Pepin et al. Contribution à la validation de la valeur moyenne d'exposition aux poussières de cobalt-métal dans l'atmosphère des locaux de travail. *Arch. Mal. Prof.* 50:248, 1989.

J. Christensen and S. Mikkelsen. Cobalt concentration in whole blood and urine from pottery plate painters exposed to cobalt paint. In: *International Conference "Heavy Metals in the Environment"*. Ed.: T. Lekkas. Vol. 2, p. 86. Athens, Sept. 1985.

R. Cornelis, A. Speeke and J. Hoste. Neutron activation analysis for bulk and trace elements in urine. *Anal. Chim. Acta* 78:317, 1975.

D. Dupas, P. Mattmann and G. Geraut. Variation de la cobalturie chez les ouvriers d'un atelier fabriquant des outils diamantés. *Arch. Mal. Prof.* 50:248, 1989.

C.-G. Elinder and L. Friberg. Cobalt. In: *Handbook on the Toxicology of Metals.* Eds.: L. Friberg, G. Nordberg, V. Vouk. Vol. II. 2nd ed. Elsevier Science Publishers, Amsterdam, 1986.

A. Ferioli, R. Roi and L. Alessio. Biological indicators for the assessment of human exposure to industrial chemicals: Cobalt. Eds.: L. Alessio, A. Berlin, M. Boni, R. Roi. Commission of the European Communities. Eur 11135 En, Luxembourg, 1987.

IARC. Monographs on the evaluation of carcinogenic risks to humans. Cobalt and cobalt compounds. Vol 52. International Agency for Research on Cancer. Lyon, 1991.

Y. Ichikawa, Y. Kusaka and S. Goto. Biological monitoring of cobalt exposure; based on cobalt concentration in blood and urine. *Int. Arch. Occup. Environ. Health* 55:269, 1985.

Y. Ichikawa. Changes in cobalt concentration in blood and urine. Abstracts. XXII International Congress on Occupational Health, Sydney, Australia, 27 Sept.-2 Oct. 1987.

Y. Ichikawa, Y. Kusaka, Y. Ogawa et al. Changes of blood and urinary levels of cobalt during single exposure to cobalt. *Jpn. J. Ind. Health* 30:208, 1988.

E. Kerfoot, W. Fredrick and E. Domeier. Cobalt metal inhalation studies on miniature swine. *Am. Ind. Hyg. Assoc. J.* 36:17, 1975.

G. Lasfargue, D. Lison and R. Lauwerys. Comparative study of the acute lung toxicity of pure cobalt powder and cobalt — tungsten carbide mixture in the rat. *Toxicol. Appl. Pharmacol.* 112:41, 1992.

C. Minoïa, E. Sabbioni, P. Apostoli et al. Trace element reference values in tissues from inhabitants of the European Community. I. A study of 46 elements in urine, blood and serum of Italian subjects. *Sci. Total Environ.* 95:89, 1990.

G. Nicolaou, R. Pietra, E. Sabbioni et al. Multielement determination of metals in biological specimens of hard metal workers: a study carried out by neutron activation analysis. *J. Trace Elem. Electrolytes Health Dis.* 1:73, 1987.

F. Pellet, A. Perdrix, M. Vincent et al. Dosage biologique du cobalt urinaire. Intérêt en médecine du travail dans la surveillance des expositions aux carbures métalliques frittés. *Arch. Mal. Prof.* 45:81, 1984.

A. Perdrix, F. Pellet, J.-F. Blatier et al. Exposition au carbure de tungstène fritté et au cobalt métallique. A propos de 64 sujets vus en consultation de pathologie professionnelle de 1969 à 1985. *Arch. Mal. Prof.* 48:173, 1987.

A. Perdrix, F. Pellet, M. Vincent et al. Cobalt and sintered metal carbides. Value of the determination of cobalt as a tracer for exposure to hard metals. *Toxicol. Eur. Res.* 5:233, 1983.

F. Posma and S. Dijstelberger. Serum and urinary cobalt levels as indicators of cobalt exposure in hard metal workers. In: *International Conference "Heavy Metals in the Environment"*. Ed.: T. Lekkas. Vol. 2, p. 89. Athens, Sept. 1985.

E. Raffn, S. Mikkelsen, D. Altman et al. Health effects due to occupational exposure to cobalt blue dye among plate painters in a porcelain factory in Denmark. *Scand. J. Work Environ. Health* 14:378, 1988.

G. Scansetti, S. Lamon, S. Talarico et al. Urinary cobalt as a measure of exposure in the hard metal industry. *Int. Arch. Occup. Environ. Health* 57:19, 1985.

E. Schumacher-Wittkopf and J. Angerer. Praxisgerechte Methode zur Kobalt Bestimmung in Harn. *Int. Arch. Occup. Environ. Health* 49:77, 1981.

A. Stebbins, S. Horstman, W. Daniell and R. Atallah. Cobalt exposure in a carbide tip grinding process. *Am. Ind. Hyg. Assoc. J.* 53:186, 1992.

J. Versieck. Trace elements in human body fluids and tissues. *CRC Crit. Rev. Clin. Lab. Sci.* 22:97, 1985.

J. Versieck and R. Cornelis. *Trace Elements in Human Plasma or Serum*. CRC Press, Boca Raton, FL, 1989.

Y. Yamada, T. Kido, R. Honda et al. Analysis of dusts and evaluation of dust exposure in a hard metal factory. *Ind. Health* 25:1, 1987.

10. COPPER

Toxicokinetics

In nonoccupationally exposed subjects, gastrointestinal absorption is the main route of copper entry to the body. This absorption is regulated by homeostatic mechanisms. In the occupational setting absorption of dust and fumes containing copper of copper compounds probably occurs via the respiratory tract but there are no data available on absorption rates of copper after inhalation. Copper intoxication has been recorded only after therapeutic application of a copper sulfate solution, but a case of skin exposure to copper particles after an explosion of copper azide suggests that it may be absorbed from the skin even if it is in the metal form (Bentur et al 1988; Grandjean 1990). Serum copper levels are subjected to a circadian rhythm. Copper is stored in the liver, heart, brain, kidney, and muscles (Triebig and Schaller 1984). Excretion mainly occurs via the bile, only a small fraction being eliminated in the urine.

Biological monitoring

The normal plasma or serum mean level lies around 0.11 mg/100 ml (Versieck and Cornelis 1989; Triebig and Schaller 1984). The mean urinary levels reported by Versieck (1985) range from 15 to 36 µg/24 h. This is in agreement with the estimate of Triebig and Schaller (1984); they reported a mean urinary daily excretion of around 20 µg and an upper limit of copper concentration in urine of 50 µg/l. These levels are influenced by physiological (such as pregnancy) or pathological conditions (such as liver diseases).

In a population of healthy, nonoccupationally exposed workers, Minoïa et al (1990) recorded as mean values: 0.12 mg/100 ml blood (n: 475; range: 0.08-0.16); 0.0985 mg/100 ml serum (n: 901; range: 0.06-0.137); 23 µg/l urine (n: 507; range: 4.2-50).

The data concerning the relationship between occupational exposure, internal dose and effect are at present too scarce to suggest reliable biological limit values.

REFERENCES

Y. Bentur, G. Koren, M. Mc Guigan et al. An unusual skin exposure to copper; clinical and pharmacokinetic evaluation. *J. Toxicol. Clin. Toxicol.* 26:371, 1988.

Ph. Grandjean. Skin Penetration. Hazardous Chemicals at Work. A report prepared for the Commission of the European Communities. Luxembourg. Taylor and Francis. London, 1990.

C. Minoïa, E. Sabbioni, P. Apostoli et al. Trace element reference values in tissues from inhabitants of the European Community. I. A study of 46 elements in urine, blood and serum of Italian subjects. *Sci. Total Environ.* 95:89, 1990.

G. Triebig and K. Schaller. Biological indicators for the assessment of human exposure to industrial chemicals: Copper. Eds.: L. Alessio, A. Berlin, M. Boni, R. Roi. Commission of the European Communities. Eur 8903 En, Luxembourg, 1984.

J. Versieck and R. Cornelis. *Trace Elements in Human Plasma or Serum.* CRC Press, Boca Raton, FL, 1989.

J. Versieck. Trace Elements in Human Body Fluids and Tissues. *CRC Crit. Rev. Clin. Lab. Sci.* 22:97, 1985.

11. FLUORIDE

Toxicokinetics

Fluoride is easily absorbed by the pulmonary and the gastrointestinal routes. The skin does not constitute a route of absorption of fluoride except in the case of burns caused by hydrofluoric acid. Fluoride appears rapidly in urine after it has been absorbed (NAS 1971). Approximately 99% of the fluoride in the body is localized in the skeleton and this retained fluoride may be slowly released and add to the levels in blood and urine (WHO 1984). Pharmacokinetic studies realized following oral absorption of sodium fluoride tablets have led to the conclusion that the half-lives of fluoride in plasma and urine are almost identical: 5.78 and 5.11 h, respectively (Ekstrand et al 1983).

Biological monitoring

Urine analysis

The urinary fluoride concentration is an excellent index of fluoride intake (Kaltreider et al 1972; Pantucek 1975). This intake, depending on fluoride contents of air, water, and food, explains the great interindividual variability of fluoride concentration in blood and in urine.

A retrospective study by Dinman et al (1976a) indicates that group postshift urinary fluoride concentrations averaging less than 8 mg/l over a long period is not associated with enhanced risk of bony fluorosis, and the same result appears to apply if preshift urinary fluoride concentrations are less than 4 mg/l. When exposure is mainly by the pulmonary route, a urinary fluoride concentration in a postshift sample of 8 mg/l corresponds to a TWA exposure of 2 mg/m^3 (Dinman et al 1976b).

Brown (1985) could not demonstrate a significant correlation between air levels and changes in urinary fluoride during the shift in workers exposed to hydrofluoric acid in a motor gasoline alkylation unit. In a survey of workers handling fluorine-containing materials, urine fluoride levels where shown to be dependent on the specific type of work (Levi et al 1986).

In workers (n: 41) employed in an aluminium plant in Sweden (average total fluoride exposure: 0.91 mg/m^3 of which 34% was gaseous fluoride) Ehrnebo and Ekstrand (1986) reported a mean fluoride plasma level before the shift of 23 ng/ml (1.2 µM/l) which increased on average to 48 ng/ml (14-151 ng/ml) at the end of the shift. They also observed that a high fluid intake during the shift increased the capacity of the kidney to excrete fluoride and decreased the levels of fluoride in the body.

In workers exposed to fluoride (mainly HF) Hogstedt (1984) observed considerable individual variations in urinary fluoride excretion. However, on the group level, the results showed a correlation between the 8-h mean air concentrations and the average urine concentrations of fluoride in postshift samples (r: 0.70), in 24-h samples (r: 0.68) and in samples collected 16 h afterwards (r: 0.63). Extrapolation from these results suggests that a TWA concentration of 2 mg/m^3 would result in urinary fluoride concentrations of about 10.6 mg/l (postshift sample), 7 mg/24 h, and 8.2 mg/l (after 16 h), respectively.

It should be mentioned that the results of a study, whose aim was to relate fluoride concentration in urine to the actual plasma fluoride concentration after intake of given doses of fluoride, suggest that plasma fluoride levels or urinary excretion rates of fluoride may give a more correct picture of occupational fluoride exposure than fluoride concentrations in urine (Ekstrand et al 1983).

In 1979, NIOSH recommended that preshift and postshift concentrations of fluoride in urine should not exceed 4 and 7 mg/l, respectively. The ACGIH (1991-1992) has adopted as biological exposure indices 3 mg/l urine prior to shift and 10 mg/l urine at the end of the shift. According to the Dutch Expert Committee for Occupational Standards, a TWA exposure to 500 µg/m^3 (as total particulate and gaseous fluoride) results in an increase of urinary fluoride of 1 mg fluoride/l during the workday (group average) (DFG 1989).

Blood analysis

A decrease in sialic acid and an increase in aminoglycans concentrations in serum have been described in subjects suffering from fluorosis (Susheela and Jha 1982).

A risk of fluorosis might exist when the ratio sialic acid/aminoglycans is below 10%.

Hair analysis

Kono et al (1990) have found a linear correlation between the concentration of fluoride in the urine and in hair of HF-exposed workers and have suggested that analysis of hair samples could present a useful monitoring method for workers exposed to HF when external contamination is minimized. This has also been proposed by Czarnowski and Krechniak (1990), who demonstrated, in workers employed in a phosphate fertilizer plant, positive correlations between the group mean concentrations of fluorides in urine and hair (r: 0.11), urine and nails (r: 0.99), and hair and nails (r: 0.70).

   In summary, it is recommended that the postshift minus preshift fluoride concentration in urine does not exceed 3-4 mg/g creatinine.

REFERENCES

ACGIH. American Conference of Governmental Industrial Hygienists. *Threshold Limit Values and Biological Exposure Indices.* Cincinnati, 1991-1992.

M. Brown. Fluoride exposure from hydrofluoric acid in a motor gasoline alkylation unit. *Am. Ind. Hyg. Assoc. J.* 46:662, 1985.

W. Czarnowski and J. Krechniak. Fluoride in the urine, hair, nails of phosphate fertiliser. *Br. J. Ind. Med.* 47:349, 1990.

B. Dinman, D. Backenstose, R. Carter et al. A five-year study of fluoride absorption and excretion. *J. Occup. Med.* 18:17, 1976a.

B. Dinman, W. Boward, T. Bonney et al. Absorption and excretion of fluoride immediately after exposure. *J. Occup. Med.* 18:7, 1976b.

Dutch Expert Committee for Occupational Standards. Directoraat-Generaal of Labour. Health-Based recommended Occupational Exposure Limits for fluorine, hydrogen fluoride and inorganic fluoride compounds. RA 1/89, 1989.

M. Ehrnebo and J. Ekstrand. Occupational fluoride exposure and plasma fluoride levels in man. *Int. Arch. Occup. Environ. Health* 58:179, 1986.

J. Ekstrand, L. Odont and M. Ehrnebo. The relationship between plasma — fluoride urinary excretion rate and urine fluoride concentrations in man. *J. Occup. Med.* 25:745, 1983.

G. Hogstedt. Fluorides. In: *Biological Monitoring and Surveillance of Workers Exposed to Chemicals.* Eds.: A. Aitio, V. Riihimaki, H. Vainio. Hemisphere Publishing Company, Washington, D.C., 1984.

N. Kaltreider, M. Elder, L. Crawlley and M. Colwell. Health survey of aluminum workers with special reference to fluoride exposure. *J. Occup. Med.* 14:531, 1972.

K. Kono, Y. Yoshida, M. Watawabe et al. Elemental analysis of hair among hydrofluoric acid exposed workers. *Int. Arch. Occup. Environ. Health* 62:85, 1990.

S. Levi, L. Zilberman, A. Frumin and M. Frydman. Exposure to fluoride in the chemical industry. *Am. J. Ind. Med.* 6:153, 1986.

NAS. Committee on Biologic Effects of Atmospheric Pollutants. Fluorides, National Academy of Sciences, Washington, D.C., 1971.

NIOSH. National Institute for Occupational Safety and Health. Recommended Standard for Occupational Exposure to Inorganic Fluoride. 657.012/313. U.S. Government Printing Office, Washington, D.C., 1979.

M. Pantucek. Hygienic evaluation of exposure to fluoride fume from basic arc-welding electrodes. *Ann. Occup. Hyg.* 18:207, 1975.

A. Susheela and M. Jha. On the significance of sialic acid and glycosaminoglycans in the serum of fluorosed human subjects. *Fluoride* 15:199, 1982.

WHO. IPCS. International Programme on Chemical Safety. Environmental Health Criteria 36: Fluorine and fluorides. World Health Organization, Geneva, 1984.

12. GERMANIUM

Toxicokinetics

Animal experimental data show that inorganic and organic germanium compounds are rapidly and almost completely absorbed from the lungs and gastrointestinal tract. In humans, ingested germanium is also well absorbed, but there are no data on the absorption rate by the pulmonary route. Absorbed germanium seems to be rapidly excreted mainly in urine (Vouk 1986; Okada et al 1989; Okuda et al 1987).

Biological monitoring

In nonoccupationally exposed subjects, germanium level in urine is below 1 μg/ g creatinine (personal observation). The available data do not permit to propose a biological exposure limit.

REFERENCES

K. Okada, K. Okagawa, K. Kawakami et al. Renal failure caused by long-term use of a germanium preparation as an elixir. *Clin. Nephrol.* 31:219, 1989.

S. Okuda, S. Kiyama, Y. Oh et al. Persistent renal failure dysfunction induced by chronic intake of germanium-containing compounds. *Curr. Ther. Res.* 41:265, 1987.

V. Vouk. Germanium. In: *Handbook on the Toxicology of Metals.* Vol. II. 2nd ed. Eds.: L. Friberg, G. Nordberg, V. Vouk. Elsevier Science Publishers, Amsterdam, 1986.

13. LEAD
13.1 Inorganic Lead

Toxicokinetics

Lead is a cumulative toxin, that is, absorbed by the lungs and the gastrointestinal tract (WHO 1978). Approximately 50% of lead deposited in the lung is absorbed,

whereas usually less than 10% of ingested lead passes into the systemic circulation. Many environmental factors (concentration, particle size, solubility, etc.) and
biological factors (age, sex, iron store, fasting state, etc.) influence the individual
rates of absorption (James et al 1985; Heard et al 1983). Some lead salts of organic
acids (e.g., lead naphthenate) can be slightly absorbed through the skin (Hine et
al 1969; Van Peteghem and DeVos 1974). It has also been found that finely
powdered lead metal or lead nitrate solution placed on skin results in rapid
absorption of lead (Lilley et al 1988).

   In blood, lead is mainly (90%) bound to erythrocytes. Kinetic studies in
man indicate that the lead body burden consists essentially of three compartments (Rabinowitz et al 1976). The first compartment, which includes the
blood mass and some rapidly exchanging tissues, contains about 2 mg of lead
in nonoccupationally exposed persons. It has a half-life of about 35 d. The
second compartment is mainly composed of soft tissues and, in control adults,
contains 0.3 to 0.9 mg of lead. Its biological half-life is about 40 d. The third
compartment (bone) contains about 90% of the total content, has a half-life of
about 20 years, and in nonoccupationally exposed subjects the amount of lead
in this compartment increases throughout life. However, it has been found that
the latter contains at least two different pools of lead with different turnover
rates. Indeed, after cessation of occupational exposure, lead is released faster
from the trabecular bone (e.g. finger, vertebrae) than from cortical bone
(Christoffersson et al 1986; Schütz et al 1987b). Moreover, the finding of an
association between vertebral and blood lead levels in retired workers indicates a possible risk of endogenous lead exposure from the skeletal pool
(Schütz et al 1987a).

   Absorbed lead is excreted mainly through the kidney (75%). Other routes of
excretion are the bile, gastrointestinal secretions, hair, nails and sweat (WHO
1978). As lead concentration in hepatic bile is about 10 times higher than in urine,
it is suggested that the majority of lead excreted into hepatic bile is probably
subjected to reabsorption in the intestine and finally excreted in the urine (Ishihara
and Matsushiro 1986).

Biological monitoring

The literature regarding the biological monitoring of exposure to lead is abundant.
The biological tests can be classified into two groups: those directly reflecting the
exposure and/or the amount stored in soft tissues (lead in blood, in urine, in hair
and in bone, and lead excretion after the administration of a chelating agent) and
those indicating the early biological effects of lead related to the intensity of
exposure (hemoglobin, hematocrit, stippled cells, coproporphyrin in urine, δ-
aminolevulinic acid in urine, porphobilinogen in urine, free erythrocyte protoporphyrin, δ-aminolevulinic acid dehydratase, and pyrimidine-5'-nucleotidase in red
blood cells). The significance, advantages, and limitations of the principal biological tests are discussed below (Lauwerys 1990).

Concentration of lead in biological media as indicator of exposure and/or body burden

Blood analysis

Lead in blood. In a steady-state situation, lead in blood is considered to be the best indicator of the concentration of lead in soft tissues and hence of recent exposure. Lead in blood does not necessarily correlate with the total body burden of lead. Balance studies performed by Kehoe (1961) have demonstrated that, in volunteers receiving various amounts of lead (0.3 to 3 mg/d) in addition to the amount normally present in their diet, the lead level in blood progressively reaches a plateau, although the lead body burden increases continuously during exposure. The level of the plateau, however, is proportional to the administered dose. Other authors have made the same observation.

It has been estimated that under low exposure conditions (environmental exposure) an increase of 1 µg lead/m^3 in air is reflected by an increase of 1 to 2 µg lead/100 ml of whole blood (Chamberlain et al 1975; Coulston et al 1972; Griffin et al 1975; Rabinowitz et al 1974). Such a relationship does not hold when the exposure exceeds 10 to 20 µg lead/m^3 of air, and only applies to well-standardized exposure conditions (e.g., experiments on volunteers) where ingestion of lead is restricted to the amount normally present in diet. In industry, there is usually no satisfactory correlation between lead in air and lead in blood (WHO 1980), mainly because the first parameter is a very crude index of the true exposure of the workers (see Chapter 1) and also because lead being a ubiquitous pollutant, uptake may also result from nonoccupational sources.

When a worker is removed from exposure, lead in blood decreases progressively with a half-time displaying considerable interindividual variation, but it may remain above the normal level because of the progressive release of lead from tissue depots. On the contrary, a worker can exhibit a level of blood lead in the normal range even though disturbances of the heme biosynthesis pathway (e.g., increased free erythrocyte protoporphyrin) are still demonstrable. According to O'Flaherty et al (1982), the average time required to reach a preestablished blood lead level in workers removed from exposure will increase with the length of the workers' previous exposure.

Since the presence of lead in the environment is ubiquitous, the normal level of lead in blood is dependent on the environmental exposure conditions (dietary and drinking habits, residence, etc.). In many countries, the progressive removal of organolead from gasoline and other preventive measures have been associated with a progressive decline of blood lead level of the general population. Presently, the 50 and 95 percentiles of blood lead values in the nonoccupationally exposed populations are usually less than 15 and 30 µg/100 ml, respectively. In male lead workers, blood lead values that do not exceed 60 to 70 µg/100 ml have usually been considered acceptable. Several authors, however, consider it necessary to lower this limit because subclinical effects of lead (reduction in nerve conduction

velocity, psychomotor disturbances, increased prevalence of subjective com-
plaints) can already be detected when lead in blood exceeds 50 µg/100 ml. We
have also some evidence that adult male workers exposed to lead for 10.7 years
on average (1-28.4) and with a mean blood lead level of 46 µg/100 ml (24-75)
have a reduced fertility (Gennart et al 1992).

The World Health Organization has recently proposed 40 µg/100 ml as the
maximal tolerable individual lead in blood value for adult male workers, and 30
µg/100 ml for women of child-bearing age (WHO 1980). The ACGIH (1991-
1992) recommends a lead concentration of 50 µg/100 ml of blood as biological
exposure indice for lead exposure.

Since more than 90% of circulating lead is bound to erythrocytes, some authors
(Grisler and Farina 1969; Van Houte and Cocle 1969; De Silva 1984) have
proposed expression of the results in µg per 100 ml of erythrocytes so that
variations in hematocrit value are taken into consideration. But up to now there
is no general agreement concerning the application of this correction factor. In
fact, blood contains lead in three forms: a major fraction (about 95%) bound to
erythrocytes, a protein-bound fraction in plasma, and a diffusible fraction that
represents the metabolic active form of circulating lead (Cavalleri et al 1978).
Lead in plasma is in equilibrium with lead in erythrocytes (De Silva 1984), but the
ratio of plasma to erythrocyte lead concentrations increases with blood lead
concentration, from 0.2% at a blood lead concentration of 10 µg/100 ml (0.5
µmol/l) up to 2% at a blood lead concentration of 100 µg/100 ml (Manton and
Cook 1984). In view of the difficulty of accurate measurement of very low
concentrations of lead in biological material, the determination of lead in plasma
cannot presently be recommended for the monitoring of lead exposure (De Silva
1984; Ong et al 1986).

Urine analysis

Lead in urine. Lead in urine reflects the amount of lead recently absorbed. Since
the analysis does not require blood sampling, it is sometimes preferred to blood
lead analysis. Although on a group basis there exists a satisfactory correlation
between lead in blood and lead in urine (Williams et al 1969), in the same
individual lead in urine fluctuates with time more than lead in blood. Skerfving
et al (1985) have confirmed the considerable interindividual variation in the
urinary lead excretion at a certain whole blood lead level. The concentration of
lead in urine is usually lower than 50 µg per gram of creatinine. A concentration
of lead in blood of 50 µg/100 ml corresponds to a concentration of lead in urine
of approximately 150 µg/g creatinine. Lead in urine will probably average 50 µg/
g creatinine for a mean blood lead level of 40 µg/100 ml.

The ACGIH (1991-1992) has proposed 150 µg/g creatinine as a biological
exposure indice. However, in view of the poor association between lead in urine
and in blood and the great risk of external contamination during sampling, we do
not recommend lead in urine for the routine assessment of lead exposure.

Urinary excretion of lead after the administration of a chelating agent. The amount of lead excreted in urine after administration of a chelating agent (EDTA, dimercaptosuccinic acid) is considered to reflect the mobilizable pool of lead (Albahary et al 1961; Cramer and Selander 1965; Markowitz and Rosen 1984). Hence, in workers currently exposed to lead, the amount of lead excreted in urine after the administration of a chelating agent should correlate with the lead level in blood. This has been confirmed by several authors (Araki 1975; Alessio et al 1979; Hansen et al 1981). Alessio et al (1979) and Hansen et al (1981) found, in workers who had been exposed to lead for at least 1 year, a correlation coefficient of 0.85-0.87 between lead in blood and the amount of lead excreted in urine within 24 h after the intravenous infusion of $CaNa_2EDTA$ (12.5-25 mg/kg). In the study by Alessio et al (1979), a level of lead in blood of 60 µg/100 ml corresponds to an average urinary excretion of 2.0 mg of lead within 24 h whereas, in the study by Hansen et al (1981), the average 24-h urinary excretion of lead for the same level of lead in blood amounts to about 1.5 mg. Different practical modalities have been proposed for the execution of this test (Gaultier et al 1973; Lahaye et al 1968; Teisinger 1971; Araka and Ushio 1982; Markowitz and Rosen 1984; Schütz et al 1987a,b).

This test may also be useful to confirm past exposure to lead in persons not currently occupationally exposed to this metal, but, in these circumstances, the correlation between chelatable lead and lead in blood is different from that found in currently exposed subjects. For an identical variation in chelatable lead, the variations in blood lead are higher in currently exposed subjects (Alessio et al 1981). In control persons, the urinary excretion of lead within 24 h after administration of 1 g EDTA does not exceed 0.6-0.7 mg (Roche et al 1960).

In summary, an excretion of lead exceeding 600 µg/24 h after EDTA (1 g IV) or dimercaptosuccinic acid (30 mg/kg PO) in subjects removed from lead exposure indicates an increased body burden of lead. The main application of this test is for diagnostic purpose; i.e., the confirmation of an increased body burden in subjects with a history of possible lead exposure and with clinical signs suggestive of lead intoxication, but without clear-cut anomalies of the other biological parameters (see below).

Hair analysis

Determination of lead in hair has been suggested as a method of evaluating the mobilizable pool of lead. A correlation between lead levels in blood and hair has been reported in some studies (Medical Research Council 1985). Grandjean (1978) has proposed to analyze the proximal segment of hair first decontaminated with freon. He found that under steady-state conditions a concentration of 70 µg lead/g of hair corresponds to a level of lead in blood of 60 µg/100 ml. In 1983, Niculescu et al deduced from their investigations a biological limit value of 3 ng lead/cm single hair for a blood value of 40 µg/100 ml and of 4 ng lead/cm for a blood level of 60 µg/100 ml.

Although hair is an easily available material and is a time-integrated index (Grandjean 1984), it has presently limited interest in occupational medicine because whatever the washing procedure selected it is highly difficult to distinguish between lead incorporated into the hair and that simply adsorbed on its surface (Renshaw et al 1972).

Feces analysis

Fecal lead excretion has been used as an index of dietary lead intake (Bruaux and Svartengren 1985). This is without interest in the occupational setting.

Bone analysis

Skeletal lead concentrations have been measured in bone biopsies; this method is of course without interest for routine biological monitoring.

Noninvasive measurements of lead in bone (finger, tibia) have been performed with an X-ray fluorescence technique (Ahlgren et al 1976, 1979, 1980; Wielopolski et al 1981, 1983a,b, 1986; Bloch and Shapiro 1986; Somervaille et al 1986, 1988, 1989; Chettle et al 1989; Chettle 1990; Christoffersson et al 1984, 1986; Landrigan 1991).

A lead urinary excretion of 600 µg/48 h (upper level of lead excretion in normal population) following the administration of Ca EDTA (1 g IV) would correspond to a lead level in bone of 70 µg/g wet weight (Wielopolski et al 1986). Somervaille et al (1986, 1988, 1989) obtained a strong correlation between bone lead measurements and the index of cumulative exposure (time-integrated blood level index).

Let us note that in children the determination of lead concentration in circumpulpal dentine of deciduous teeth has been proposed to estimate exposure to lead (Rabinowitz et al 1991).

Biological effects of lead related to the internal dose

The biological tests that have been proposed for the detection of excessive exposure to lead are mainly based on the interference of lead with several stages of the heme synthesis pathway, illustrated by Figure 7. The enzymes that have been reported to be inhibited by lead are underlined. *In vitro*, lead can also inhibit the enzyme δ-aminolevulinic acid synthetase, but *in vivo* it seems that an induction of the activity of this enzyme is more likely to be produced (Meredith et al 1978). This action could result from a derepression of the enzyme synthesis due to the decreased amount of heme.

Blood analyses

Hemoglobin, hematocrit, stippled cells. It is now well accepted that hemoglobin, hematocrit, and stippled cell determinations do not have much value in the early diagnosis of lead exposure. In a group of workers, it may be possible to detect a slight reduction of the hemoglobin concentration when the level of lead in blood

Krebs cycle
↓
Succinyl CoA + Glycine
Pyridoxal-PO$_4$ ↓ ALA synthetase
δ-Aminolevulinic acid (ALA)
↓   ALA dehydratase
Porphobilinogen (PBG)
↓   PGB deaminase, isomerase
Uroporphyrinogen III
↓   Decarboxylase
Coproporphyrinogen III
↓   Decarboxylase
Protoporphyrinogen III
↓   Oxidase
Fe^{++} + Protoporphyrin IX
↓   Chelatase
Heme

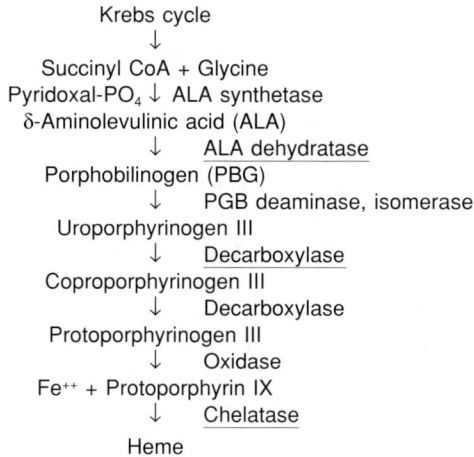

FIGURE 7 Biosynthesis of heme.

exceeds 50 µg/100 ml (Tola et al 1973; Hammond et al 1980). On an individual basis the sensitivity of this test is too low.

The basophilic granulations in red cells are ribonucleic acid aggregates probably resulting from the inhibition of the erythrocyte enzyme pyrimidine-5'-nucleotidase by lead (Paglia et al 1975). Numerous observations have demonstrated that the stippled cell count is not a satisfactory reflection of the amount of lead absorbed. Their increase occurs much later than other biological changes, which are discussed below. Furthermore, many factors (time of blood drying, anticoagulant, etc.) markedly influence the stippled cell count.

Free erythrocyte protoporphyrins. The inhibition of ferro-chelatase (Figure 7) in erythroblasts leads to an accumulation of protoporphyrin IX in red blood cells. Protoporphyrin IX constitutes more than 95% of the noniron-bound porphyrins present in red blood cells (free erythrocyte porphyrins, FEP). In fact, it has been demonstrated that the protoporphyrin that accumulates in the erythrocytes during lead intoxication is not "free", but exists as zinc protoporphyrin (Lamola and Yamane 1974). Since the accumulation of protoporphyrin in erythrocytes results from the action of lead in the bone marrow and since the average life span of the erythrocytes is about 120 d, in newly exposed workers there is a time lag in the increase of erythrocyte protoporphyrin as compared to the rise of lead in blood. Under steady-state conditions, there is, however, a relationship between the concentration of lead in blood and the erythrocyte protoporphyrin content (Alessio et al 1976a,b; Roels et al 1975, 1979; Sassa et al 1973; Tomokuni et al 1975). FEP starts to increase significantly at levels of lead in blood of about 35 µg/100 ml in males and 25 µg/100 ml in females (Roels et al 1979; Roels and Lauwerys 1987). With increasing lead levels in blood, the slope of the increase in FEP is significantly steeper in females than in males (Roels et al 1975, 1979). After the termination of occupational exposure to lead, the erythrocyte protoporphyrin remains elevated out of proportion to current levels of lead in blood. In this case, the protoporphyrin level is better correlated with the amount of chelatable lead excreted in urine than with lead in blood (Alessio et al 1976c).

The FEP concentration measured by an extraction method (Roels et al 1975; Sassa et al 1973) is usually below 75 µg/100 ml red blood cells (mean of about 50) in males and below 85 µg/100 ml red blood cells (mean of about 60) in females (Roels et al 1979). Horiguchi (1990) reports in occupationally nonexposed subjects a geometric mean of FEP concentration of 36.4 µg/100 ml red blood cells (22-72). Under steady-state conditions, a level of lead in blood of 60 µg/100 ml corresponds to an average free protoporphyrin level of around 300 µg/100 ml erythrocytes in males and 350-400 in females. When the mean lead level in blood does not exceed 40 µg/100 ml, the average free protoporphyrin level will probably not exceed 80 µg/100 ml erythrocytes.

Zinc protoporphyrin (ZPP) can also be determined in dilute whole blood by front surface fluorometry (hematofluorometer) (Blumberg et al 1977). Its concentration is usually less than 2.5 µg/g hemoglobin or 40 µg/100 ml whole blood. In a group of 1599 referents, Wildt et al (1987) found a normal mean ZPP value of 25 µg/100 ml whole blood (values rarely exceeding 45 µg/100 ml). In male workers, under steady-state conditions, the corresponding average values for blood lead of 40 and 60 µg/100 ml are 3 µg/g hemoglobin (40 µg/100 ml whole blood) and 12.5 µg/g hemoglobin (200 µg/100 ml whole blood), respectively.

It is important to keep in mind that slight iron deficiency, which is not an uncommon disorder, will also cause an elevated protoporphyrin concentration in red blood cells. When positive, this test should therefore be supplemented by other analyses (e.g., lead in blood, serum ferritin, urinary excretion of lead after administration of a chelating agent) to distinguish between the two conditions.

Values of 100 µg ZPP/100 ml blood or 250 µg/100 ml red blood cells are recommended as biological exposure indices by the ACGIH (1991-1992), these values corresponding to a TWA exposure of 0.15 mg/m^3.

δ-Aminolevulinic acid dehydratase in red blood cells. This enzyme is highly sensitive to inhibition by lead. Within the range of blood lead levels of 5 to 40 µg/100 ml, Hernberg et al (1970) found a close negative correlation between δ-aminolevulinic acid dehydratase and blood lead. The correlation is even better when the enzyme activity is expressed in percent of its activity found in the presence of the reactivator dithiothreitol rather than in absolute values (Lauwerys et al 1978). Although this enzyme can be influenced by substances other than lead (inorganic mercury, zinc, ethanol, carbon monoxide), in practice it can be considered as a highly specific index of lead exposure.

However, at blood levels below 40 µg/100 ml, a poor correlation between blood lead values and δ-aminolevulinic acid dehydratase was found by Boudène et al (1984) and later confirmed by the work of Witting et al (1987). In a group of lead workers with blood lead <40 µg/100 ml the correlation coefficient between blood lead and δ-aminolevulinic acid dehydratase was greater than that between blood lead and pyrimidine-5'-nucleotidase (Ichiba and Tomokuni 1988).

Contrary to the response of FEP, depression of δ-aminolevulinic acid dehydratase following lead exposure occurs with no demonstrable time lag (Tola et al 1973).

Pyrimidine-5'-nucleotidase in erythrocytes. This enzyme is also very sensitive to the inhibitory action of lead. A reduced activity can already be evidenced in persons whose level of lead in blood does not exceed 40 µg/100 ml (Angle and McIntire 1978; Fendler et al 1978; Paglia et al 1975).

Several studies have confirmed the usefulness of erythrocyte pyrimidine-5'-nucleotidase test as a sensitive biological marker of exposure to lead (Mohammed-Brahim et al 1985; Sakai and Ushio 1986; Ichiba and Tomokuni 1988, 1990; Tomokuni et al 1988). Cook et al (1986) reported "normal values" ± SD for pyrimidine-5'-nucleotidase activity as 12.0 µmol uridine/h/g Hb ± 0.71 (n: 14) (mean blood lead ± SD: 7.6 ± 4 µg/100 g), Sakai et al (1988) as 16.2 ± 2.5 (n: 72), Mohammed-Brahim et al (1985) as 8.3 ± 2.3 (n: 100) (mean blood lead ± SD: 14.3 ± 6.7 µg/100 ml), Sato et al (1981) as 13.9 (n: 209) (mean blood lead: 7.34 µg/100 g ± 3.7), Sakai and Ushio (1986) as 17.7 ± 2.95 (n: 35) (mean blood lead ± 6.58 ± 1.6 µg/100 g) and Ichiba and Tomokuni (1990) as 12.6 ± 2.7 (n: 48) (blood lead: 6 ± 2 µg/100 ml); Cook et al (1987) as 12.4 ± 2.4. In practice, pyrimidine-5'-nucleotidase activity is usually higher than 10 µmol uridine/h/g Hb when blood lead does not exceed 10 µg/100 ml.

Urine analysis

Coproporphyrinuria. The urinary excretion of coproporphyrins, mostly the III isomer, starts to increase when the concentration of lead in blood reaches a value of about 40 µg/100 ml (Omae et al 1988; Tola et al 1973). Above this value, it is proportional to the amount of circulating lead. According to Abe and Hori (1987) in an occupationally nonexposed population, the concentration of coproporphyrin I ranges from 22.1 to 57.4 µg/g creatinine, the concentration of coproporphyrin III from 22.1 to 109.1 µg/g creatinine, and the concentration ratio of III/I varies between 0.71 and 3.73. A concentration of coproporphyrin of 100 and 250 µg/g urinary creatinine corresponds to a concentration of lead in blood of about 40 and 60 µg/100 ml, respectively. It is important to recognize the limitations of the coproporphyrinuria test. Firstly, coproporphyrinuria is not specific to lead exposure, since it also occurs in cirrhosis, hepatitis, hemolytic anemia, malign hemopathies, infectious diseases, various intoxications, and even after alcohol consumption. Secondly, it is not only possible to find false positive results, but also false negative ones, i.e., persons overexposed to lead, but without change in urinary coproporphyrins, although this situation is much rarer than the former. If one keeps in mind the preceding limitations, urinary coproporphyrin determination remains an interesting screening method for assessing lead exposure (Williams et al 1968; Omae et al 1988).

Porphobilinogen. The determination of porphobilinogen in urine is not useful for monitoring workers exposed to lead because its increase usually does not precede the development of clinical signs of intoxication (Gibson et al 1968). This parameter is therefore not sensitive enough.

δ-Aminolevulinic acid. Because of the inhibition of δ-aminolevulinic dehydratase by lead, δ-aminolevulinic acid accumulates in tissues and is excreted in greater amounts. When the levels of lead in blood exceed 40 µg/100 ml, there is under steady-state conditions a correlation (r in the range of 0.5 to 0.7) between δ-aminolevulinic acid excretion in urine and the level of lead in blood. A significant increase in urinary δ-aminolevulinic acid excretion occurs at approximately the same concentration of lead in blood as coproporphyrinuria, i.e., 40 µg/100 ml (Lauwerys et al 1974). The specificity of the δ-aminolevulinic acid test is nevertheless better than that of coproporphyrinuria. In control individuals the concentration of δ-aminolevulinic acid in urine does not exceed 4.5 mg/g creatinine. A concentration of 10 mg/g creatinine corresponds on average to a blood lead level of 60 µg/100 ml. When lead in blood level is maintained below 40 µg/100 ml it is likely that δ-aminolevulinic acid concentration in urine will not exceed 5 mg/g creatinine in the majority of workers.

However, by using a new sensitive high-performance liquid chromatography (HPLC) method, Tabuchi et al (1989) concluded that δ-aminolevulinic acid concentration adjusted to urinary creatinine could still be a useful indicator of exposure to lead at low blood lead levels (<40 µg/100 ml). They observed a curvilinear relationship between blood lead and urinary δ-aminolevulinic levels; the latter increasing slowly in the blood lead range from 15 to 60 µg/100 ml and more rapidly above 60 µg/100 ml. Letourneau et al (1988) have suggested 5 mg/g creatinine as a threshold to prescreen workers who should have their blood lead levels measured.

On the basis of a study carried out on battery workers, Wang et al (1985) have proposed as symptomatic (abdominal and joint pain, fatigue, peripheral nerve conduction) threshold values: 30 µg Pb/100 ml blood, 45 µg Pb/l urine, and 40 µg ZPP/100 ml blood. In a further study concerning the validation of screening indicators for early detection of overexposure to lead, Liang and Wang (1991) stated that blood lead at 35 µg/100 ml, urinary lead at 35 µg/l, ZPP at 70 µg/100 ml blood, and erythrocyte protoporphyrin at 45 µg/100 ml have adequate sensitivity and specificity.

The DFG (1991) biological tolerance values are 70 µg Pb/100 ml whole blood and 15 mg δ-aminolevulinic acid/l urine for men, 30 µg Pb/100 ml whole blood and 6 mg δ-aminolevulinic acid/l urine for women <45 years. The ACGIH (1991-1992) BEI are 50 µg Pb/100 ml blood, 150 µg Pb/g creatinine, 250 µg ZPP/100 ml erythrocytes or 100 µg ZPP/100 ml whole blood after 1 month exposure.

The Commission of the European Communities (CEC 1982) directive can be summarized as follows: lead in blood >40 µg/100 ml: information of workers; lead in blood from 40 µg/100 ml to 50 µg/100 ml: periodic biological surveillance; lead in blood >50 µg/100 ml: application of a surveillance program. The limit values of the biological parameters are lead in blood 70 µg/100 ml or 80 µg/100 ml if the urinary δ-aminolevulinic acid level remains lower than 20 mg/g creatinine or the ZPP remains lower than 20 μg/g Hb or the δ-aminolevulinic acid dehydratase remains greater than 6 European units.

It is evident that not all the tests described above can be applied routinely for the monitoring of workers exposed to lead. The choice depends on the internal dose of lead considered as acceptable. In practice, we recommend three tests for the routine surveillance of lead workers: determination of lead in blood, δ-aminolevulinic acid in urine and FEP (or ZPP). The first reflects the absorption of lead, the other two its biological action.

In view of the recent studies on the threshold effect level of lead in adult workers, we propose more stringent guidelines than those currently recommended by the CEC (see Chapter 4, Summary of Recommendations). It has also been recommended that one should prevent any significant increase in lead body burden in child-bearing age women.

13.2 Tetraalkyllead

In some countries, tetraethyllead and tetramethyllead are still used as antiknock agents. They may be absorbed by all the usual routes. Lead in blood is not a good index of exposure to tetraalkyllead, whereas lead in urine seems to be useful for evaluating the risk of overexposure. Signs of poisoning by tetraethyllead are usually associated with urinary lead levels exceeding 200 μg/l (Foa et al 1970).

Linch et al (1970) found a correlation between the airborne concentration of tetraalkyllead (plus inorganic lead) measured with personal air samplers and the level of lead in urine, at least when the weekly average values of both parameters were compared. They found that the current ACGIH TLVs for tetraethyllead and tetramethyllead (0.075 mg/m³ calculated as lead) were exceeded by factors of two and four, respectively, before urinary lead excretion exceeded 0.1 mg/l. Cope et al (1979) also carried out personnel monitoring for alkyllead and inorganic lead in air for 6 weeks on five workers in an alkyl manufacturing plant. No correlation was found between lead in blood, lead in urine, and lead in air values. They are also of the opinion that the determination of lead in urine is more effective for detecting overexposure than personnel air monitoring.

Lauwerys (1990) proposes a maximum permissible concentration of 100 μg/g creatinine.

REFERENCES

K. Abe and T. Hori. Determination of coproporphyrins I and III in urine by high-performance liquid chromatography. *Jpn. J. Clin. Chem.* 16:33, 1987.

ACGIH. American Conference of Governmental Industrial Hygienists. *Threshold Limit Values and Biological Exposure Indices.* Cincinnati, 1991-1992.

L. Ahlgren, K. Liden, S. Mattsson and S. Tejning. X-ray fluorescence analysis of lead in human skeleton in vivo. *Scand. J. Work Environ. Health* 2:82, 1976.

L. Ahlgren and S. Mattsson. An X-ray fluorescence technique for in vivo determination of lead concentration in bone matrix. *Phys. Biol.* 24:136, 1979.

L. Ahlgren, B. Haeger-Aronson, S. Mattsson and A. Schütz. In vivo determination of lead in the skeleton after occupational exposure to lead. *Br. J. Ind. Med.* 37:109, 1980.

G. Albahary, R. Truhaut, C. Boudène and H. Desoille. Le dépistage de l'imprégnation saturnine par un test de mobilisation du plomb. *Presse Méd.* 69:2121, 1961.

L. Alessio, P. Bertazzi, F. Toffoletto and V. Foa. Free erythrocyte protoporphyrin as an indicator of the biological effect of lead in adult males. I. Relationship between FEP and indicators of internal dose of lead. *Int. Arch. Occup. Environ. Health* 37:73, 1976a.

L. Alessio, P. Bertazzi, O. Monelli and V. Foa. FEP as an indicator of the biological effect of lead in adult males. II. Comparison between FEP and other indicators of effect. *Int. Arch. Occup. Environ. Health* 37:89, 1976b.

L. Alessio, P. Bertazzi, O. Monelli and F. Toffoletto. FEP as an indicator of the biological effect of lead in adult males. III. Behavior of FEP in workers with past lead exposure. *Int. Arch. Occup. Environ. Health* 38:77, 1976c.

L. Alessio, M. Castoldi, O. Monelli et al. Indicators of internal dose in current and past exposure to lead. *Int. Arch. Occup. Environ. Health* 44:127, 1979.

L. Alessio, M. Castoldi, P. Odone and I. Franchini. Behavior of indicators of exposure and effect after cessation of occupational exposure to lead. *Br. J. Ind. Med.* 38:262, 1981.

C. Angle and M. McIntire. Low level lead and inhibition of erythrocyte pyridine nucleotidase. *Environ. Res.* 17:296, 1978.

S. Araka and K. Ushio. Assessment of the body burden of chelatable lead: a model and its application to lead workers. *Br. J. Ind. Med.* 39:157, 1982.

S. Araki. Evaluation of lead mobilization test with intravenous infusion of Ca EDTA in workers occupationally exposed to lead. *Ind. Health* 13:179, 1975.

P. Bloch and I. Shapiro. An X-ray fluorescence technique to measure in situ the heavy metal burdens of persons exposed to these elements in the workplace. *J. Occup. Med.* 28:609, 1986.

W. Blumberg, J. Eisinger, A. Lamola and D. Zuckerman. Zinc protoporphyrin level in blood determined by a portable hematofluorometer: a screening device for lead poisoning. *J. Lab. Clin. Med.* 89:712, 1977.

C. Boudène, N. Despaux-Pages, E. Comoy and C. Bohuon. Immunological and enzymatic studies of erythrocytic and delta aminolevulinate dehydratase. *Int. Arch. Occup. Environ. Health* 55:87, 1984.

P. Bruaux and M. Svartengren. Assessment of human exposure to lead: comparison between Belgium, Malta, Mexico and Sweden. Karolinska Institute, Stockholm, 1985.

A. Cavalleri, C. Minoia, L. Pozzoli and A. Baruffini. Determination of plasma lead levels in normal subjects and in lead exposed workers. *Br. J. Ind. Med.* 35:21, 1978.

CEC. Council Directive of 28 July 1982 on the protection of workers from the risks related to exposure to metallic lead and its ionic compounds at work (first individual Directive within the meaning of Article 8 of Directive 80/1107/EEC). Official Journal of the European Communities. L 247. Vol. 25, 1982.

A. Chamberlain, W. Clough, M. Heard et al. Uptake of lead by inhalation of motor exhaust. *Proc. R. Soc. London* B192:77, 1975.

D. Chettle, M. Scott and L. Somervaille. Improvements in the precision of in vivo bone lead measurements. *Phys. Med. Biol.* 34:1295, 1989.

D. Chettle. Photoelectron Bremsstrahlung — Analytical possibilities. *Phys. Med. Biol.* 35:259, 1990.

J. Chisolm. Lead in red blood cells and plasma. *J. Pediatr.* 84:163, 1974.

J. Christoffersson, A. Schütz, L. Ahlgren et al. Lead in finger-bone analysed in vivo in active and retired lead workers. *Am. J. Ind. Med.* 6:447, 1984.

J. Christoffersson, L. Ahlgren, A. Schütz et al. Decrease of skeletal lead in man after the end of occupational exposure. *Arch. Environ. Health* 41:312, 1986.

L. Cook, C. Angle and S. Stohs. Erythrocyte arginase, pyrimidine 5'nucleotidase and deoxypyrimidine 5'nucleotidase as indices of lead exposure. *Br. J. Ind. Med.* 43:387, 1986.

L. Cook, C. Angle, C. Kubitschek and S. Stohs. Prediction of blood lead by HPLC assay of erythrocyte pyrimidine 5'nucleotidase. *J. Anal. Toxicol.* 11:39, 1987.

R. Cope, B. Pancamo, W. Rinehart and G. Ter Haar. Personnel monitoring for tetraalkyllead in the work place. *Am. Ind. Hyg. Assoc. J.* 40:372, 1979.

F. Coulston, L. Goldberg, T. Griffin and J. Russel. The effects of continuous exposure to airborne lead. 2. Exposure of man to particulate lead at a level of 10.9 $\mu g/m^3$. Final Report to the U.S. Environmental Protection Agency, 1972.

K. Cramer and S. Selander. Studies in lead poisoning. Comparison between different laboratory tests. *Br. J. Ind. Med.* 22:311, 1965.

P. De Silva. Blood lead levels and the haematocrit correction. *Ann. Occup. Hyg.* 28:417, 1984.

DFG. Deutsche Forschungsgemeinschaft. Maximum Concentrations at the Workplace and Biological Tolerance Values for Working Materials. Report No. XXVII. Commission for the Investigation of Health Hazards of Chemical Compounds in the Work Area. VCH, Weinheim, 1991.

J. Fendler, G. Dongradi and H. Buc. Le diagnostic infraclinique du saturnisme. *Ann. Med. Int.* 129:473, 1978.

V. Foa, G. Cavagna and M. Manfredi. Valutazione della piomburia nella diagnosi di intossicazione da plombo tetraetile. *Med. Lav.* 61:491, 1970.

M. Gaultier, E. Fournier, P. Gervais et al. Critères diagnostiques du saturnisme dans une consultation de maladies professionnelles. *Arch. Mal. Prof.* 34:613, 1973.

J.-Ph. Gennart, J.-P. Buchet, H. Roels et al. Fertility of male workers exposed to cadmium, lead or manganese. *Am. J. Epidemiol.* 135:1028, 1992.

S. Gibson, J. MacKenzie and A. Goldberg. The diagnosis of industrial lead poisoning. *Br. J. Ind. Med.* 25:40, 1968.

Ph. Grandjean. Lead concentration in single hairs as a monitor of occupational lead exposure. *Int. Arch. Occup. Environ. Health* 42:69, 1978.

Ph. Grandjean. Lead poisoning: hair analysis shows the calendar of events. *Hum. Toxicol.* 3:223, 1984.

T. Griffin, F. Coulston, H. Wills and J. Russel. In: *Environmental Quality and Safety.* Eds.: F. Coulston, K. Korte. Vol. II, Suppl. George Thieme, Stuttgart, p. 221, 1975.

R. Grisler and G. Farina. Validita clinica per la diagnosi di saturnisma di un nuovo indice ematologia: la concentrazione eritrocitaria media del plombo. *Med. Lav.* 60:360, 1969.

P. Hammond, S. Lerner, P. Gartside et al. The relationship of biological indices of lead exposure to the health status of workers in a secondary lead smelter. *J. Occup. Med.* 22:475, 1980.

J. Hansen, M. Dossing, and P. Paulev. Chelatable lead body burden (by calcium-disodium EDTA) and blood lead concentration in man. *J. Occup. Med.* 23:39, 1981.

M. Heard, A. Chamberlain and J. Sherlock. Uptake of lead by humans and effect of minerals and food. *Sci. Total Environ.* 30:245, 1983.

S. Hernberg, J. Nikkanen, G. Mellin and H. Lilius. δ-Aminolevulinic acid dehydrase as a measure of lead exposure. *Arch. Environ. Health* 21:140, 1970.

C. Hine, R. Cavalli and S. Beltran. Percutaneous absorption of lead from industrial lubricants. *J. Occup. Med.* 11:568, 1969.

S. Horiguchi. Biological monitoring of lead. In: *Biological Monitoring of Exposure to Industrial Chemicals. Proceedings of the U.S.A.-Japan Cooperative Seminar on Biological Monitoring.* Eds.: V. Fiserova-Bergerova, M. Ogata. ACGIH, Cincinnati, 1990.

M. Ichiba and K. Tomokuni. Response of erythrocyte pyrimidine 5'nucleotidase activity in workers exposed to lead. *Br. J. Ind. Med.* 45:718, 1988.

M. Ichiba and K. Tomokuni. Studies on erythrocyte pyrimidine 5'-nucleotidase test and its evaluation in workers occupationally exposed to lead. *Int. Arch. Occup. Environ. Health* 62: 305, 1990.

N. Ishihara and T. Matsushiro. Biliary and urinary excretion of metals in humans. *Arch. Environ. Health* 41:324, 1986.

H. James, M. Hilburn and J. Blair. Effects on meals and meal times on uptake of lead from the gastrointestinal tract in humans. *Hum. Toxicol.* 4:401, 1985.

R. Kehoe. The metabolism of lead in man in health and disease. Lecture II. *J. R. Inst. Public Health Hyg.* 24:120, 1961.

D. Lahaye, D. Roosels and R. Verwilghen. Diagnostic sodium calcium edetate mobilization test in ambulant patients. *Br. J. Ind. Med.* 25:148, 1968.

A. Lamola and T. Yamane. Zinc protoporphyrin in the erythrocytes of patients with lead intoxication and iron deficiency anemia. *Science* 186:936, 1974.

Ph. Landrigan. Strategies for epidemiologic studies of lead in bone in occupationally exposed populations. *Environ. Health Perspect.* 91:81, 1991.

R. Lauwerys, J.-P. Buchet, H. Roels and D. Materne. Relationship between urine δ-aminolevulinic acid excretion and the inhibition of red cell δ-aminolevulinic dehydrase by lead. *Clin. Toxicol.* 7:383, 1974.

R. Lauwerys, J.-P. Buchet, H. Roels and G. Hubermont. Placental transfer of lead, mercury, cadmium and carbon monoxide in women. I. Comparison of the frequency distributions of the biological indices in maternal and umbilical cord blood. *Environ. Res.* 15:278, 1978.

R. Lauwerys. *Toxicologie Industrielle et Intoxications Professionnelles.* 3rd ed. Masson, Paris, 1990.

G. Letourneau, R. Plante and J.-Ph. Weber. Blood lead and maximal urinary excretion of δ-aminolevulinic acid. *Am. Ind. Hyg. Assoc. J.* 49:342, 1988.

Y. Liang and Y. Wang. Inorganic lead hazard and its control in small scale industries: exploration of screening indicators. *J. Occup. Med. Singapore* 3:65, 1991.

S. Lilley, T. Florence and J. Stauber. The use of sweat to monitor lead absorption through the skin. *Sci. Total Environ.* 76:267, 1988.

A. Linch, E. Wiest and M. Carter. Evaluation of tetraalkyl lead exposure by personnel monitoring survey. *Am. Ind. Hyg. Assoc. J.* 31:170, 1970.

W. Manton and J. Cook. High accuracy (stable isotope dilution) measurements of lead in serum and cerebrospinal fluid. *Br. J. Ind. Med.* 41:313, 1984.

M. Markowitz and J. Rosen. Assessment of lead stores in children validation of an 8-hour CaNa$_2$EDTA provocative test. *J. Pediatr.* 104:327, 1984.

Medical Research Council. The neuropsychological effects of lead in children. A review of recent research 1979-1983. London, 1985.

P. Meredith, M. Moore, B. Campbell et al. δ-aminolevulinic acid metabolism in normal and lead exposed humans. *Toxicology* 9:1, 1978.

B. Mohammed-Brahim, J.-P. Buchet and R. Lauwerys. Erythrocyte pyrimidine 5′nucleoti-
dase activity in workers exposed to lead, mercury or cadmium. *Int. Arch. Occup.
Environ. Health* 55:247, 1985.

T. Niculescu, R. Dumitru, V. Botha et al. Relationship between the lead concentration in
hair and occupational exposure. *Br. J. Ind. Med.* 40:67, 1983.

E. O'Flaherty, P. Hammond and S. Lerner. Dependence of apparent blood lead half-life on
the length of previous lead exposure in humans. *Fund. Appl. Toxicol.* 2:49, 1982.

K. Omae, H. Sakurai, T. Higashi et al. Reevaluation of urinary excretion of coproporphyrins
in lead-exposed workers. *Int. Arch. Occup. Environ. Health* 60:107, 1988.

C. Ong, W. Phoon, B. Lee et al. Lead in plasma and its relationships to other biological
indicators. *Ann. Occup. Hyg.* 30:219, 1986.

D. Paglia, W. Valentine and J. Dahlgren. Effect of low-level lead exposure on pyrimidine
5′-nucleotidase and other erythrocyte enzymes. *J. Clin. Invest.* 56:1164, 1975.

M. Rabinowitz, G. Wetherhill and J. Kopple. Studies of human lead metabolism by use of
stable isotope tracers. *Environ. Health Perspect.* 7:145, 1974.

M. Rabinowitz, G. Wetherhill and J. Kopple. Kinetic analysis of lead metabolism in
healthy humans. *J. Clin. Invest.* 58:260, 1976.

M. Rabinowitz, J. Wang and W. Soong. Dentine lead and child intelligence in Taiwan.
Arch. Environ. Health 46:351, 1991.

G. Renshaw, C. Pounds and E. Pearson. Variation in lead concentration along single hairs as
measured by non-flame atomic absorption spectrophotometry. *Nature* 238:162, 1972.

L. Roche, A. Badinand and E. Lejeune. Les chélateurs dans le saturnisme. *Arch. Mal. Prof.*
21:1, 1960.

H. Roels, R. Lauwerys, J.-P. Buchet and M. Vrelust. Response of free erythrocyte porphyrin and urinary δ-aminolevulinic acid in men and women moderately exposed to
lead. *Int. Arch. Arbeitsmed.* 34:97, 1975.

H. Roels, M. Balis-Jacques, J.-P. Buchet and R. Lauwerys. The influence of sex and
chelation therapy on erythrocyte protoporphyrin and urinary δ-aminolevulinic acid in
lead exposed workers. *J. Occup. Med.* 21:527, 1979.

H. Roels and R. Lauwerys. Evaluation of dose-effect and dose-response relationships for
lead exposure in different belgian population groups (fetus, child, adult men and
women). *Trace Elem. Med.* 4:80, 1987.

T. Sakai and K. Ushio. A simplified method for determining erythrocyte pyrimidine
5′nucleotidase (P5N) activity by HPLC and its value in monitoring lead exposure. *Br.
J. Ind. Med.* 43:839, 1986.

T. Sakai, T. Araki and K. Ushio. Determination of pyrimidine 5′nucleotidase (P5N) activity
in whole blood as an index of lead exposure. *Br. J. Ind. Med.* 45:420, 1988.

S. Sassa, J. Granick, S. Granick et al. Studies in lead poisoning. 1. Microanalysis of
erythrocyte protoporphyrin levels by spectrofluorometry in the detection of chronic
lead intoxication in the subclinical ranges. *Biochem. Med.* 8:135, 1973.

Y. Sato, T. Sasaki, N. Taniguchi and K. Saito. Normal pyrimidine 5′nucleotidase activity
level of Japanese subjects and a significance as a marker for low concentration of lead
in blood. *Jpn. J. Hyg.* 36:518, 1981.

A. Schütz, S. Skerfving, S. Mattson et al. Lead in vertebral bone biopsies from active and
retired lead workers. *Arch. Environ. Health* 42:340, 1987a.

A. Schütz, S. Skerfving, J. Christoffersson and I. Tell. Chelatable lead versus lead in human
trabecular and compact bone. *Sci. Total Environ.* 61:201, 1987b.

S. Skerfving, L. Ahlgren, J. Christoffersson et al. Metabolism of inorganic lead in man.
Nutr. Res. Suppl. 1:601, 1985.

L. Somervaille, D. Chettle, M. Scott et al. Comparison of two in vitro methods of bone lead analysis and the implications for in vivo measurements. *Phys. Med. Biol.* 31:1267, 1986.

L. Somervaille, D. Chettle, M. Scott et al. In vivo tibia lead measurements as an index of cumulative exposure in occupationally exposed subjects. *Br. J. Ind. Med.* 45:174, 1988.

L. Somervaille, U. Nilsson, D. Chettle et al. In vivo measurements of bone lead. A comparison of two X-ray fluorescence techniques used at three different bone sites. *Phys. Med. Biol.* 34:1833, 1989.

T. Tabuchi, A. Okayama, K. Miyajima et al. A new HPLC fluorimetric method to monitor urinary γ-aminolevulinic acid (Ala-U) levels in workers exposed to lead. *Int. Arch. Occup. Environ. Health* 61:297, 1989.

J. Teisinger. Biochemical responses to provocative chelation by edetate disodium calcium. *Arch. Environ. Health* 23:280, 1971.

S. Tola, S. Hernberg, S. Asp and J. Nikkanen. Parameters indicative of absorption and biological effect in new lead exposure: a prospective study. *Br. J. Ind. Med.* 30:134, 1973.

K. Tomokuni, I. Osaka and M. Ogata. Erythrocyte protoporphyrin test for occupational lead exposure. *Arch. Environ. Health* 30:588, 1975.

K. Tomokuni, M. Ichiba and Y. Hirai. Relationship between inhibition of erythrocyte pyrimidine 5′nucleotidase activity and biological response for porphyrin metabolism in workers occupationally exposed to lead. *Int. Arch. Occup. Environ. Health* 60:431, 1988.

G. Van Houte and P. Cocle. Proposition d'un nouveau test simple plus fidèle de dépistage du saturnisme. *Arch. Mal. Prof.* 30:675, 1969.

Th. Van Peteghem and H. De Vos. Toxicity study of lead naphtenate. *Br. J. Ind. Med.* 31:233, 1974.

Y. Wang, P. Lu, Z. Chen et al. Effects of occupational lead exposure. *Scand. J. Work Environ. Health* 11:20, 1985.

WHO. Task Group on Environmental Health Criteria for Lead. Environmental Health Criteria 3. Lead. World Health Organization, Geneva, 1978.

WHO. World Health Organization. Report of a study group: Recommended Health-based Limits in Occupational Exposure to Heavy Metals. Technical Report Series 647. Geneva, 1980.

L. Wielopolski, D. Slatkin, D. Vartsky et al. Feasibility study for the in vivo L X-rays fluorescence. *IEEE Trans. Nucl. Sci.* 28:114, 1981.

L. Wielopolski, J. Rosen, D. Slatkin et al. Feasibility of non invasive analysis of lead in the human tibia by soft X-ray fluorescence. *Med. Phys.* 10:248, 1983a.

L. Wielopolski, D. Vartsky, S. Yasumura and S. Cohn. Application of XRF to measure strontium in human bone in vivo. *Adv. X-ray Anal.* 26:415, 1983b.

L. Wielopolski, K. Ellis, A. Vaswani et al. In vivo bone lead measurements: a rapid monitoring method for cumulative lead exposure. *Am. J. Ind. Med.* 9:221, 1986.

K. Wildt, M. Berlin and P. Isberg. Monitoring of zinc protoporphyrin levels in blood following occupational lead exposure. *Am. J. Ind. Med.* 12:395, 1987.

M. Williams, E. King and J. Walford. Method for estimating objectively the comparative merits of biological tests of lead exposure. *Br. Med. J.* 1:618, 1968.

M. Williams, E. King and J. Walford. An investigation of lead absorption in an electric accumulator factory with the use of personal samplers. *Br. J. Ind. Med.* 26:202, 1969.

U. Witting, N. Binding and G. Müller. Evaluation of a new specific analysis of urinary δ-aminolevulinic acid in man. *Int. Arch. Occup. Environ. Health* 59:375, 1987.

14. MANGANESE

Toxicokinetics

In industry, workers absorb manganese mainly through the lungs; the gastrointestinal absorption is low (3% on the average); it is controlled by homeostatic mechanisms and reduced by concomitant ingestion of calcium (Freeland-Graves and Lin 1991). In blood, manganese is mainly present in red blood cells, in which the manganese concentration is about 25-fold higher than in serum (Versieck et al 1974a). Excretion occurs mainly through the bile. Very small amounts are eliminated with hair, perspiration, and urine.

Biological monitoring

The normal concentration of manganese in urine is usually less than 3 µg/l (Buchet et al 1976; Watanabe et al 1978). Minoïa et al (1990) suggest a mean reference value of 1.02 µg/l (n: 777; range: 0.12-1.9) and Järvisalo et al (1992) 0.3 µg/l. For Oberdoerster and Cherian (1988) the normal value is less than 1 µg/l. Normally, the concentration of manganese in whole blood and in plasma is less than 1 µg/100 ml (Lauwerys 1990) and 0.1 µg/100 ml (Versieck et al 1974a,b), respectively.

The mean blood and serum reference values reported by Minoïa et al (1990) are 0.88 µg/100 ml (n: 88; range: 0.71-1.05) and 0.06 µg/100 ml (n: 414; range: 0.03-0.09), respectively. This is in agreement with the normal concentration of less than 0.05 µg/100 ml plasma or serum reported by Versieck and Cornelis (1989), the mean manganese concentrations of 0.059 µg/100 ml serum and 1.04 µg/100 ml blood found by Nève and Leclercq (1991) and Järvisalo et al (1992), respectively. The values reported by Oberdoerster and Cherian (1988) also range between 0.7-1.2 µg/100 ml blood.

In the hair, normal concentrations based on old studies have been reported to be below 4 µg/g (Oberdoerster and Cherian 1988; EPA 1984), but, using a more sensitive method, Guillard et al (1984) found mean (±SE) normal values of 0.26 ± 0.05 µg/g hair.

Urine analysis

Manganese in urine. Although the excretion in urine is low (about 1% of the absorbed dose), the determination of manganese in urine has been proposed for estimating recent exposure. Tanaka and Lieben (1969) found that, on a group basis, the urinary manganese concentration of exposed workers shows a rough correlation to the average air concentration. Similar results were obtained by Roels et al (see below). Smyth et al (1973) only found a slight correlation between the airborne concentration of manganese and its concentration in urine of exposed workers.

Urinary excretion of manganese after the administration of a chelating agent. Since EDTA has been shown to mobilize manganese and to increase its urinary excretion, analysis of urine after EDTA administration might be useful to confirm the existence of an increased manganese body burden. Petkova and Kostova (1992) have proposed 60 µg/l as a reference value. Further work is required to confirm the validity of this test (Lauwerys 1975).

Blood analysis

Smyth et al (1973) and Tsalev et al (1977) found no significant correlation between manganese exposure and the blood manganese level in workers, whereas Ulrich et al (1979) reported a relationship between both parameters in monkeys and rats exposed by inhalation to manganese oxide aerosol. In view of the efficient homeostatic mechanisms controlling the metabolism of manganese and its extremely short half-life in the blood compartment, blood concentration changes very little with exposure. Following Oberdoerster and Cherian (1988), one would expect that blood levels will be increased during continuous exposure, but decreased rapidly after termination of exposure.

Feces analysis

The determination of manganese in stool has also been recommended as a group test for the evaluation of the level of occupational exposure to manganese (Jindrichova 1969).

Hair analysis

Attempts have also been made to measure the concentration of manganese in hair of exposed workers, but the studies do not permit any final judgment as to the value of this material as a measure of increased manganese exposure (Valentin and Schiele 1983). The problem of external contamination must be taken into account. Stauber et al (1987) have found no correlation between individual hair levels and the severity of neurological effects in manganese-exposed persons.

Two cross-sectional epidemiological studies undertaken among workers exposed to inorganic manganese dust in a manganese oxide- and salt-producing plant (Roels et al 1987a,b), and workers exposed to manganese oxide dust in a dry alkaline battery plant (Roels et al 1992) suggest that, on an individual basis, neither manganese in blood nor manganese in urine are correlated to external exposure parameters (duration of exposure, current exposure, lifetime-integrated exposure to airborne manganese). There is no relationship between blood and urine manganese. On a group basis, manganese in urine seems partly influenced by recent exposure, and the authors suggest that its periodic determination in a group of workers may detect changes in environmental pollution and/or a time trend in the risk of overexposure.

The study of Järvisalo et al (1992) on manual metal arc welders of mild steel also indicates that the measurement of manganese in urine or blood may be used for monitoring manganese exposure at the group level only.

In summary, the possibilities for monitoring exposure to manganese by a biological method are still very limited. Most authors have concluded that there is no direct relation between manganese concentration in biological material and the severity of chronic manganese poisoning, probably because individual susceptibility to the disease plays an important role (Valentin and Schiele 1983). The measurement of manganese in urine can probably be recommended to confirm the absorption of manganese, but no biological threshold limit value can yet be proposed.

REFERENCES

J.-P. Buchet, R. Lauwerys and H. Roels. Determination of manganese in blood and urine by flameless atomic absorption spectrophotometry. *Clin. Chim. Acta* 73:481, 1976.

EPA. U.S. Environmental Protection Agency. Health Assessment Document for Manganese. Final Report. EPA 600/8-83-013F, 1984.

J. Freeland-Graves and P. Lin. Plasma uptake of manganese as affected by oral loads of manganese, calcium, milk, phosphorus, copper and zinc. *J. Am. Coll. Nutr.* 10:38, 1991.

O. Guillard, J. Brugier, A. Piriou et al. Improved determination of manganese in hair by use of a mini autoclave and flameless atomic absorption spectrometry with zeeman background correction: an evaluation in unexposed subjects. *Clin. Chem.* 30:1642, 1984.

J. Järvisalo, M. Olkinuora, M. Kiilunen et al. Urinary and blood manganese in occupationally nonexposed populations and in manual metal arc welders of mild steel. *Int. Arch. Occup. Environ. Health* 63:495, 1992.

J. Jindrichova. Anwendungsmoeglichkeiten der Mangan Bestimmung im Stuhl als Expositionstest. *Int. Arch. Gewerbepathol. Gewerbehyg.* 25:347, 1969.

R. Lauwerys. Biological criteria for selected toxic chemicals: a review. *Scand. J. Work Environ. Health* 1:139, 1975.

R. Lauwerys. *Toxicologie Industrielle et Intoxications Professionnelles.* 3rd ed. Masson, Paris, 1990.

C. Minoïa, E. Sabbioni, P. Apostoli et al. Trace element reference values in tissues from inhabitants of the European Community. I. A study of 46 elements in urine, blood and serum of Italian subjects. *Sci. Total Environ.* 95:89, 1990.

J. Nève and N. Leclercq. Factors affecting determinations of manganese in serum by atomic absorption spectrometry. *Clin. Chem.* 37:723, 1991.

G. Oberdoerster and G. Cherian. Manganese. In: *Biological Monitoring of Toxic Metals. Rochester Series on Environmental Toxicity.* Eds.: Th. Clarkson, L. Friberg, G. Nordberg, P. Sager. Plenum Press, New York, 1988.

V. Petkova and V. Kostova. Syndrome parkinsonien, traduction d'intoxications professionnelles chroniques. *Arch. Mal. Prof.* 53:99, 1992.

H. Roels, R. Lauwerys, P. Genet et al. Relationship between external and internal parameters of exposure to manganese in workers from a manganese oxide and salt producing plant. *Am. J. Ind. Med.* 11:297, 1987a.

H. Roels, R. Lauwerys, J.-P. Buchet et al. Epidemiological survey among workers exposed to manganese: effects on lung, central nervous system, and some biological indices. *Am. J. Ind. Med.* 11:307, 1987b.

H. Roels, P. Ghyselen, J.-P. Buchet et al. Assessment of the permissible exposure level to manganese in workers exposed to manganese dioxide dust. *Br. J. Ind. Med.* 49:25, 1992.

L. Smyth, R. Ruhe, N. Whitman and T. Dugan. Clinical manganism and exposure to manganese in the production and processing of ferromanganese alloy. *J. Occup. Med.* 15:101, 1973.

J. Stauber, T. Florence and W. Webster. The use of scalp hair to monitor manganese in aborigines from Groote Eylandt. *Neurotoxicology* 8:431, 1987.

S. Tanaka and J. Lieben. Manganese poisoning and exposure in Pennsylvania. *Arch. Environ. Health* 19:674, 1969.

D. Tsalev, J. Langmyhr and N. Gunderson. Direct atomic absorption spectrometric determination of manganese in whole blood of unexposed individuals and exposed workers in Norwegian manganese alloy plant. *Bull. Environ. Contam. Toxicol.* 17:660, 1977.

C. Ulrich, W. Rinehart and M. Brandt. Evaluation of the chronic inhalation toxicity of a manganese oxide aerosol. III. Pulmonary function, electromyograms, limb tremor and tissue manganese data. *Am. Ind. Hyg. Assoc. J.* 40:349, 1979.

H. Valentin and R. Schiele. *Human Biological Monitoring of Industrial Chemicals: Manganese.* Eds.: L. Alessio, A. Berlin, R. Roi, M. Boni. Commission of the European Communities. Eur 8476 En, Luxembourg, 1983.

J. Versieck, F. Barbier, A. Speecke and J. Hoste. Manganese, copper and zinc concentrations in serum and packed blood cells during acute hepatitis, chronic hepatitis and posthepatic cirrhosis. *Clin. Chem.* 20:1141, 1974a.

J. Versieck, F. Barbier, A. Speecke and J. Hoste. Normal manganese concentration in human serum. *Acta Endocrinol.* 76:783, 1974b.

J. Versieck and R. Cornelis. *Trace Elements in Human Plasma or Serum.* CRC Press, Boca Raton, FL, 1989.

T. Watanabe, R. Tokunaga, T. Iwahana et al. Determination of urinary manganese by the direct chelation-extraction method and flameless atomic absorption spectrophotometry. *Br. J. Ind. Med.* 35:73, 1978.

15. MERCURY
15.1 Metallic Mercury and Its Inorganic Salts

Toxicokinetics

Inhalation represents the main route of uptake of metallic mercury in industry. Approximately 80% of the amount of mercury inhaled as a vapor is absorbed at the alveolar level. The gastrointestinal absorption of metallic mercury is negligible. Inorganic mercury salts can be absorbed through the lungs (inhalation of inorganic mercury aerosol) and the gastrointestinal tract (accidental or intentional intake). The cutaneous absorption of metallic mercury and its inorganic salts is possible in industry, but its importance is limited by comparison with the pulmonary route.

In blood, inorganic mercury is equally distributed between plasma and red blood cells. The principal sites of deposition of mercury are the kidney and the brain after exposure to mercury vapor and the kidney after exposure to inorganic mercury salts. Inorganic mercury is excreted mainly via the feces and urine. The urinary route dominates when exposure is high. In urine, also, a very small percentage of mercury (<0.1%) is excreted as elemental mercury (Yoshida 1985). Small quantities are also excreted through salivary, lacrimal, and sweat glands. However, in the case of heavy perspiration, the amount of mercury eliminated in the sweat may be important (Lovejoy et al 1973). Mercury can also be detected in the expired air during the few hours following exposure to mercury vapor. After exposure to mercury vapor, the decrease in blood mercury can be described by two half-times; one of 2-4 days, accounting for about 90% of the absorbed mercury, and another of 15-30 days, accounting for most of the remainder; the existence of a third compartment with a still longer half-time, however, cannot be excluded (Clarkson et al 1988). After 3 days of high exposure (>100 $\mu g/m^3$) to metallic mercury vapor, Barregard (1991) observed a first phase of blood mercury decrease with a half-time of 3.1 days and a slower decrease with a half-time of 18 days. Peak mercury in urine was not seen until 2-3 weeks after exposure. A mean urinary half-time of 50 days (median 40 days) was thereafter observed. Excretion in feces is initially higher than in urine, but seems to be lower after about 40 days. The biological half-time of mercury is on the order of 2 months in kidney, but is much longer in the central nervous system (for a review on mercury metabolism see WHO 1991).

Biological monitoring

Several studies have demonstrated that on a group basis there is a correlation between the intensity of recent exposure to mercury vapor and the concentration of mercury in blood, urine, and saliva. Such a relationship holds only when exposure has lasted for at least 1 year. Under chronic exposure conditions there is also a relationship between blood or urinary concentration of mercury and the occurrence of clinical and biological signs of intoxication by mercury vapor.

In persons nonoccupationally exposed to mercury, the concentration of mercury in urine is usually less than 5 $\mu g/g$ creatinine. Minoïa et al (1990) found a mean value of 3.5 $\mu g/l$ (range: 0.1-6.9) in the urine of 380 healthy Italians. Since mercury in fish is mainly methylmercury, which is not excreted through the kidney, the dietary habits may well influence the concentration of mercury in blood, but not in urine. It should, however, be pointed out that dental restorations (amalgam fillings) and the use of local disinfectants containing mercury will transiently increase the excretion of mercury through urine (Desbaumes 1973; Lauwerys et al 1977; Zander et al 1990; Langworth et al 1991). In control persons, the concentration of mercury in blood is probably below 1 $\mu g/100$ ml. Minoïa et al (1990) observed mean blood and serum values of 0.53 $\mu g/100$ ml (n: 368; range: 0.17-0.99) and 0.21 $\mu g/100$ ml (n: 349; range: 0.06-0.38), respectively.

On the basis of a critical review of the literature, Brune et al (1991) have established tentative mean reference values according to the amount of fish consumed. Category I (no fish consumption): 0.2 $\mu g/100$ ml whole blood (SD: 0.18; n: 223) or 0.13 $\mu g/100$ ml plasma (SD: 0.08; n: 24); category II (<2 fish meals/week): 0.48 $\mu g/100$ ml whole blood (SD: 0.19; n: 339) and 0.25 $\mu g/100$ ml plasma (SD: 0.14; n: 131); category III (≥2-4 fish meals/week): 0.84 µg/100 ml whole blood (SD: 0.45; n: 658); category IV (>4 fish meals/week): 4.44 µg/100 ml whole blood (SD: 2.99; n: 613); category V (unknown fish consumption): 0.58 µg/100 ml whole blood (SD: 0.36; n: 3182) and 0.21 µg/100 ml plasma (SD: 0.11; n: 370); category III-IV: 0.47 µg/100 ml plasma (SD: 0.26; n: 118).

Urine analysis

In newly exposed workers, urinary excretion does not immediately follow the onset of exposure; there is a latent period while the body accumulates a certain quantity of mercury, mainly in the kidney (Kobal and Stegner cited by Clarkson et al 1988). It is often stated that the determination of mercury concentration in urine can only be used to estimate the exposure of a group of workers because the great variability of urinary excretion of mercury limits the usefulness of this index in individual cases (Berlin 1979). Piotrowski et al (1975), however, have demonstrated that the urinary mercury excretion in a given subject may be assessed precisely enough provided the urine sampling is always performed at the same time, the concentration is corrected for specific gravity, the subject has been exposed for a sufficiently long time to reach steady-state, and has not been off work for several days. It is also possible that the individual variability in mercury excretion is a true reflection of the variability of the external micro-environment of the workers, which is not reflected by measuring the airborne concentration of mercury in the general work area (Stopford et al 1978). Indeed, studies in which external exposure to mercury vapor had been monitored with personal samplers indicate that breathing zone samples may average severalfold higher in concentration than concurrent area samples. Furthermore, this microenvironmental exposure to mercury vapor, presumably from contaminated clothing and hands, may continue after cessation of work. Authors who used personal samplers to monitor environmental exposure to mercury vapor found that the ratio between mercury in air (in µg/m³) and mercury in urine (µg/l) is closer to 1:1 (Bell et al 1973) or 1:1.6 (Lindstedt et al 1979) or 1 (µg/m³): 1.22 (µg/g creatinine) (Roels et al 1987) than to 1:2.5, a ratio found when the airborne concentration of mercury was measured with static samplers (Smith et al 1970).

It should be noted that with static air sampling, Mattiussi et al (1982) also found a ratio of 1:1.3; a mean urinary value of 0.065 mg/l corresponding to a mean airborne mercury concentration of 0.05 mg/m³.

It is interesting to note that interruption of the daily rhythm of exposure on weekends should not lead to a marked deviation from the excretory "plateau" of the steady state (Piotrowski et al 1975).

A correlation has been demonstrated between the urinary excretion of mercury, the prevalence of neuropsychomotor effects (Langolf et al 1978; Miller et al 1975; Roels et al 1982) and the prevalence of increased excretion of urinary proteins (Buchet et al 1980). Clinical signs of poisoning usually do not occur when the concentration of mercury in urine is kept below 300 µg/l. According to Langolf et al (1978), there is no change in tremor, electromyography (EMG), and psycho-motor tests when urinary mercury does not exceed 500 µg/l in the previous years. Both urinary *N*-acetyl β-D-glucosaminidase (NAG) levels and the prevalence of neuropsychological symptoms were greater among workers with mercury levels above 100 µg/l (Rosenman et al 1986). However, other authors have reported an increased prevalence of slight tremor and of biological signs of renal dysfunction in workers excreting more than 50 µg Hg/l of urine (standardized for a urinary creatinine of 1 g/l) (Buchet et al 1980; Roels et al 1982, 1985; Gennart et al 1986). Other data seem to indicate that exposure to metallic mercury <50 µg/m³ which leads to a urinary excretion below 20 µmol Hg/mol creatinine may still increase finger tremor (Verberk et al 1986). The value of 50 µg Hg/g creatinine has been proposed as a biological threshold limit value for chronic exposure to mercury vapor (Buchet et al 1980) and has been endorsed by WHO (1980). It should, however, be stressed that on the basis of the results of a survey on the effects of low exposure to inorganic mercury on psychological performance, Soleo et al (1990) have suggested that the TLV-TWA for mercury should be lowered to 0.025 mg/m³ and that the biological urinary exposure indicator for biological monitoring should be 25 µg/l.

It should be noted that this threshold effect level of mercury in urine only applies during the period of exposure. As pointed out by Foa and Bertelli (1986), if mercury in urine can be considered as an exposure indicator, it cannot be used to assess the dose in the brain because the average biological half-life of mercury in brain is longer than in other organs, which implies that high quantities can remain in the brain after cessation of exposure.

Yamamura (1990) considers inorganic mercury concentrations in urine above 50 µg/l as a sign of increased absorption and above 100 µg/l as a warning level. Lauwerys (1990) estimates that a urinary concentration above 50 µg/l is a warning sign and a value above 100 µg/l justifies removal from exposure.

According to WHO (1991), when exposure is above 80 µg/m³ corresponding to a urinary mercury level of 100 µg/g creatinine, the probability of developing neurological signs and proteinuria is high. Exposure in the range of 25 to 80 µg/m³ corresponding to a level of 30 to 100 µg/g creatinine increases the risk of certain less severe toxic effects that do not lead to overt clinical impairment (psychomotor performances, tremor, fatigue, irritability).

Urinary excretion of mercury after the administration of a chelating agent. In view of the results of animal experiments it has been suggested that meso-2,3-dimercaptosuccinic acid (DMSA), a less toxic analog of British anti-lewisite which has been shown to be effective as an antidote for intoxication by heavy metal (Aposhian 1983), chiefly removes mercury from its main peripheral site of deposition: the kidney (Buchet and Lauwerys 1989).

Data on workers seem to confirm that after a few days of cessation of occupational exposure to mercury vapor the difference between the concentration of mercury in urine before and after the administration of DMSA mainly reflects the amount of mercury stored in the kidney (Roels et al 1991).

Blood analysis

The concentration of mercury in blood may be influenced by dietary habits, in particular by the consumption of fish containing methylmercury (see above). Blood concentration of mercury is mainly influenced by exposure to mercury during recent days. In workers exposed to mercury vapor, and who are not consuming fish regularly, there is a significant correlation between mercury in blood and mercury in urine (Buchet et al 1980; Smith et al 1970; Yoshida 1985; Brune et al 1991 (see above)).

Suspicion of temporary high exposure to mercury vapor can be confirmed by immediate blood sampling, whereas urinary mercury level seems less informative (Barregard 1991).

Lindstedt et al (1979) also found a good correlation between mercury in blood and mercury in air, even better than that between urine mercury and mercury in air. According to these authors, an average airborne mercury concentration of 50 $\mu g/m^3$ corresponds to a mercury concentration in blood of about 3-3.5 $\mu g/100$ ml. Buchet et al (1980) have found that, in workers from chloralkali plants, the likelihood of increased urinary excretion of high-molecular weight proteins and of some enzymes (e.g., β-galactosidase) is greater when the mercury concentration in blood exceeds 3 $\mu g/100$ ml.

According to Schaller and Triebig (1984), a mean atmospheric concentration of mercury vapor of 0.05 mg/m^3 and 0.1 mg/m^3 (personal samplers) corresponds to a blood concentration of 2 $\mu g/100$ ml and 5 $\mu g/100$ ml and a urinary concentration of 50 $\mu g/g$ creatinine and 150 $\mu g/g$ creatinine, respectively. Under steady-state exposure conditions, strong correlations were observed by Roels and co-workers (1987), between the daily intensity of exposure to mercury vapor and the daily level of mercury in blood (end of shift) and in urine (following morning); an atmospheric concentration of 1 ($\mu g/m^3$) corresponding to a level in blood of 0.45 ($\mu g/l$ whole blood) and in urine of 1.22 ($\mu g/g$ creatinine). Since a biological TLV of 50 $\mu g/g$ creatinine has been proposed for mercury in urine, this implies corresponding threshold values of 1.8 μg mercury/100 ml whole blood and 40 μg mercury/m^3 air.

The biological exposure indices recommended by the ACGIH (1991-1992) are 35 μg mercury/g creatinine (urine specimens collected preferably in the morning, prior to the shift at the end of the workweek) and 1.5 $\mu g/100$ ml blood (blood specimens collected at the end of the shift at the end of the workweek). The biological tolerance values established by the DFG (1991) are 5 $\mu g/100$ ml blood and 200 $\mu g/l$ urine.

Saliva analysis

Joselow et al (1968, 1969) have found a good correlation between mercury in saliva and mercury in blood or in urine. They have established the relationship of the mercury concentration of parotid fluid to that of blood and urine to be about 1:4 and 1:10, respectively. Data are, however, too limited to propose a biological threshold limit value for this parameter.

Hair analysis

In view of the risk of external contamination by mercury, hair does not seem to be a satisfactory biological material for evaluating exposure to mercury vapor.

15.2 Organic Mercury Compounds

The organic mercury compounds are easily absorbed by all the usual routes. In blood, they will be found mainly in red blood cells (around 90%). A distinction must be made, however, between the short-chain alkyl compounds (mainly methylmercury), which are very stable and are resistant to biotransformation, and the aryl or alkoxyalkyl derivatives, which more easily liberate inorganic mercury *in vivo*. For the latter compounds, the concentration of mercury in blood as well as in urine is probably indicative of the exposure intensity.

Methylmercury is mainly excreted through feces and determination of mercury in urine has no practical value. It has been demonstrated that, under steady-state conditions, mercury in whole blood and in hair (Bakir et al 1973; Berlin 1979) correlates with methylmercury body burden and with the risk of signs of methylmercury poisoning. Since contaminated fish containing methylmercury probably represents the major source of methylmercury in food, it is not surprising that in the general population a correlation has been found between fish consumption and mercury hair concentration (Airey 1983; Renzoni 1987). In practice, hair analysis is mainly useful for the biological monitoring in subjects suspected of exposure to methylmercury through seafood (WHO 1990). It has been estimated that in persons chronically exposed to alkylmercury, the earliest signs of intoxication (paresthesia, sensory disturbances) may occur when the level of mercury in blood and in hair exceeds 200 µg/l and 50 µg/g, respectively (WHO 1976). It has therefore been suggested that, in persons exposed to alkylmercury, the concentration of mercury in whole blood should not exceed 10 µg/100 ml.

REFERENCES

ACGIH. American Conference of Governmental Industrial Hygienists. *Threshold Limit Values and Biological Exposure Indices*. Cincinnati, 1991-1992.

D. Airey. Total mercury concentrations in human hair from 13 countries in relation to fish consumption and location. *Sci. Total Environ*. 31:157, 1983.

H. Aposhian. DMSA and DMPS — water soluble antidotes for heavy metal poisoning. *Annu. Rev. Pharmacol. Toxicol*. 23:193, 1983.

F. Bakir, S. Damluji, L. Amin-aki et al. Methylmercury poisoning in Iraq. An interuniversity report. *Science* 181:230, 1973.

L. Barregard. Occupational exposure to inorganic mercury in chloralkali workers. Studies on metabolism and health effect. Thesis: University of Göteborg. Department of Occupational Medicine. Sweden, 1991.

Z. Bell, H. Lovejoy and T. Vizena. Mercury exposure evaluations and their correlation with urine mercury excretion. 3. Time-weighted average (TWA) mercury exposure and urine mercury levels. *J. Occup. Med.* 15:501, 1973.

M. Berlin. Mercury. In: *Handbook on the Toxicology of Metals.* Vol. II. 2nd ed. Eds.: L. Friberg, G. Nordberg, V.B. Vouk. Elsevier Science Publishers. Amsterdam, 1979.

D. Brune, G. Nordberg, O. Versterberg et al. A review of normal concentration of mercury in human blood. *Sci. Total Environ.* 100:235, 1991.

J.-P. Buchet, H. Roels, A. Bernard and R. Lauwerys. Assessment of renal function of workers exposed to inorganic lead, cadmium or mercury vapor. *J. Occup. Med.* 22:741, 1980.

J.-P. Buchet and R. Lauwerys. Influence of 2,3-dimercaptopropane-1-sulfonate and dimercaptosuccinic acid on the mobilization of mercury from tissues of rats pretreated with mercuric chloride, phenylmercury acetate or mercury vapors. *Toxicology* 54:323, 1989.

Th. Clarkson, J. Hursh, P. Sager and T. Syversen. Mercury. In: *Biological Monitoring of Toxic Metals. Rochester Series on Environmental Toxicity.* Eds.: Th. Clarkson, L. Friberg, G. Nordberg, P. Sager. Plenum Press, New York, 1988.

P. Desbaumes. Pseudo-hydrargyrisme dû à certains médicaments (enquêtes toxicologiques parmi le personnel exposé et erreur de diagnostic). *Arch. Mal. Prof.* 34:372, 1973.

DFG. Deutsche Forschungsgemeinschaft. Maximum Concentrations at the Workplace and Biological Tolerance Values for Working Materials. Report No. XXVII. Commission for the Investigation of Health Hazards of Chemical Compounds in the Work Area. VCH, Weinheim, 1991.

V. Foa and G. Bertelli. Biological indicators for the assessment of human exposure to industrial chemicals: mercury. Eds.: L. Alessio, A. Berlin, M. Boni, R. Roi. Commission of the European Communities, Luxembourg. Eur 10704 En, 1986.

J.-Ph. Gennart, H. Roels, J.-P. Buchet et al. Synthèse de trois études épidémiologiques parmi des travailleurs exposés aux vapeurs de mercure. *LARC Médical* 6:317, 1986.

M. Joselow, R. Ruiz and L. Goldwater. Absorption and excretion of mercury in man. XIV. Salivary excretion of mercury and its relationship to blood and urine mercury. *Arch. Environ. Health* 17:35, 1968.

M. Joselow, R. Ruiz and L. Goldwater. The use of salivary (parotid) fluid in biochemical monitoring. *Am. Ind. Hyg. Assoc. J.* 30:77, 1969.

G. Langolf, D. Chaffin, R. Henderson and H. Whittle. Evaluation of workers exposed to elemental mercury using quantitative tests of tremor and neuromuscular functions. *Am. Ind. Hyg. Assoc. J.* 39:976, 1978.

S. Langworth, C.-G. Elinder, C. Göthe and O. Vesterberg. Biological monitoring of environmental and occupational exposure to mercury. *Int. Arch. Occup. Environ. Health* 63:161, 1991.

R. Lauwerys, H. Roels, J.-P. Buchet and A. Bernard. Non-job related increased urinary excretion of mercury. *Int. Arch. Occup. Environ. Health* 39:33, 1977.

R. Lauwerys. *Toxicologie Industrielle et Intoxications Professionnelles.* 3rd ed. Masson, Paris, 1990.

G. Lindstedt, I. Gottberg, B. Holmgren et al. Individual mercury exposure of chloralkali workers and its relation to blood and urinary mercury levels. *Scand. J. Work Environ. Health* 5:59, 1979.

H. Lovejoy, Z. Bell and T. Vizena. Mercury exposure evaluations and their correlation with urine mercury excretion. 4. Elimination of mercury by sweating. *J. Occup. Med.* 15:590, 1973.

R. Mattiussi, G. Armelli and V. Bareggi. Statistical study of the correlation between mercury exposure (TWA) and urinary concentrations in chloralkali workers. *Am. J. Ind. Med.* 3:335, 1982.

J. Miller, D. Chaffin and R. Smith. Subclinical psychomotor and neuromuscular changes in workers exposed to inorganic mercury. *Am. Ind. Hyg. Assoc. J.* 36:725, 1975.

C. Minoïa, E. Sabbioni, P. Apostoli et al. Trace element reference values in tissues from inhabitants of the European Community. I. A study of 46 elements in urine, blood and serum of Italian subjects. *Sci. Total Environ.* 95:89, 1990.

J. Piotrowski, B. Trojanowska and E. Mogilnicka. Excretion kinetics and variability of urinary mercury in workers exposed to mercury vapour. *Int. Arch. Occup. Environ. Health* 35:245, 1975.

A. Renzoni. Mercury levels in human hair and their relevance to health. In: *International Conference "Heavy Metals in the Environment"*. Vol. 2. Eds.: S. Lindberg, T. Hutchinson. New Orleans, p. 80, Sept 1987. CEP Consultants Ltd., Edinburgh.

H. Roels, R. Lauwerys, J.-P. Buchet et al. Comparison of renal function and psychomotor performance in workers exposed to elemental mercury. *Int. Arch. Occup. Environ. Health* 50:77, 1982.

H. Roels, J.-Ph. Gennart, R. Lauwerys et al. Surveillance of workers exposed to mercury vapour. Validation of a previously proposed biological threshold limit value for mercury concentration in urine. *Am. J. Ind. Med.* 7:45, 1985.

H. Roels, S. Abdeladim, E. Ceulemans and R. Lauwerys. Relationships between the concentrations of mercury in air and in blood or urine in workers exposed to mercury vapour. *Ann. Occup. Hyg.* 31:135, 1987.

H. Roels, M. Boeckx, E. Ceulemans and R. Lauwerys. Urinary excretion of mercury after occupational exposure to mercury vapour and influence of the chelating agent meso-2,3-dimercaptosuccinic acid (DMSA). *Br. J. Ind. Med.* 48:247, 1991.

K. Rosenman, J. Valciukas and L. Glickman. Sensitive indicators of inorganic mercury toxicity. *Arch. Environ. Health* 41:208, 1986.

K. Schaller and G. Triebig. Personenbezogene Probennahme von Quecksilberdämpfen am Arbeitsplatz — ein vergleich zwischen externer und interner Quecksilberexposition. *Arbeitsmed Sozialmed. Präventivmed.* 19:289, 1984.

R. Smith, A. Vorwald, L. Patil and T. Mooney, Jr. Effect of exposure to mercury in the manufacture of chlorine. *Am. Ind. Hyg. Assoc. J.* 31:687, 1970.

L. Soleo, M. Urbano, V. Petrera and L. Ambrosi. Effects of low exposure to inorganic mercury on psychological performance. *Br. J. Ind. Med.* 47:105, 1990.

W. Stopford, S. Bundy, L. Goldwater and J. Bittikofer. Microenvironmental exposure to mercury vapor. *Am. Ind. Hyg. Assoc. J.* 39:378, 1978.

M. Verberk, H. Salle and C. Kemper. Tremor in workers with low exposure to metallic mercury. *Am. Ind. Hyg. Assoc. J.* 47:559, 1986.

WHO. World Health Organization. Environmental Health Criteria 1. Mercury. Geneva, 1976.

WHO. World Health Organization. Report of a study group: Recommended Health-Based Limits in Occupational Exposure to Heavy Metals. Technical report series 647. Geneva, 1980.

WHO. IPCS. International Programme on Chemical Safety. Environmental health criteria 101: methylmercury. World Health Organization, Geneva 1990.

WHO. IPCS. International Programme on Chemical Safety. Environmental Health Criteria 118: inorganic mercury. World Health Organization, Geneva, 1991.

Y. Yamamura. Mercury concentrations in blood and urine as indicators of exposure to mercury vapor. In: *Biological Monitoring of Exposure to Industrial Chemicals*. Eds.:

V. Fiserova-Bergerova, M. Ogata. Proceedings of the US-Japan cooperative seminar on biological monitoring. ACGIH, Cincinnati, 1990.

M. Yoshida. Relation of mercury exposure to elementary mercury levels in the urine and blood. *Scand. J. Work Environ. Health* 11:33, 1985.

D. Zander, U. Ewers, I. Freier et al. Studies on human exposure to mercury. II. Mercury concentrations in urine in relation to the number of amalgam fillings. *Zentralbl. Hyg. Umweltmed.* 190:325, 1990.

16. NICKEL

Toxicokinetics

Nickel is not a cumulative toxin. In industry, absorption mainly occurs through the respiratory tract, but possibly also through the gastrointestinal tract and the skin. Inhalation involves dust (relatively insoluble nickel compounds), aerosols derived from solutions (soluble nickel), or gaseous nickel (usually nickel carbonyl). The absorption rate is greatly dependent on the solubility of the compound. Hence, if levels of nickel in biological media markedly increase following inhalation of soluble compounds (such as nickel chloride, sulfate, nitrate), poorly soluble compounds (such as nickel carbonate, sulfide, oxide) result in lesser, but more prolonged, elevation. Sparingly soluble compounds are highly retained in the lung and lymph nodes. Sunderman et al (1989) have also shown that dietary constituents profoundly reduce the bioavailability of nickel for gastrointestinal absorption, e.g., about 25% of nickel ($NiSO_4$) ingested in drinking water is absorbed and excreted in urine, compared with only 1% of nickel ingested in food.

In post-mortem tissues from adults without known exposure to nickel compounds, the highest concentrations are found in bone, followed by lung, kidney, liver, and heart (Sunderman 1988).

The excretion of nickel occurs predominantly via the urine (Tossavainen et al 1980; Sunderman et al 1989). Investigations on welders suggest at least one pool with fast elimination (biological half-life one or a couple of days) and one pool with slower elimination (biological half-life one to several months) (Akesson and Skerfving 1985).

Biological monitoring

In adults nonoccupationally exposed to nickel, the urinary concentration of nickel is usually less than 5 µg/l (Ader and Stoeppler 1977; McNeely et al 1972; Mikac-Devic et al 1977) and the concentration in plasma is much below 1 µg/100 ml (Hogetveit et al 1978; Nomoto and Sunderman 1970; Torjussen and Andersen 1979; Sunderman et al 1989; Linden et al 1985). According to Nixon et al (1989), who have taken great care to prevent any external contamination, the normal concentration of nickel in serum averages 0.014 µg/100 ml (range: 0.005-0.023). The reference mean values established by Minoïa et al (1990), in a population of healthy Italians, are 0.12 µg/100 ml serum (range: 0.024-0.28; n: 385), 0.23 µg/100 ml blood (range: 0.13-0.33; n: 36) but Versieck and Cornelis (1989) have suggested that the mean normal plasma value lies somewhere in the vicinity of 0.03 to 0.06 µg/100 ml. Recent studies indicate that in nonoccupationally exposed subjects, the concentration of nickel in urine is usually below 2 µg/g creatinine (Minoïa et al 1990; personal observations). A large number of workers exposed to various nickel compounds have been found to have elevated levels of nickel in body fluids.

16.1 Soluble Nickel Compounds

Urine and blood analyses

Several studies have demonstrated that the concentrations of nickel in plasma and urine are indicators of recent exposure to soluble nickel compounds. Tola et al (1979) have studied the relationship between atmospheric exposure to nickel and urine and plasma nickel concentrations by following four workers from an electroplating shop for 1 week. The urinary and plasma nickel were higher in the samples taken after than before the workshift, and a close correlation was found between the air nickel concentrations and the urine (r: 0.82) and plasma (r: 0.83) nickel concentrations, respectively. There was also a close correlation (r: 0.82) between urinary and plasma nickel concentrations. Torjussen and Andersen (1979) also found a significant correlation (r: 0.69) between plasma and urine nickel concentration in 318 workers from a nickel-refining factory. In electroplating workers, Bernacki et al (1980) reported a correlation coefficient of 0.70 between nickel concentrations in the air samples and in end-shift urine specimens. A TWA exposure of 0.1 mg/m^3 of soluble nickel compounds corresponds approximately to a concentration of nickel in plasma and in urine collected at the end of the workshift of 0.7 µg/100 ml and 70 µg/l (corrected for a specific gravity of 1.018), respectively (Tola et al 1979). Angerer and Lehnert (1990) recently reported that they could not establish a limit value for nickel in urine on the basis of their investigation on stainless steel welders, but speculate that the limit value might be situated somewhere between 30 and 50 µg/l. Hogetveit et al (1978, 1980) recommend the determination of nickel in plasma as a routine biological monitoring method. This parameter fluctuates less than urinary nickel. They have proposed 1 µg/100 ml plasma as the critical concentration. Employees with plasma nickel measurements higher than that should be closely supervised, required to wear masks (which efficiency has still recently been demonstrated by Morgan and Rouge 1984; Delabarre 1989), or suspended until the blood level has dropped. According to Grandjean et al (1988), in the absence of detailed dose-response relationships, exposure levels of nickel and inorganic nickel compounds should be kept as low as reasonably achievable; they have suggested as a preliminary guideline that nickel levels should be kept below 0.5 µg/100 ml serum or plasma.

With the exception of nickel carbonyl overexposure, where nickel in urine has been successfully used as an indicator of the risk of the poisoning, there is currently no health-based biological limit values for nickel in blood and in urine.

Therefore, biological monitoring of nickel may not be used as a means of risk estimation, but only as an indicator of exposure (Aitio 1984).

16.2 Sparingly Soluble Nickel Compounds

Urine and blood analyses

Sparingly soluble nickel compounds (e.g., stainless steel welding fumes) are poorly absorbed from the lung. In this case, significant amounts of nickel may be deposited in the respiratory tract without rapid and important change in plasma or urinary nickel concentration (Akesson and Skerfving 1985; Kalliomaki et al 1981). Nevertheless, nickel will be slowly released from the lungs and the serum and urine levels may, to certain extent, reflect lung burden (Grandjean et al 1988). This may explain the elevated concentration of nickel in plasma and in urine of nickel refinery workers, 3-4 years after cessation of exposure (Boysen et al 1984). When workers are exposed simultaneously to insoluble and readily soluble nickel compounds, the concentration of nickel in plasma and in urine will mainly reflect the exposure to the soluble compounds.

Determination of nickel in nasal mucosa

Torjussen and Andersen (1979) have proposed to determine the concentration of nickel in the nasal mucosa to reflect the local nickel exposure of the upper respiratory tract, a major site of action of nickel. This measurement may be useful when exposure is to insoluble nickel compounds or in retired workers who have developed nasal carcinoma to confirm their previous excessive exposure to the metal. Among 57 control subjects, 95% had mucosal nickel concentration below 53 µg/100 g wet weight. Nevertheless, practical problems preclude the routine use of these measurements.

Hair analysis

Bencko et al (1986) found increased concentrations of nickel in hair samples from welders (1.6 to 3.5 mg/kg; mean: 2.39 mg/kg) and nickel smelters (42.7 to 2140 mg/kg; mean: 22.5 mg/kg). The control values ranged from 0 to 1.3 mg/kg (mean: 0.29 mg/kg). However, the use of hair as an internal exposure index is not reliable because of the risk of external contamination.

REFERENCES

D. Ader and M. Stoeppler. Radiochemical and methodological studies on the recovery and analysis of nickel in urine. *J. Anal. Toxicol.* 1:252, 1977.
A. Aitio. Biological monitoring of occupational exposure to nickel. In: *Nickel in the Human Environment.* Ed.: F. Sunderman. IARC Scientific Publication No. 53. Lyon, 1984.

B. Akesson and S. Skerfving. Exposure in welding of high nickel alloy. *Int. Arch. Occup. Environ. Health* 56:111, 1985.

J. Angerer and G. Lehnert. Occupational chronic exposure to metals. II. Nickel exposure of stainless steel welders. Biological monitoring. *Int. Arch. Occup. Environ. Health* 62:7, 1990.

V. Bencko, T. Geist, D. Arbetova et al. Biological monitoring of environmental pollution and human exposure to some trace element. *J. Hyg. Epidemiol. Microbiol. Immunol.* 30:1, 1986.

E. Bernacki, E. Zygowicz and F. Sunderman, Jr. Fluctuations of nickel concentrations in urine of electroplating workers. *Ann. Clin. Lab. Sci.* 10:33, 1980.

M. Boysen, L. Solberg, W. Torjussen et al. Histological changes, rhinoscopical findings and nickel concentration in plasma and urine in retired nickel workers. *Acta Otolaryngol. (Stockholm)* 97:105, 1984.

P. Delabarre. Occupational exposures during the production of catalysts containing inorganic nickel. *Polish J. Occup. Med.* 2:357, 1989.

Ph. Grandjean, O. Andersen and G. Nielsen. Biological indicators for the assessment of human exposure to industrial chemicals: Nickel. Eds.: L. Alessio, A. Berlin, M. Roni, R. Roi. Commission of the European Communities, Luxembourg. Eur 11478 En, 1988.

A. Hogetveit, R. Barton and C. Kostol. Plasma nickel as a primary index of exposure in nickel refining. *Ann. Occup. Hyg.* 21:113, 1978.

A. Hogetveit, R. Barton and I. Anderson. Variations of nickel in plasma and urine during the work period. *J. Occup. Med.* 22:597, 1980.

P. Kalliomaki, E. Rahkonen, V. Vaaranen et al. Lung-retained contaminants, urinary chromium and nickel among stainless steel workers. *Int. Arch. Occup. Environ. Health* 49:67, 1981.

J. Linden, S. Hopper, H. Gossling and F. Sunderman. Blood nickel concentrations in patients with stainless-steel hip prostheses. *Ann. Clin. Lab. Sci.* 15:459, 1985.

M. McNeely, M. Nechay and F. Sunderman, Jr. Measurement of nickel in serum and urine as indices of environmental exposure to nickel. *Clin. Chem.* 18:992, 1972.

C. Minoïa, E. Sabbioni, P. Apostoli et al. Trace element reference values in tissues from inhabitants of the European Community. I. A study of 46 elements in urine, blood and serum of Italian subjects. *Sci. Total Environ.* 95:89, 1990.

D. Mikac-Devic, F. Sunderman, Jr. and S. Nomoto. Furildioxime method for nickel analysis in serum and urine by electrothermal atomic absorption spectrometry. *Clin. Chem.* 23:948, 1977.

L. Morgan and P. Rouge. Biological monitoring in nickel refinery workers. In: *Nickel in the Human Environment.* Ed.: F. Sunderman. IARC Scientific Publication No. 53. Lyon, 1984.

D. Nixon, T. Moyer, D. Squillace and J. McCarthy. Determination of serum nickel by graphite furnace atomic absorption spectrometry with Zeeman-effect background correction: value in a normal population undergoing dialysis. *Analyst* 114:1671, 1989.

S. Nomoto and F. Sunderman, Jr. Atomic absorption spectrometry of nickel in serum, urine, and other biological materials. *Clin. Chem.* 16:477, 1970.

F. Sunderman, S. Hopper, K. Sweeney et al. Nickel absorption and kinetics in human volunteers. *Proc. Soc. Exp. Biol. Med.* 191:5, 1989.

F. Sunderman. Nickel. In: *Biological Monitoring of Toxic Metals. Rochester Series on Environmental Toxicity.* Eds.: Th. Clarkson, L. Friberg, G. Nordberg, P. Sager. Plenum Press, New York, p. 265, 1988.

S. Tola, J. Kilpio and M. Vitarmo. Urinary and plasma concentrations of nickel as
    indicators of exposure to nickel in an electroplating shop. *J. Occup. Med.* 21:184,
    1979.
W. Torjussen and I. Andersen. Nickel concentrations in nasal mucosa, plasma and urine in
    active and retired nickel workers. *Ann. Clin. Lab. Sci.* 9:289, 1979.
A. Tossavainen, M. Nurminen, P. Mutanen and S. Tola. Application of mathematical
    modelling for assessing the biological half-times of chromium and nickel in field
    studies. *Br. J. Ind. Med.* 37:285, 1980.
J. Versieck and R. Cornelis. *Trace Elements in Human Plasma or Serum.* CRC Press, Boca
    Raton, FL, 1989.

17. SELENIUM

Toxicokinetics

Selenium is an essential nutrient. Soluble selenium compounds seem to be easily
absorbed through the lungs and the gastrointestinal tract. Selenium compounds,
including those used in some dandruff shampoos, are not well absorbed through the
skin. However, under special circumstances, direct contact may be of importance,
i.e., when cutaneous absorption is facilitated by the local irritation and skin damage
caused by vesicant selenium compounds (WHO 1987). Under normal conditions,
levels of selenium are higher in the kidney and liver than in the other body tissues.
Muscle selenium levels are lower; however, being the tissue present in greatest
amount in the body, it accounts for the highest proportion of the total body selenium
(WHO 1987). In the body, selenium is methylated to trimethylselenonium ion and
eliminated via the kidney. In nonoccupationally exposed subjects, it seems that
trimethylselenonium constitutes only a minor fraction of selenium in urine. The
reported relative concentration of trimethylselenonium is, however, highly variable,
ranging from less than 1% to 10% or even 47% of total content (Sun et al 1987; Nève
1991). Urinary excretion is rapid and represents the most important route of selenium elimination. It normally accounts for approximately half of the daily intake of
the element in food (Robinson et al 1985). However, urine selenium (24-h collection) was reported to amount up to 70% of selenium intake in 44 subjects
nonoccupationally exposed to selenium (Swanson et al 1990). The urinary selenium
excretion effectively represents the principal known process of regulation of selenium metabolism (Dubois and Belleville 1988). Guidi et al (1988), who administered increasing oral doses of sodium selenite, observed that urinary excretion of
selenium quite perfectly paralleled intake during the first 4-6 weeks supplementation and continued to increase while the plasma selenium level plateaued. Above a
certain intake level, the relation did not hold any more, the urinary selenium
excretion remaining stable despite increasing intake levels. Nève et al (1989a) noted
that a twofold increase in the selenium intake by healthy subjects (100-200 µg/d
additional selenium) had no significant effect on urinary excretion of selenium,
suggesting that under high exposure conditions, other excretory routes might be
preferred (feces, sweat, breath).

Only under conditions of high exposure is a volatile metabolite, dimethylselenide, also eliminated through the lungs. The normal concentrations of selenium in biological media vary considerably, depending on the dietary intake. Besides the variation in soil content, the most probable explanation for the difference observed in the dietary intake of selenium, many factors have been reported to influence the selenium levels of body fluids: e.g., age, sex, pregnancy, lactation, smoking habits, alcohol or coffee consumption, race, etc. However, according to Van Cauwenbergh et al (1990) the absolute differences between various groups are small.

Biological monitoring

The biological significance of selenium in blood and urine is not yet fully established. It seems, however, that the concentration in plasma (or serum) and urine mainly reflects short-term exposure, whereas the selenium content of erythrocytes reflects more long-term exposure (Valentine et al 1978; Suzuki et al 1989; Van Cauwenbergh et al 1990; Schaller and Schiele 1989).

Since, under "normal" exposure conditions, selenium is more concentrated in erythrocytes than in plasma (Magos and Berg 1988; Lederer 1986), whole blood selenium may also reflect long-term exposure.

In nonoccupationally exposed subjects, the concentration of selenium in serum or plasma has been reported to range from 3.3 to 20 µg/100 ml in various part of the world (Versieck 1985; Versieck and Cornelis 1989; Van Caillie-Bertrand et al 1986; Jacobson and Lockitch 1988; Morisi et al 1988; Nève et al 1989b; Schaller and Schiele 1989; Lauwerys 1990; Burguera et al 1990; Lederer 1986; Robberecht et al 1990; Verlinden et al 1983; Dubois et al 1990; Swanson et al 1990; Hongo et al 1985; Zachara et al 1986; Bukkens et al 1990). In 441 healthy Italian subjects, Minoïa et al (1990) found a mean value of 8.1 µg/100 ml (range: 5.6-10.5).

Given the wide range of selenium intake and blood levels, a classification in three categories has been proposed. Countries with inhabitants having a mean plasma selenium of less than 5-6 µg/100 ml are called "poor"- or "low"-selenium areas, while those with values higher than 10-12 µg/100 ml are called "high"-selenium areas or selenium-"rich" countries. Those between the two are "intermediates" (Nève 1991).

In whole blood, values ranging from 2 to 18 µg/100 ml have been found (Schaller and Schiele 1989; Thomson and Robinson 1980; Jaakkola et al 1983; Verlinden et al 1983; Minoïa et al 1990; Robberecht et al 1990; Van Cauwenbergh et al 1990; Lederer 1986; Zachara et al 1986). Particularly high mean values have been described in South Dakota: 40 µg/100 ml (range: 22.1-64) (Whanger et al 1988) and among a group of South Dakota and Wyoming residents: 25 µg/100 ml (range: 18.7-56) (Swanson et al 1990). In 564 Belgian subjects aged between 20 and 80 years a mean value of 9.9 µg/100 ml was found with a range from 2.7 to 18.1. There was no sex difference (unpublished data).

A wide range of mean selenium concentrations in erythrocytes in control subjects has been reported: from 0.24 to 0.59 µg/g Hb (Hongo et al 1985; Suzuki et al 1989; Imai et al 1990; Bukkens et al 1990) or from 7.3 to 17.4 µg/100 ml (Rea et al 1979; Robinson et al 1985; Verlinden et al 1983; Zachara et al 1986).

Depending on the oral intake, the urinary concentrations of selenium found in nonoccupationally exposed persons may vary from 1 to 200 µg/l (Sterner and Lidfeldt 1941; Hopkins et al 1967; Glover 1967; Hadjimarkos 1973; Young and Christian 1977; Valentine et al 1978; Lauwerys 1990; Minoïa et al 1990; Hongo et al 1985; Swanson et al 1990; Schaller and Schiele 1989). According to Robberecht and Deelstra (1984), the normal selenium values for urine are generally below 30 µg/l. Among 484 Italians, Minoïa et al (1990) found a mean urinary concentration of 22.1 µg/l (range: 2.1-30.9). Similar results were obtained in Belgium (Lauwerys 1990).

Very limited data are available on the relationships between selenium concentration in air at the workplaces, the levels in blood or urine, and the risk of adverse effects. There is no indication that selenium level in hair may be used to assess body burden (Cheng et al 1988). Some authors have measured selenium in toenails and obtained mean ($\pm$ SE) values of 0.78 $\pm$ 0.17 ppm (Bukkens et al 1990) and 1.2 µg/g (Swanson et al 1990).

Urine analysis

A biological threshold limit value for selenium concentration in urine of 100 µg/l has been proposed (Lacasse and Richer 1976), but the scientific basis of this proposal is not evident.

REFERENCES

S. Bukkens, N. de Vos, F. Kok et al. Selenium status and cardiovascular risk factors in healthy Dutch subjects. *J. Am. Coll. Nutr.* 9:128, 1990.

J. Burguera, M. Burguera, M. Gallignani et al. Blood serum selenium in the province of Merida, Venezuela, related to sex, cancer incidence and soil selenium content. *J. Trace Elem. Electrolytes Health Dis.* 4:73, 1990.

Y. Cheng, G. Zhuang, M. Tan et al. Preliminary study of correlation of Se content in human hair and tissues. *J. Trace Element Exp. Med.* 1:19, 1988.

F. Dubois and F. Belleville. Selenium: Physiology and human medical implications. *Pathol. Biol.* 36:1017, 1988.

F. Dubois, A. Teby, F. Belleville et al. Selenium status in eastern France. *Ann. Biol. Clin.* 48:28, 1990.

J. Glover. Selenium in human urine: a tentative maximum allowable concentration for industrial and rural population. *Ann. Occup. Hyg.* 10:3, 1967.

G. Guidi, G. Bellisola, R. Schiavon et al. Effects of increasing dietary selenium intake in humans. Preliminary results of a longitudinal study. In: *Trace Element Analytical Chemistry in Medicine and Biology.* Vol. 5. Eds.: P. Brätter, P. W. Schramel, de Gruyter, Berlin, p. 377, 1988.

D. Hadjimarkos. Selenium in relation to dental carries. *Food Cosmet. Toxicol.* 11:1083, 1973.

T. Hongo, Ch. Watawabe, S. Himeno and T. Suzuki. Relationship between erythrocyte mercury and selenium in erythrocyte, plasma, and urine. *Nutr. Res.* 5:1285, 1985.

L. Hopkins, A. Majaj, O. Muth et al. Selenium in human nutrition. In: *Selenium in Biomedicine*. AVI Publishing, Westport, CT, 1967.

H. Imai, T. Suzuki, H. Kashiwazaki et al. Dietary habit and selenium concentrations in erythrocyte and serum in a group of middle aged and elderly Japanese. *Nutr. Res.* 10:1205, 1990.

K. Jaakkola, J. Tummavuori, A. Pirinen et al. Selenium levels in whole blood of Finnish volunteers before and during organic and inorganic selenium supplementation. *Scand. J. Clin. Lab. Invest.* 43:473, 1983.

B. Jacobson and G. Lockitch. Direct determination of selenium in serum by graphite-furnace atomic absorption spectrometry with deuterium background correction and a reduced palladium modifier: age specific reference ranges. *Clin. Chem.* 34:709, 1988.

Y. Lacasse and C. Richer. Toxicité du sélénium et de ses dérivés. *Union Méd. Can.* 105:1192, 1976.

R. Lauwerys. *Toxicologie Industrielle et Intoxications Professionnelles.* 3rd ed. Masson, Paris, 1990.

J. Lederer. *Sélénium et Vitamine E.* Maloine, Paris, 1986.

L. Magos and G. Berg. Selenium. In: *Biological Monitoring of Toxic Metals*. Eds.: Th. Clarkson, L. Friberg, G. Nordberg, P. Sager. Plenum Press, New York, 1988.

C. Minoïa, E. Sabbioni, P. Apostoli et al. Trace element reference values in tissues from inhabitants of the European Community. I. A study of 46 elements in urine, blood and serum of Italian subjects. *Sci. Total Environ.* 95:89, 1990.

G. Morisi, M. Patriarca and A. Menotti. Improved determination of selenium in serum by zeeman atomic absorption spectrometry. *Clin. Chem.* 34:127, 1988.

J. Nève. Methods in determination of selenium states. *J. Trace Elem. Electrolytes Health Dis.* 5:1, 1991.

J. Nève, S. Chamart, S. Van Erum et al. Selenium status in humans as investigated by the effects of supplementation with selenium enriched yeast tablets. In: *Selenium in Medicine and Biology*. Eds.: J. Nève, A. Favier, W. De Gruyter, Berlin, p. 315, 1989a.

J. Nève, F. Vertongen, A. Peretz and Y. Carpentier. Valeurs usuelles du sélénium et de la glutathion peroxydase dans une population belge. *Ann. Biol. Clin.* 47:138, 1989b.

H. Rea, Ch. Thomson, D. Campbell and M. Robinson. Relation between erythrocyte selenium concentrations and glutathione peroxidase activities of New Zealand residents and visitors to New Zealand. *Br. J. Nutr.* 42:201, 1979.

H. Robberecht, H. Deelstra and R. Van Grieken. Determination of selenium in blood components by X-ray emission spectrometry procedures , concentration levels, and health implications. *Biol. Trace Element Res.* 25:149, 1990.

H. Robberecht and H. Deelstra. Selenium in human urine. Determination, speciation and concentration levels. *Talanta* 31: 497, 1984.

J. Robinson, M. Robinson, O. Levander and Ch. Thomson. Urinary excretion of selenium by New Zealand and North American human subjects on differing intakes. *Am. J. Clin. Nutr.* 41:1023, 1985.

K. Schaller and R. Schiele. Biological indicators for the assessment of human exposure to industrial chemicals: selenium. Eds.: L. Alessio, A. Berlin, M. Boni, R. Roi. Commission of the European Communities, Luxembourg. Eur 12174 En, 1989.

J. Sterner and V. Lidfeldt. The selenium content of "normal urine". *J. Pharmacol. Exp. Ther.* 73:205, 1941.

X. Sun, B. Ting and M. Janghorbani. Excretion of trimethylselenonium ion in human urine. *Anal. Biochem.* 167:304, 1987.

T. Suzuki, T. Hongo, T. Ohba et al. The relation of dietary selenium to erythrocyte and plasma selenium concentrations in Japanese college women. *Nutr. Res.* 9:839, 1989.

Ch. Swanson, M. Longnecker, Cl. Veillon et al. Selenium intake age, gender, and smoking
in relation to indices of selenium status of adults residing in a seleniferous area. *Am.
J. Clin. Nutr.* 52:858, 1990.

Ch. Thomson and M. Robinson. Selenium in human health and disease with emphasis on
those aspects peculiar to New Zealand. *Am. J. Clin. Nutr.* 33:303, 1980.

H. Uchida, Y. Shimoishi and K. Toei. Rapid determination of trace amounts of selenium
in biological samples by gas chromatography with electron-capture detection. *Analyst*
106:757, 1981.

J. Valentine, H. Kang and G. Spivey. Selenium levels in human blood, urine and hair in
response to exposure via drinking water. *Environ. Res.* 17:347, 1978.

M. Van Caillie-Bertrand, H. Degenhart and J. Fernandes. Influence of age on the selenium
status in Belgium and the Netherlands. *Pediatr. Res.* 20:574, 1986.

R. Van Cauwenbergh, H. Robberecht and H. Deelstra. Selenium concentration levels in
whole blood of Belgian blood blank donors, as determined by direct graphite furnace
atomic absorption spectrometry. *J. Trace Elem. Electrolytes Health Dis.* 4:215, 1990.

M. Verlinden, M. Van Sprundel, J. Van Der Auwera and W Eylenbosch. The selenium status
of Belgian population groups. I. Healthy adults. *Biol. Trace Element Res.* 5:91, 1983.

J. Versieck and R. Cornelis. Normal levels of trace elements in human blood plasma or
serum. *Anal. Chim. Acta* 116:217, 1980.

J. Versieck. Trace elements in human body fluids and tissues. *CRC Crit. Rev. Clin. Lab.
Sci.* 22: 97, 1985.

J. Versieck and R. Cornelis. *Trace Elements in Human Plasma or Serum.* CRC Press, Boca
Raton, FL, 1989.

P. Whanger, M. Beilstein, Ch. Thomson et al. Blood selenium and glutathione peroxidase
activity of populations in New Zealand, Oregon, and South Dakota. *FASEB J.* 2:2996,
1988.

WHO. IPCS International Programme on Chemical Safety. Environmental Health Criteria
58: selenium. World Health Organization. Geneva, 1987.

J. Young and G. Christian. Gas-chromatographic determination of selenium. *Anal. Chim.
Acta* 65:127, 1977.

B. Zachara, W. Wazowicz, J. Gromadzinska et al. Glutathione peroxidase activity, selenium, and lipid peroxide concentrations in blood from a healthy Polish population. I.
Maternal and cord blood. *Biol. Trace Element Res.* 10:175, 1986.

18. SILVER

Toxicokinetics

Silver may enter the body via the gastrointestinal tract, the respiratory tract, and
the skin. It is likely that the predominant routes of exposure in the workplace are
pulmonary and dermal; however, little is known about the absorption rate. Silver
is primarily eliminated through feces, by biliary excretion; relatively small amounts
are eliminated in the urine.

Biological monitoring

According to Versieck and Cornelis (1989), 0.1 µg/100 ml may be a reliable
indicative value of normal concentration of silver in plasma or serum. In a popula-

tion of healthy Italians, Minoïa et al (1990) found mean serum, blood, and urinary values of 0.018 μg/100 ml (range: 0.006-0.03; n: 394), 0.037 μg/100 ml (range: 0.013-0.061; n: 437), and 0.46 μg/l (range: 0.04-0.88; n: 472), respectively.

Urine, blood, hair, feces analyses

Di Vincenzo et al (1985) have measured silver concentrations in blood, urine, feces, and hair samples collected from smelting and refining silver workers (insoluble silver was considered to be the primary form of airborne silver). Significantly higher silver levels were found in the hair of the workers compared to the controls. However, the possibility of hair contamination with exogenous silver makes the value of measuring silver in hair as an index of exposure questionable.

Since silver is mainly eliminated through feces, these authors suggest fecal measurements as an index of exposure and as a means of calculating body burden. According to their results, occupational human exposure to metallic silver at 0.1 mg/m^3 is expected to lead to a fecal excretion of about 1 mg of silver per day.

Other authors found higher urinary or blood levels of silver in occupationally exposed workers (Minoïa et al 1985; Starkey et al 1987; Rosenman et al 1987), but the data are too scanty to propose any biological threshold limit value.

REFERENCES

G. Di Vincenzo, C. Giordano and L. Schriever. Biological monitoring of workers exposed to silver. *Int. Arch. Occup. Environ. Health* 56:207, 1985.

C. Minoïa, E. Sabbioni, P. Apostoli et al. Trace element reference values in tissues from inhabitants of the European Community. I. A study of 46 elements in urine, blood and serum of Italian subjects. *Sci. Total Environ.* 95:89, 1990.

C. Minoïa, M. Oppezzo, L. Pozzoli et al. Environmental and biological monitoring of subjects occupationally exposed to precious metals. *G. Ital. Med. Lav.* 7:65, 1985.

K. Rosenman, N. Seixas and I. Jacobs. Potential nephrotoxic effects of exposure to silver. *Br. J. Ind. Med.* 44:267, 1987.

B. Starkey, A. Taylor and A. Walker. Measurement of silver in blood by electrothermal atomic absorption spectrometry. *Ann. Clin. Biochem.* 24:SI 91, 1987.

J. Versieck and R. Cornelis. *Trace Elements in Human Plasma or Serum.* CRC Press, Boca Raton, FL, 1989.

19. TELLURIUM

Toxicokinetics

The metabolism of tellurium has not been extensively investigated in man. Tellurium can be absorbed following ingestion, but in occupational exposure the respiratory tract predominates.

Some tellurium compounds, after causing skin burns, are absorbed through skin. Organic tellurium esters are absorbed through intact skin. The highest concentrations of tellurium are found in the blood, liver, kidneys, lungs, thyroid, and spleen. Following long-term exposure, some accumulation occurs also in bone (Gerhardsson et al 1986). It is partly excreted through the kidney, the gastrointestinal tract, and the lungs (as dimethyltelluride, producing a garliclike odor), but the relative importance of each route of excretion has not been established.

Biological monitoring

Minoïa et al (1990) report a mean value of less than 1 µg/l urine in 20 non-occupationally exposed subjects.

Urine analysis

The concentration of tellurium in urine is probably a reflection of the amount absorbed, but the relationship between the internal dose and the amount excreted in urine is not yet known. It has been suggested that, to prevent the garlic odor of the breath, the concentration of tellurium in urine should be kept below 1 µg/l (Amdur 1947; Steinberg et al 1942).

REFERENCES

M. Amdur. Tellurium. *J. Occup. Med.* 3:386, 1947.
L. Gerhardsson, J. Glover, G. Nordberg and V. Vouk. Tellurium. In: *Handbook on the Toxicology of Metals.* Vol II. 2nd ed. Eds.: L. Friberg, G. Nordberg, V. Vouk. Elsevier Science Publishers, Amsterdam, 1986.
C. Minoïa, E. Sabbioni, P. Apostoli et al. Trace element reference values in tissues from inhabitants of the European Community. I. A study of 46 elements in urine, blood and serum of Italian subjects. *Sci. Total Environ.* 95:89, 1990.
H. Steinberg, S. Massari, A. Miner and R. Rink. Industrial exposures to tellurium. *J. Ind. Hyg.* 29:183, 1942.

20. THALLIUM

Toxicokinetics

Thallium is easily absorbed by the pulmonary, gastrointestinal, and cutaneous routes. In blood, approximately 70% of thallium is present in erythrocytes (Rauws 1974). Soluble thallium salts are widely distributed in the body, the highest concentration being found initially in the kidneys (Kazantzis 1986). It is mainly excreted by the kidney and to a lesser extent by the intestine, hair, and saliva. The half-life for the urinary excretion rate of thallium is between 15-30 days (Barclay et al 1953; Innis and Moses 1978).

Biological monitoring

The concentration of thallium in the urine of normal persons is usually less than 1.5 µg/l or 1 µg/g creatinine (Goenechea and Sellier 1967; Schaller et al 1980; Weinig and Zink 1967). In a population of healthy Italians, Minoïa et al (1990) reported mean reference values of 0.42 µg/l in urine (range: 0.07-0.7) (n: 496); 0.039 µg/100 ml (range: 0.015-0.063) in whole blood (n: 418), and 0.018 µg/100 ml (range: 0.002-0.034) in serum (n: 360).

Urine analysis

Very little is known about the relationship between the exposure to thallium and its concentration in human biological media. Most of the limited information concerns cases of clinical poisoning. As a summary statement, it can be said that the determination of thallium in urine is probably a more reliable indicator of exposure than its determination in blood (Schaller et al 1980). On the basis of the results of a study carried out on workers manufacturing thallium-containing seawater batteries, Marcus (cited by Kazantzis 1986) estimated that a 40-h weekly exposure to a TWA exposure of 0.1 mg/m^3 would correspond to a concentration of thallium in urine of about 100 µg/l. He considered this level to be acceptable, but proposed an alerting level requiring action of 50 µg/l urine.

REFERENCES

R. Barclay, W. Peacock and D. Karnofsky. Distribution and excretion of radioactive thallium in the chick embryo, rat and man. *J. Pharm. Expl. Ther.* 107:178, 1953.

S. Goenechea and K. Sellier. Ueber die natuerlichen Thalliumgehalt des menschlichen Koerpers. *Dtsch. Z. Gesamte Gerichtl. Med.* 60:135, 1967.

R. Innis and H. Moses. Thallium poisoning. *Johns Hopkins Med. J.* 142:27, 1978.

G. Kazantzis. Thallium. In: *Handbook on the Toxicology of Metals.* Vol. II. 2nd ed. Eds.: L. Friberg, G. Nordberg, V. Vouk. Elsevier Science Publishers. Amsterdam, 1986.

C. Minoïa, E. Sabbioni, P. Apostoli et al. Trace element reference values in tissues from inhabitants of the European Community. I. A study of 46 elements in urine, blood and serum of Italian subjects. *Sci. Total Environ.* 95:89, 1990.

A. Rauws. Thallium pharmacokinetics and its modification by Prussian blue. *Naunyn-Schmiedeberg's Arch. Pharmacol.* 284:295, 1974.

K. Schaller, G. Manke, H. Raithel et al. Investigation of thallium-exposed workers in cement factories. *Int. Arch. Occup. Environ. Health* 47:223, 1980.

E. Weinig and P. Zink. Ueber die quantitative massenspektrometrische Bestimmung des normalen Thalliumgehalts im menschlichen Organisms. *Arch. Toxicol.* 22:255, 1967.

21. URANIUM

Toxicokinetics

Soluble uranium compounds are likely to be partly absorbed after inhalation; insoluble compounds deposited in the respiratory tract do not easily enter the systemic circulation, but can remain in the lung tissue for a relatively long time.

Absorption of uranium compounds following oral exposure is generally considered quite low; studies in animals estimate a rate of absorption of approximately 1% (more soluble compounds absorbed somewhat more readily than insoluble compounds) (ATSDR 1990). Uranium occurs in the body in soluble form only as tetravalent uranium or hexavalent uranium in uranyl complexes. Oxidation of tetravalent uranium to hexavalent uranium is likely to occur in the body (Berlin and Rudell 1986). Upon autopsy, uranium has been found primarily in the lungs, kidneys, and bones (ATSDR 1990).

After entering the organism, the soluble uranium compounds are rapidly eliminated through the kidney. One must also keep in mind that insoluble uranium particles that are retained in the lungs constitute a local radiological hazard.

Biological monitoring

In nonoccupationally exposed persons, Welford et al (1960) have found urinary levels ranging from 0.03 to 0.3 µg/l. According to Minoïa et al (1990), the mean values are below 0.1 µg/l in urine and below 0.01 µg/100 ml in whole blood.

Urine analysis

The determination of uranium in urine can be used to evaluate recent exposure to soluble uranium salts. It has been proposed that to prevent renal damage, the postshift urine concentration of uranium should not exceed 250 µg/l (Heid et al 1975).

REFERENCES

ATSDR. Agency for Toxic Substances and Disease Registry. U.S. Department of Health and Human Services. Public Health Service. ATSDR/TP-90/29, 1990.

M. Berlin and B. Rudell. Uranium. In: *Handbook on the Toxicology of Metals*. Vol. II. 2nd ed. Eds.: L. Friberg, G. Nordberg, V. Vouk. Elsevier Science Publishers, Amsterdam, 1986.

K. Heid, W. Walsh and J. Houston. Conference on Occupational Health Experience with Uranium. U.S. Energy Research and Development Administration, Washington D.C., pp. 297-322, 1975.

C. Minoïa, E. Sabbioni, P. Apostoli et al. Trace element reference values in tissues from inhabitants of the European Community. I. A study of 46 elements in urine, blood and serum of Italian subjects. *Sci. Total Environ.* 95:89, 1990.

G. Welford, R. Morse and J. Alercio. Urinary uranium levels in non-exposed individuals. *Am. Ind. Hyg. Assoc. J.* 21:68, 1960.

22. VANADIUM

Toxicokinetics

In industry, vanadium is mainly absorbed by the pulmonary route. It has been estimated that about 25% of soluble compounds may be absorbed (WHO 1988). The oral absorption rate seems low (less than 1%). The dermal absorption of vanadium compounds is likely to be extremely small. Vanadium concentrations in human tissues are generally low, but are higher in the liver, kidney, and lung than in other tissues (WHO 1988). Some data suggest the possibility of a slow accumulation in the body in the course of occupational or environmental exposure (Schaller and Triebig 1987). Approximately 90% of circulating vanadium is bound to plasma transferrin. It is excreted in urine with a biological half-life of about 20-40 h, and to a minor degree in feces.

Biological monitoring

The normal concentrations of vanadium in serum, whole blood or urine reported in the literature vary widely (Byrne and Kosta 1978; Gylseth et al 1979; Holzhauser and Schaller 1977; Buchet et al 1982; Schaller and Triebig 1987; WHO 1988), but this is probably due to analytical errors rather than to different background exposure levels. For Versieck and Cornelis (1989), it is likely that the value in whole blood lies below 0.1 µg/100 ml.

The normal urinary concentration is probably less than 1 µg/g creatinine. This is in agreement with the results recently reported by Minoïa et al (1990). The mean reference values in healthy Italians are 0.035 µg/100 ml whole blood (range: 0.009-0.075; n: 65), 0.062 µg/100 ml serum (range: 0.007-0.11; n: 415), and 0.8 µg/l urine (range: 0.2-1; n: 382).

The determination of vanadium in blood and in urine has been proposed for evaluating recent exposure to the metal. Up to now only very limited data are available on the relationship between uptake and the concentration of vanadium in blood and urine.

Urine analysis

Urinary vanadium excretion is probably a more suitable indicator of exposure than blood vanadium (Maroni et al 1983, 1984, 1987; Stonard et al 1984; Alessio et al 1988; Schaller et al 1987; Buchet et al 1985; White et al 1987; Kawai et al 1989). In highly exposed workers, urinary vanadium levels may increase up to 30 times over a workshift. However, the majority of studies have obtained poor correlation between vanadium concentrations in air and the amounts excreted in urine (WHO 1988).

It has been suggested that assessment of exposure during the workday is best carried out by determining the difference between the concentrations of vanadium in urine collected at the end and at the beginning of the shift, whereas determination of urinary vanadium levels 2 days after cessation of exposure (Monday morning) might reflect accumulation of the metal in the body (Alessio et al 1988).

A tentative biological threshold limit value of 50 µg/g creatinine has been proposed for urinary vanadium (postshift sample), but investigations on occupationally exposed persons are needed to test the validity of this proposal (Lauwerys et
al 1980).

Hair analysis

According to Mountain et al (1955), exposure to vanadium reduces the cystine
content of fingernail. The latter analysis has been proposed as a biological
monitoring method. It seems, however, that this effect is not constant (Kiviluoto
et al 1980) or occurs only after long-term exposure to vanadium.

REFERENCES

L. Alessio, M. Maroni and A. Dell' Orto. Biological monitoring of vanadium. In: *Biological Monitoring of Toxic Metals*. Eds.: Th. Clarkson, L. Friberg, G. Nordberg. Plenum
Press. New York, 1988.

J.-P. Buchet, E. Knepper and R. Lauwerys. An electrothermal atomic absorption spectrometric technique for the measurement of vanadium in urine. *Anal. Chim. Acta* 136:243,
1982.

J.-P. Buchet, R. Lauwerys, E. De Maere et al. Evaluation de l'intensité d'exposition au
vanadium de trois groupes de travailleurs par le dosage du vanadium urinaire. *Cah.
Méd. Trav.* 22:247, 1985.

A. Byrne and L. Kosta. Vanadium in foods and in human body fluids and tissues. *Sci. Total
Environ.* 10:17, 1978.

B. Gylseth, H. Leira, E. Steinnes and Y. Thomassen. Vanadium in the blood and urine of
workers in a ferroalloy plant. *Scand. J. Work Environ. Health* 5:188, 1979.

K. Holzhauser and K. Schaller. *Arbeitsmedizinische Untersuchungen bei Schornsteinfegern*.
Thieme Verlag, Stuttgart, 1977.

T. Kawaï, K. Seiki, T. Watanabe et al. Urinary vanadium as a biological indicator of
exposure to vanadium. *Int. Arch. Occup. Environ. Health* 61:283, 1989.

M. Kiviluoto, L. Pyy and A. Pakarinen. Fingernail cystine of vanadium workers. *Int. Arch.
Occup. Environ. Health* 46:179, 1980.

R. Lauwerys, J.-P. Buchet and H. Roels. Les méthodes biologiques de surveillance des
travailleurs exposés à divers toxiques industriels. *Cah. Méd. Trav.* 17:91, 1980.

M. Maroni, A. Colombi, M. Buratti et al. Urinary elimination of vanadium in boiler
cleaners. *International Conference "Heavy Metals in the Environment"*. Heidelberg,
1983.

M. Maroni, A. Colombi, M. Buratti et al. Assessment of occupational exposure to vanadium at fossil-fuel power plants. Communication: XXI International Congress on
Occupational Health. Dublin, Sept. 9-14, 1984.

M. Maroni, A. Colombi, M. Buratti et al. Human exposure to vanadium and nickel from
fuel-oil combustion residues. *International Conference "Heavy Metals in the Environment"*. Vol. 2. Eds.: S. Lindberg, T. Hutchinson. New Orleans, Sept. 1987.

C. Minoïa, E. Sabbioni, P. Apostoli et al. Trace element reference values in tissues from
inhabitants of the European Community. I. A study of 46 elements in urine, blood and
serum of Italian subjects. *Sci. Total Environ.* 95:89, 1990.

J. Mountain, F. Stockell and H. Stokinger. Fingernail cystine as an early indicator of metabolic changes in vanadium workers. *Arch. Ind. Health* 12:494, 1955.

K. Schaller and G. Triebig. Biological indicators for the assessment of human exposure to industrial chemicals: Vanadium. Eds.: L. Alessio, A. Berlin, M. Boni, R. Roi. Commission of the European Communities, Luxembourg. Eur 11135 En, 1987.

M. Stonard, J. O'Sullivan and D. Duffield. Absorption and respiratory effects of vanadium in individuals exposed to vanadium pentoxide. Communication: XXI International Congress on Occupational Health. Dublin, Sept. 9-14, 1984.

J. Versieck and R. Cornelis. *Trace Elements in Human Plasma or Serum.* CRC Press, Boca Raton, FL, 1989.

M. White, G. Reeves, S. Moore et al. Sensitive determination of urinary vanadium as a measure of occupational exposure during cleaning of oil fired boilers. *Ann. Occup. Hyg.* 31:339, 1987.

WHO. IPCS. International Programme on Chemical Safety. Environmental health criteria 81: vanadium. World Health Organization. Geneva, 1988.

23. ZINC

Toxicokinetics

Zinc can be absorbed via the respiratory and gastrointestinal routes. Inhalation (principally ZnO fumes) is the main route of exposure in occupational situations, but no data are available on the percentage absorbed. Skin absorption may occur only in the case of certain soluble compounds. Homeostatic mechanisms exist for the gastrointestinal absorption and excretion of zinc. Excretion takes place mainly via the feces, nonabsorbed zinc probably accounting for the major part, urinary excretion, being the second route of elimination (Bertelli et al 1984; Wastney et al 1986; Elinder 1986). A circadian rhythm of urinary zinc excretion has been observed (Franco et al 1987).

Biological monitoring

The mean blood, plasma, serum, and urinary levels reported by Bertelli et al (1984) in their review vary between 0.108 and 1.116 mg/100 ml, 0.084 and 0.21 mg/100 ml, 0.09 and 0.167 mg/100 ml, and 354 and 887 µg/l, respectively.

In a population of healthy, nonoccupationally exposed Italians, Minoïa et al (1990) report the following mean values: 456 µg/l urine (n: 683; range: 266-846); 0.634 mg/100 ml blood (n: 502; range: 0.4076-0.7594); 0.0922 mg/100 ml plasma (n: 682; range: 0.0587-0.1215).

English and Hambidge (1988) have stressed the effect of time between collection and separation of blood samples with respect to plasma or serum zinc level. In summary the levels of zinc in serum and plasma are, in nonoccupationally exposed subjects, probably around 0.1 mg/100 ml (Versieck and Cornelis 1989; Elinder 1986). The urinary excretion lies between 0.1 and 1.2 mg/24 hours (mean 0.3-0.4 mg/d) (Versieck 1985).

Blood and urine analyses

Zinc in blood (whole blood, plasma, serum) and urine have been used as biological indicators for the assessment of occupational exposure (Cirla et al 1978; D'Andrea et al 1981; Bruzzone et al 1983; Trevisan et al 1983). Although the levels observed are generally significantly higher compared to the controls, no correlation between these values and the levels of exposure or toxic effect has been established yet, and no biological threshold can be proposed.

REFERENCES

G. Bertelli, G. Cortona, P. Odone and L. Alessio. Biological indicators for the assessment of human exposure to industrial chemicals: Zinc. Eds.: L. Alessio, A. Berlin, M. Boni, R. Roi. Commission of the European Communities, Luxembourg. Eur 8903 En, 1984.

M. Bruzzone, C. Calabresi, G. Figari et al. Indagine su saldatori esposti a fumi di zinco. Prevenzione ambientale, indicazioni per il monitoraggio biologico degli esposti. *Med. Lav.* 73:619, 1983.

A. Cirla, G. Pisati, C. Sala and S. Zedda. Biological evaluation of zinc retention. Normal values in adult subjects. *Med. Lav.* 3:244, 1978.

F. D'Andrea, P. Apostoli, F. Brugnone et al. Monitoraggio ambientale e biologico in lavoratori esposti a Pb, Zn, e Cu in fonderie artistiche di bronzo. *Ann. Ist. Super. Sanita* 17:475, 1981.

C.-G. Elinder. Zinc. In: *Handbook on the Toxicology of Metals*. Vol II. 2nd ed. Eds.: L. Friberg, G. Nordberg, V. Vouk. Elsevier Science Publishers, Amsterdam, 1986.

J. English and K. Hambidge. Plasma and serum zinc concentrations: effect of time between collection and separation. *Clin. Chim. Acta* 175:211, 1988.

G. Franco, L. Fidanza, P. Bacchi et al. Circadian rhythm of urinary zinc excretion and biomonitoring occupational risk. *Trace Elem. Med.* 4:163, 1987.

C. Minoïa, E. Sabbioni, P. Apostoli et al. Trace element reference values in tissues from inhabitants of the European Community. I. A study of 46 elements in urine, blood and serum of Italian subjects. *Sci. Total Environ.* 95:89, 1990.

A. Trevisan, A. Buzzo and G. Gori. Indici biologici nell' esposizione professionale a basse concentrazioni di zinco. *Med. Lav.* 73:614, 1983.

J. Versieck. Trace elements in human body fluids and tissues. *CRC Crit. Rev. Clin. Lab. Sci.* 22:97, 1985.

J. Versieck and R. Cornelis. *Trace Elements in Human Plasma or Serum*. CRC Press, Boca Raton, FL, 1989.

M. Wastney, R. Aamodt, W. Rumble et al. Kinetic analysis of zinc metabolism and its regulation in normal humans. *Am. J. Physiol.* 251:R398, 1986.

24. NITROUS OXIDE

Toxicokinetics

Nitrous oxide (N_2O) is an anaesthetic gas, easily absorbed through the lungs. There is no tendency for N_2O to accumulate in the body in case of repeated exposure.

Biological monitoring

Urine analysis

In 1983 (a,b), Sonander et al suggested that urine sampling could be a practical method for routine control of the uptake of N_2O during anaesthetic work, reflecting the TWA exposure over the period during which the urine is produced. Their results indicated that a TWA exposure to 100 ppm would lead to a mean urinary concentration of 66 µg/l.

Stevens et al (1987) studied the exposure to N_2O in operating theatres of four hospitals and measured its urinary concentration. The ambient air levels ranged from 9 to 481 ppm (median 47 ppm). The mean preshift and postshift values for N_2O in urine were 7.7 µg/g creatinine (median: 6.9; SD: 1.6) and 59.6 µg/g creatinine (median: 36.9; SD: 17.6) on Monday; 12-6 µg/g creatinine (median: 10.1; SD: 5.7) and 73 µg/g creatinine (median: 26.3; SD: 41.8) on Friday.

The observations of Imbriani et al (1988) confirmed that for the assessment of individual exposure to N_2O the method of choice is the measurement of its concentration in urine. They found a significant correlation between the N_2O concentration in urine produced during the shift and N_2O environmental concentration. The mean urinary concentrations associated with TWA exposures of 25, 50, and 100 ppm were 20, 35, and 65 µg/l, respectively. The corresponding biological exposure indices (95% lower limit confidence of the regression line) proposed by the authors and based on 4 h of exposure are 13, 21, and 55 µg/l, respectively.

REFERENCES

M. Imbriani, S. Ghittori, G. Pezzagno et al. Nitrous oxide (N_2O) in urine as biological index of exposure in operating room personnel. *Appl. Ind. Hyg.* 3:223, 1988.

H. Sonander, O. Stenqvist and K. Nilsson. Urinary N_2O as a measure of biologic exposure to nitrous oxide anesthetic contamination. *Ann. Occup. Hyg.* 27:73, 1983a.

H. Sonander, O. Stenqvist and K. Nilsson. Exposure to trace amounts of nitrous oxide. Evaluation of urinary gas content monitoring in anaesthetic pratice. *Br. J. Anaesth.* 55:1225, 1983b.

M.-P. Stevens, J. Walrand, J.-P. Buchet and R. Lauwerys. Evaluation de l'exposition à l'halothane et au protoxyde d'azote en salle d'opération par des mesures d'ambiance et des mesures biologiques. *Cah. Méd. Trav.* 34:41, 1987.

3 BIOLOGICAL MONITORING OF EXPOSURE TO ORGANIC SUBSTANCES

1. UNSUBSTITUTED ALIPHATIC AND ALICYCLIC HYDROCARBONS

Like all volatile solvents, the low-molecular weight (less than 16 carbons) aliphatic and alicyclic hydrocarbons are rapidly eliminated, partly unchanged in the expired air and partly as water-soluble metabolites in urine. Since they are not cumulative chemicals, only recent exposure can be evaluated by a biological parameter. Experiments on volunteers have demonstrated the validity of several tests (analysis of the parent compounds in blood and expired air or measurement of urinary metabolites) for estimating the intensity of recent exposure, but the practical application of these measurements has not yet been sufficiently investigated to propose biological threshold limit values for many solvents.

1.1. n-Hexane

Toxicokinetics

Hexane is mainly taken up through the lungs. Retention is approximately 15% (Brugnone et al 1978; Jorgensen and Cohr 1981; Veulemans et al 1982; Mutti et al 1984). In agreement with the volunteer study of Veulemans et al (1982), Mutti et al (1984) found that the postexposure alveolar excretion was biphasic; the median half-lives of the fast and slow phases being 11 min and 99 min, respectively. The simulation of an exposure to n-hexane repeated 5 d a week has suggested that the solvent accumulates in the fat tissue. The half-life of n-hexane in fat tissue equalled 64 h (Perbellini et al 1986). The metabolic pathways of n-hexane and methyl-n-butyl ketone are closely related (Figure 8); 2,5-hexanedione, 2-hexanol, γ-valerolactone, and 2,5-dimethylfuran have been found in urine of workers exposed to n-hexane (Perbellini et al 1980). It seems, however, that 2-hexanol cannot be detected in urine when exposure is moderate (<15 ppm) (Iwata et al 1983); it is therefore a less sensitive indicator of exposure than 2,5-hexanedione. Recent studies in both rat and man seem to indicate that 4,5-dihydroxy-2-hexanone may also be a metabolite of n-hexane (Fedtke and Bolt 1987). Animal experiments have shown that the urinary excretion of n-hexane metabolites significantly decreases in case of simultaneous exposure to methylethylketone or toluene (Iwata et al 1984).

Biological monitoring

Recent exposure to n-hexane can be monitored by measuring its concentration in blood, in alveolar air and possibly also in urine or by determining the concentration of its metabolites in urine.

N-hexane is a ubiquitous environmental pollutant (De Bortoli et al 1985; Krause et al 1987; Brugnone et al 1991) and therefore it is not surprising that it has been detected in blood and urine of the general population. Brugnone et al (1991) reported a normal median value of n-hexane in blood of 36.5 ng/100 ml, 95% of the results being below 147.5 ng/100 ml. In control urines, 95% of the values were below 5900 ng/l with a median concentration of 549 ng/l. According to Fedtke and Bolt (1986) the small amount of 2,5-hexanedione (from 0.12 to 0.78 mg/l, mean 0.45 ± 0.20 mg/l) detected in the urine samples from persons (n: 12) not occupationally exposed to n-hexane might be the result of an endogenous production of n-hexane (e.g., by lipid peroxidation). It might also result from the biotransformation of absorbed n-hexane since, as indicated above, the latter is normally present in the general environment.

Urine analyses

2-Hexanol, 2,5-hexanedione. A correlation has been found between n-hexane exposure and the concentration of the urinary metabolites, mainly 2-hexanol and 2,5-hexanedione. According to Perbellini et al (1981), a time-weighted average (TWA) exposure to n-hexane of approximately 180 mg/m³ (50 ppm) corresponds to an average postshift urine sample concentration for 2-hexanol, 2,5-hexanedione, and total n-hexane metabolites of approximately 0.20, 5.3, and 12.5 mg/l, respectively. The same authors have estimated that, at the 95% confidence level, an n-hexane exposure of 90 mg/m³ (25 ppm) corresponds to a urinary excretion of 5.5-9.7 mg/l of total n-hexane metabolites in postshift urine sample. For each of the individual n-hexane metabolites, the urinary excretion corresponding to an n-hexane exposure of 90 mg/m³ (25 ppm) is: 0-1.3 mg/l for 2-hexanol, 0.8-3.7 mg/l for 2,5-dimethylfuran, 1.5-3.5 mg/l for γ-valerolactone, and 1.5-4.4 mg/l for 2,5-hexanedione. Additional investigations are required, however, to validate these proposals because only grab samples were used to estimate the environmental exposure to n-hexane. Perbellini et al (1981) consider that, from the practical point of view, 2,5-hexanedione, which shows higher urinary concentrations than 2-hexanol, is more reliable. Comparing different methods for the analysis of 2,5-hexanedione, Kawai et al (1990, 1991) have observed that acid hydrolysis results in an elevated background level in the urine. This is in agreement with the findings by Fedtke and Bolt (1986); 2,5-hexanedione concentrations without any hydrolysis correlated best with the intensity of exposure to n-hexane.

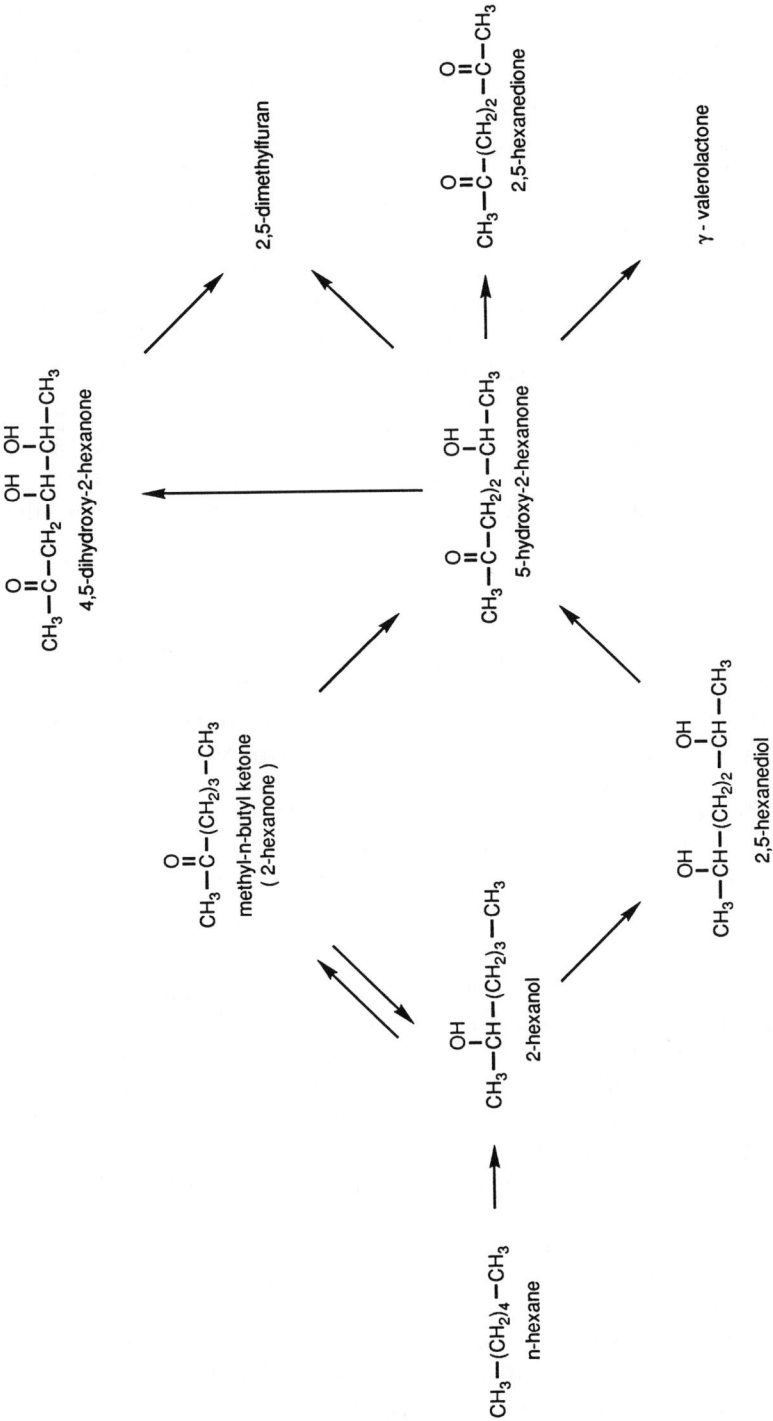

FIGURE 8 Main metabolic pathways of *n*-hexane and methyl-*n*-butyl ketone.

Other authors have found a close relationship between the mean daily exposure to *n*-hexane and the urinary excretion of 2,5-hexanedione in end-of-shift samples; a TWA exposure of 50 ppm would correspond to a urinary concentration of 1.838 mg/l according to Iwata et al (1983), 2.039 mmol/mol creatinine (2.057 mg/g creatinine) according to Mutti et al (1984), 4.95 mg/l according to Perbellini et al (1987), 14.17 µmol/l (1.617 mg/l) according to Ahonen and Schimberg (1988), and 4.214 mg/l according to De Rosa et al (1988). Ahonen and Schimberg (1988) have also estimated that, at this exposure level, the difference between postshift and preshift concentrations amounts to 9.81 µmol/l (1.120 mg/l).

The biological exposure indice adopted by the ACGIH (1991-1992) is 5 mg 2,5-hexanedione/g creatinine (end-of-shift sample), corresponding to a TWA of 50 ppm (180 mg/m^3). The German biological tolerance value is 9 mg/l (DFG 1991).

n-hexane. N-hexane itself can also be measured in the urine of workers. According to Imbriani et al (1984) and Ghittori et al (1987), in workers exposed to 180 mg/m^3, the urine collected at the end of the workshift has a mean *n*-hexane concentration of 13 µg/l. Ghittori et al recommend a biological equivalent exposure limit of 9 µg/l.

Breath analysis

n-hexane. According to Brugnone et al (1978), at a TWA exposure of 180 mg/m^3 (50 ppm) the *n*-hexane concentration in expired air collected during exposure should not exceed 200 mg/m^3 (95% upper confidence limit). This is in agreement with Perbellini et al (1990), who found that the mean alveolar concentration related to an occupational exposure of 180 µg/l corresponds to 153 µg/l (confidence limits between 140 and 167 µg/l). The ACGIH (1991-1992) recommends 40 ppm *n*-hexane in end-exhaled air (during shift) as biological exposure indice (BEI) corresponding to a TWA exposure of 50 ppm (180 mg/m^3).

Blood analysis

n-Hexane. For an exposure level of 50 ppm, the concentration of *n*-hexane in blood collected during exposure should not exceed 15 µg/100 ml (Brugnone et al 1978). Veulemans et al (1982) have exposed volunteers to *n*-hexane and have found that, after the start of exposure, *n*-hexane levels in venous blood tend to quickly attain a maximum. An exposure to 360 mg/m^3 (100 ppm) at rest leads to a blood concentration of approximately 20 µg/100 ml. At the same exposure level, the blood concentration of *n*-hexane increases proportionally to the workload, i.e., to 23 µg/100 ml for a workload of 20 W, to 26 µg/100 ml for a workload of 40 W, and to 29 µg/100 ml for a workload of 60 W. This value is quite close to that proposed by Perbellini et al (1986), who observed that an environmental exposure to 360 mg/m^3 of *n*-hexane corresponded to a mean blood concentration of 17.8 µg/100 ml (confidence limits of 15.9-19.3 µg/100 ml) at the end of the shift. These authors have established a physiologically based pharmacokinetic model which predicts a blood concentration of 18.5-22 µg/100 ml for an environmental exposure of 360 mg/m^3 (Perbellini et al 1990).

We propose the following maximum biological concentrations corresponding to a TWA of 50 ppm (180 mg/m^3): 0.2 mg 2-hexanol/g creatinine; 2 mg 2,5-hexanedione/g creatinine (end-of-shift, first day of the week); 4 mg 2,5-hexanedione/g creatinine (end-of-shift, last day of the working week); 15 µg *n*-hexane/100 ml blood (during exposure); 50 ppm in exhaled air (during exposure).

1.2. 2-Methylpentane and 3-Methylpentane

Biological monitoring

The determination in urine of 2-methyl-2-pentanol and 3-methyl-2-pentanol can be used to monitor the environmental exposure to 2- and 3-methylpentane (Perbellini et al 1981). Unfortunately, the concentration of these metabolites in urine samples collected at the end of a shift, during which the airborne concentration of the solvents is in the order of 1800 mg/m^3 (1991 ACGIH TLV), is not yet known. Alveolar air concentration during exposure is directly related to environmental level, the ratio between the concentrations being approximately 0.9 (Brugnone et al 1980).

These authors suggest as alveolar and blood BEI (calculated on the basis of an acceptable TWA of 1800 mg/m^3) 1512 µg/l alveolar air and 37.8 µg/100 ml blood for 2-methylpentane and 1530 µg/l alveolar air and 34.2 µg/100 ml blood for 3-methylpentane (Brugnone and Perbellini 1985).

1.3. White Spirit

Biological monitoring

White spirit, a widely used solvent, is a mixture of hydrocarbons, the proportion of which varies, but consists of approximately 80% aliphatic and 20% aromatic hydrocarbons. Measurement of the concentration of white spirit components in blood has been suggested as a biological check on exposure (Astrand et al 1975).

The determination of the urinary concentration of dimethylbenzoic acid has been suggested to assess white spirit exposure. According to Pfäffli et al (1985) a urinary concentration of 50 and 200 mg/g creatinine corresponds to an atmospheric concentration of 100 and 500 mg/m^3, respectively.

1.4. Cyclohexane

Toxicokinetics

The main route of absorption is via inhalation. The alveolar excretion after respiratory uptake of cyclohexane is initially rapid, but reaches a slower phase after approximately 1 hour. The half-times are, respectively, 11.2 and 115.3 min. The alveolar excretion does not exceed 10% of the total uptake (9 h postexposure)

(Mutti et al 1981). Cyclohexanol and cyclohexanone have been identified as urinary metabolites of cyclohexane in man (Perbellini et al 1980; Mutti et al 1981).

Biological monitoring

Breath analysis

Cyclohexane. The cyclohexane concentration in alveolar air during exposure is directly related to environmental level. The ratio between these parameters is 0.7 on the average (Perbellini et al 1980). An environmental cyclohexane exposure to 1030 mg/m^3 (300 ppm), which is the TLV proposed by the ACGIH (1991-1992), corresponds to an alveolar cyclohexane concentration of 780-880 mg/m^3 (0.78-0.88 mg/l) (95% confidence levels) (Perbellini and Brugnone 1980). In 1985, these authors proposed an alveolar concentration of cyclohexane of 0.83 mg/l as biological indice (end-of-shift sample) (Brugnone and Perbellini 1985). The prediction from the correlation established by Mutti et al (1981) supports this data. However, these authors consider alveolar analysis not appropriate for biological monitoring, since it is too sensitive to environmental fluctuations.

Blood analyscs

Cyclohexane. The blood concentrations corresponding to a cyclohexane exposure of 300 ppm were 46-52 µg/100 ml (95% confidence limits) in the study of Perbellini and Brugnone (1980). These authors later advised 44 µg/100 ml at the end of the shift as the biological limit value (Brugnone and Perbellini 1985).

Urine analysis

Cyclohexanol. Mutti et al (1981) have observed a poor relation between urinary excretion of cyclohexanol and the level of exposure and concluded that this metabolite cannot be used for biological monitoring. On the contrary, Perbellini and Brugnone (1980) recommended the determination of cyclohexanol in urine, since it is not normally excreted in man. They have suggested that a TWA exposure of 300 ppm entails a cyclohexanol urine concentration (adjusted to a specific gravity of 1024) ranging between 3.2 and 5.5 mg/l urine (95% confidence limit interval).

Cyclohexane. From the observations of Ghittori et al (1987), it emerges that a cyclohexane TWA exposure of 300 ppm leads to a mean urinary cyclohexane concentration of 65 µg/l with a 95% lower confidence limit of 57 µg/l.

On the basis of the above summarized studies, the following values can be suggested as maximum permissible concentrations: 3.2 mg cyclohexanol/g creatinine (end-of-shift), 45 µg cyclohexane/100 ml blood, 0.76 mg/l (220 ppm) in expired air (during shift).

REFERENCES

ACGIH. American Conference of Governmental Industrial Hygienists. Threshold Limit Values for Chemical Substances and Physical Agents and Biological Exposure Indices. Cincinnati, 1991-1992.

I. Ahonen and R. Schimberg. 2,5-Hexanedione excretion after occupational exposure to n-hexane. Br. J. Ind. Med. 45:133, 1988.

I. Astrand, A. Kilbom and P. Ovrum. Exposure to white spirit: 1. Concentration in alveolar air and blood during rest and exercise. Scand. J. Work Environ. Health 1:15, 1975.

F. Brugnone, L. Perbellini, L. Grigolini and P. Apostoli. Solvent exposure in a shoe upper factory. 1. n-Hexane and acetone concentration in alveolar and environmental air and in blood. Int. Arch. Occup. Environ. Health 42:51, 1978.

F. Brugnone, L. Perbellini, E. Gaffuri and P. Apostoli. Biomonitoring of industrial solvent exposures in workers' alveolar air. Int. Arch. Occup. Environ. Health 47:245, 1980.

F. Brugnone and L. Perbellini. Biological monitoring of occupational exposure to solvents by analysis of alveolar air and blood. In: Environmental Health 5. Organic Solvents and the Central Nervous System. WHO, Regional Office for Europe — Copenhagen, Nordic Council of Ministers. Oslo, 1985, p. 57.

F. Brugnone, G. Maranelli, L. Romeo et al. Ubiquitous pollution by n-hexane and reference biological levels in the general population. Int. Arch. Occup. Environ. Health 63:157, 1991.

M. De Bortoli, H. Knoppel, E. Pecchio et al. Measurements of indoor air quality and comparison with ambient air: a study on 15 homes in northern Italy. Commission of the European Communities, Luxembourg. Eur 9656 En VI, 1985.

E. De Rosa, G. Bartolucci, L. Perbellini et al. Environmental and biological monitoring of exposure to toluene, styrene, n-hexane. Appl. Ind. Hyg. 3:332, 1988.

DFG. Deutsche Forschungsgemeinschaft. Maximum Concentrations at the Workplace and Biological Tolerance Values for Working Materials. Report No. XXVII. Commission for the Investigation of Health Hazards of Chemical Compounds in the Work Area. VCH, Weiheim, 1991.

N. Fedtke and H. Bolt. Detection of 2,5-hexanedione in the urine of persons not exposed to n-hexane. Int. Arch. Occup. Environ. Health 57:143, 1986.

N. Fedtke and H. Bolt. The relevance of 4,5-dihydroxy-2-hexanone in the excretion kinetics of n-hexane metabolites in rat and man. Arch. Toxicol. 61:131, 1987.

S. Ghittori, M. Imbriani, G. Pezzagno and E. Capodaglio. The urinary concentration of solvents as a biological indicator of exposure: proposal for the biological equivalent exposure limit for nine solvents. Am. Ind. Hyg. Assoc. J. 48:186, 1987.

M. Imbriani, S. Ghittori, G. Pezzagno and E. Capodaglio. n-Hexane urine elimination and weighted exposure concentration. Int. Arch. Occup. Environ. Health 55:33, 1984.

M. Iwata, Y. Takeuchi, N. Hisanagia and Y. Ono. A study on biological monitoring of n-hexane exposure. Int. Arch. Occup. Environ. Health 51:253, 1983.

M. Iwata, Y. Takeuchi, N. Hisanaga and Y. Ono. Changes of n-hexane neurotoxicity and its urinary metabolites by long term co-exposure with methyl ethyl ketone or toluene. Int. Arch. Occup. Environ. Health 54:273, 1984.

N. Jorgensen and K. Cohr. n-Hexane and its toxicologic effects. Scand. J. Work Environ. Health 7:157, 1981.

T. Kawai, K. Mizunuma, T. Yasugi et al. The method of choice for the determination of 2,5-hexanedione as an indicator of occupational exposure to n-hexane. *Int. Arch. Occup. Environ. Health* 62:403, 1990.

T. Kawai, T. Yasugi, K. Mizunuma et al. Dose-dependent increase in 2,5-hexanedione in the urine of workers exposed to *n*-hexane. *Int. Arch. Occup. Environ. Health* 63:285, 1991.

C. Krause, W. Mailhan, R. Nagel et al. Occurrence of volatile organic compounds in the air of 500 homes in the Federal Republic of Germany. Proceedings 4th International Conference on indoor air quality and climate. Vol.1. Berlin, August 17-21, p. 102, 1987.

A. Mutti, M. Falzoi, S. Lucertini et al. Absorption and alveolar excretion of cyclohexane in workers in a shoe factory. *J. Appl. Toxicol.* 1:220, 1981.

A. Mutti, M. Falzoi, S. Lucertini et al. N-hexane metabolism in occupationally exposed workers. *Br. J. Ind. Med.* 41:533, 1984.

L. Perbellini and F. Brugnone. Lung uptake and metabolism of cyclohexane in shoe factory workers. *Int. Arch. Occup. Environ. Health* 45:261, 1980.

L. Perbellini, F. Brugnone and I. Pavan. Identification of the metabolism of *n*-hexane, cyclohexane and their isomers in men's urine. *Toxicol. Appl. Pharmacol.* 53:220, 1980.

L. Perbellini, F. Brugnone and G. Faggionato. Urinary excretion of the metabolites of *n*-hexane and its isomers during occupational exposure. *Br. J. Ind. Med.* 38:20, 1981.

L. Perbellini, P. Mozzo, F. Brugnone and A. Zedde. Physiologico-mathematical model for studying human exposure to organic solvents: kinetics of blood/tissue n-hexane concentrations and of 2,5-hexanedione in urine. *Br. J. Ind. Med.* 43:760, 1986.

L. Perbellini, F. Brugnone and E. Gaffuri. Urinary metabolite excretion in the exposure to technical hexane. In: *Biological Monitoring of Exposure to Chemicals Organic Compounds.* Eds.: M. Ho, H. Dillon. Wiley Interscience, New York, 1987, p. 197.

L. Perbellini, P. Mozzo, D. Olivato and F. Brugnone. "Dynamic" biological exposure indices for n-hexane and 2,5-hexanedione, suggested by a physiologically based pharmacokinetic model. *Am. Ind. Hyg. Assoc. J.* 51:356, 1990.

O. Pfäffli, H. Harkonen and H. Savolainen. Urinary dimethylbenzoic acid excretion as an indicator of occupational exposure to white spirit. *J. Chromatogr.* 337:146, 1985.

H. Veulemans, E. Van Vlem, H. Janssens et al. Experimental human exposure to *n*-hexane. Study of the respiratory uptake and elimination, and of n-hexane concentrations in peripheral venous blood. *Int. Arch. Occup. Environ. Health* 49:251, 1982.

2. UNSUBSTITUTED AROMATIC HYDROCARBONS
2.1. Benzene

Toxicokinetics

Numerous reports on the indoor air quality indicate that benzene is a ubiquitous environmental pollutant (De Bortoli et al 1985; Wester et al 1986; Krause et al 1987; Holmberg and Lundberg 1985; Berlin 1985; Wallace and Clayton 1987; Wallace et al 1987; Brugnone et al 1987; Perbellini et al 1988). Wester et al (1986) even suggest the possibility of an *in vivo* source of benzene production.

Absorption of benzene occurs mainly through inhalation of vapors and through skin contact. Blank and McAuliffe (1985) have calculated from their *in vitro* studies on human skin that contact of 100 cm^2 of glabrous skin with gasoline containing 5% benzene would lead to skin absorption comparable to the respiratory uptake from an airborne benzene concentration of 10 ppm. Fiserova-Bergerova and Pierce (1989) have estimated that exposure of a skin area of 360 cm^2 to benzene would cause an absorption more than 30-fold above inhalation of 10 ppm benzene for the same period of time. According to Susten et al (1985) in the occupational setting, the dermal absorption could contribute to 20-40% of the total dose of benzene absorbed. Thus, benzene should be regarded as a skin exposure hazard (Grandjean 1990).

A fraction of the absorbed benzene is excreted unchanged in the exhaled air. Several authors have found that, in man, the fraction eliminated in the exhaled air varies between 10 and 50% depending on the metabolic activities and the quantity of body fat. The remaining fraction is metabolized. Pathways of benzene metabolism are summarized in Figure 9.

Assuming that about 50% of inhaled benzene is absorbed through the lungs and that the rates of respiration and urine excretion are 15 l/min and 1 ml/min, respectively, Inoue et al (1986) calculated that in workers occupationally exposed to a TWA concentration of 100 ppm, 13.2% of absorbed benzene is excreted as phenol, 1.6% as catechol, 10.2% as quinol, 0.5% as 1,2,4-benzenetriol, and 1.9% as t-t-muconic acid at the end of a shift of a workday.

Biological monitoring

In practice, phenol (free or conjugated) constitutes the main urinary metabolite of benzene. Excretion of the metabolites is usually completed within 24-48 h after a single exposure, which represents a biological half-life of less than 12 h. The determination of the ratio between inorganic and total sulfates in urine (normally more than 85%, but reduced following significant uptake of benzene) was the first biological test proposed for monitoring exposure. It is now completely abandoned because of its lack of sensitivity and specificity. The biological tests that are currently used are the measurement of total (free and conjugated) phenol in urine, muconic acid in urine, and the measurement of benzene in blood and exhaled air.

In nonoccupationally exposed subjects, Brugnone et al (1989) found a linear correlation between alveolar and blood concentration of different solvents, among which benzene, and very low correlations between blood and environmental concentrations of those solvents. Other markers have been proposed (e.g. catechol, 1,2,4-benzenetriol, S-phenyl-N-acetylcysteine in urine), but their possible application must wait further validation.

Urine analyses

Phenol. In persons nonoccupationally exposed to benzene, the phenol concentration in urine does not exceed 20 mg/l. On a group basis there is a significant correlation between exposure to benzene and the concentration of phenol found in a urine sample collected at the end of the exposure. Following a 6-h exposure to a benzene level of approximately 10 ppm, the average phenol concentration in

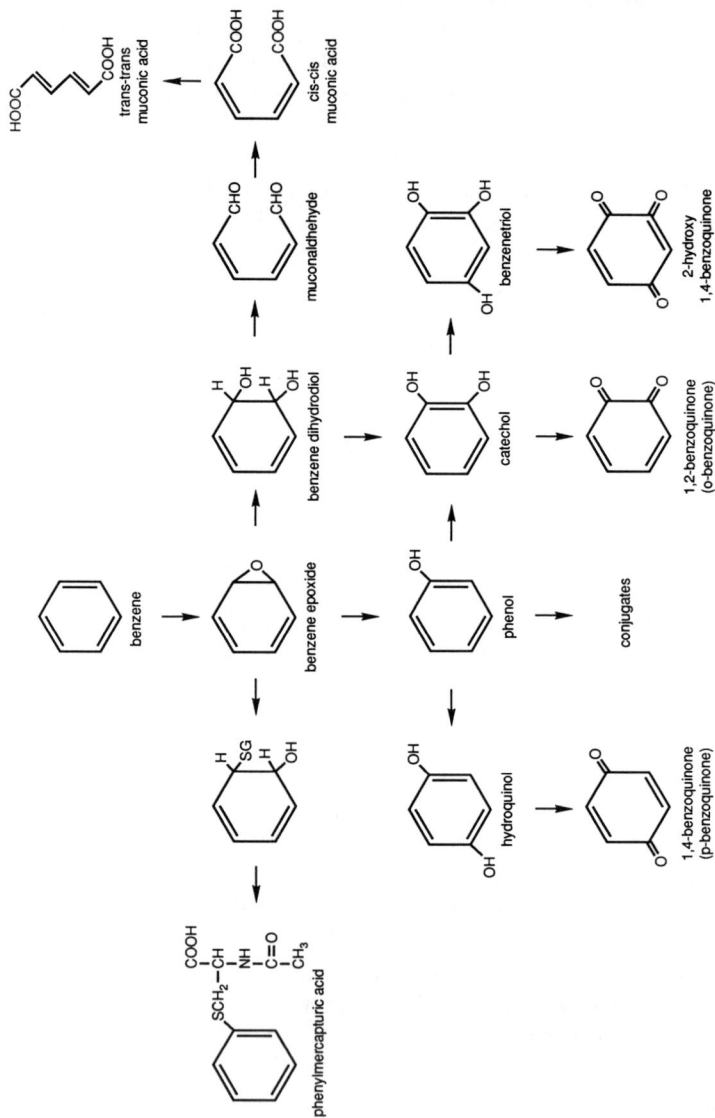

FIGURE 9 Main metabolic pathways of benzene.

the postshift urine sample is around 40 mg/l (corrected for a specific gravity of 1.016), provided a specific method of analysis is used (e.g., gas chromatography). However, in using 20 mg/l as the threshold, very slight exposure to benzene (e.g., exposure to 1 ppm benzene) may be overlooked, since the pre-exposure level of phenol may be much lower than 20 mg/l. Drummond et al (1988) have confirmed that at exposures of about 1 ppm benzene, a single determination of urinary phenol at the end of the shift is of no value as an indicator of uptake/exposure.

In that case, comparison of phenol concentrations between pre- and postshift samples may be useful, since it seems that in control individuals there is no significant difference in phenol concentration between morning and afternoon urine samples. Determination of the change of urinary phenol over the shift may also be useful to detect exposure to benzene resulting from the use of industrial substances (solvents, glues, paints) containing benzene as a nondeclared constituent (Karacic et al 1987). It is, however, likely that at a level of benzene exposure as low as 1 ppm only the determination of benzene in breath or in blood and possibly muconic acid in urine may be useful as a biological monitoring method. The factors which may influence phenol excretion should also be kept in mind when interpreting the results. Dermal application of phenol-containing preparations, exposure to phenol itself, some gastrointestinal disorders favoring the bacterial degradation of tyrosine and phenylalanine, and ingestion of phenylsalicylate-containing drugs increase urinary phenol concentration.

In summary, on a group basis, the relationship between benzene exposure and phenol excretion shown in Table 2 can be proposed (Lauwerys 1979).

These estimates have been confirmed by Inoue et al (1986), who surveyed 152 workers exposed to benzene in shoe factories. They found that the mean phenol concentration in urine (samples collected in the afternoon during the second moiety of the week) corresponding to a TWA exposure of 10 ppm (i.e. 80 ppm-hours) was 46.6 mg/l (SG 1.016) or 57.9 mg/g creatinine.

Muconic acid. t,t-Muconic acid has been measured in the urine of workers exposed to benzene (Inoue et al 1989a; Ducos et al 1990; Bechtold et al 1991). It is a more reliable indicator of exposure to low levels of benzene than phenol,

Table 2
Relationship Between Benzene Exposure and Phenol Excretion

Benzene Exposure (Concentration in ppm times duration of exposure in h)	Phenol Concentration in Urine Collected at the End of the Working Day (mg/l)*
10	20
40	30-35
80	45-50
100	50-55
200	85-90

* Corrected for a specific gravity of 1.016.

provided a sensitive method of analysis is used (Ducos et al 1990). These authors found that in nonoccupationally exposed persons, the mean and median values of t,t-muconic acid in urine are about 0.2 and <0.1 mg/l (range <0.1-0.5 mg/l). These results are in agreement with our observations. We, however, found that the geometric mean value of urinary muconic acid (afternoon sample) is about twice as high in smokers (0.13 mg/g creatinine) than in nonsmokers. In nonsmokers nonoccupationally exposed to benzene, the morning and afternoon muconic acid concentrations are not significantly different, whereas in smokers the concentration is significantly higher in the afternoon, reflecting exposure to benzene present in tobacco smoke.

Sorbitol, which is present in certain foods, may be partly transformed to muconic acid (Ducos et al 1990). This factor and environmental exposure to benzene may be responsible for the different background levels of muconic acid found in nonsmokers nonoccupationally exposed to benzene. The latter rarely exceeds 0.5 mg/g creatinine (smokers and nonsmokers combined). In workers moderately exposed to benzene (<2 ppm), there is a good relation between the TWA exposure to benzene and muconic acid concentration in postshift urine. Our results indicate that the mean postshift muconic acid concentration in subjects exposed to 0.5 or 1 ppm benzene (8-h TWA) amounts to 0.8 or 1.4 mg/g creatinine, respectively. Bechtold et al (1991) also found a linear correlation between the TWA benzene exposure and muconic acid in urine. The average values for unexposed and exposed workers at a TWA of 4.4 ppm for 8 h were 0.27 and 6.2 mg muconic acid/g creatinine, respectively. There was no overlapping between the two data sets.

Other markers. Inoue et al (1988b, 1989a,b) have measured the concentrations of quinol, catechol, t,t-muconic acid, and 1,2,4-benzenetriol in the urine of workers exposed to benzene. The analysis of 1,2,4-benzenetriol seems to present the advantage that this metabolite is not detectable in nonexposed subjects. Unfortunately, the analytical method is rather elaborated. Catechol concentration did not permit to separate those exposed to benzene at 10 ppm from those without exposure. However, the same authors (Inoue et al 1988a) have observed that, on the contrary to the other metabolites of benzene, catechol formation does not seem to decrease in case of coexposure to toluene.

Jongeneelen et al (1987) have detected *S*-phenyl-*N*-acetyl-cysteine in urine of workers exposed to benzene. However, its concentration seems too low to present any practical interest for assessing exposure to benzene. The hemoglobin adduct *S*-phenylcysteine has been detected in blood of rats and mice exposed to benzene, and an exposure dose-response relationship was found. So far, the adduct has not yet been demonstrated to be present in the globin of humans occupationally exposed to benzene (Bechtold et al 1992).

On the basis of animal experiments, some authors (Mueller et al 1987; Norpoth et al 1988) have suggested that phenylguanine in urine might represent a relevant indicator of benzene exposure because it might be related to the critical effect of benzene.

Breath analysis

Benzene. Benzene concentration in expired air during the work period reflects the exposure at the time of sampling, whereas, as for many other solvents, the content of benzene in exhaled air on the morning following exposure best reflects the integrated exposure during the preceding day. Berlin et al (1980) have found that, in nonoccupationally exposed persons, the breath concentration of benzene is in the range of 0.001 to 0.002 ppm in nonsmokers and 0.002 to 0.006 ppm in smokers. Higher levels of benzene in the breath of smokers in comparison with nonsmokers have since been confirmed by many authors. Berlin et al (1980) reported that 16 h after a daily exposure to about 80 ppm-hours (i.e., a TWA over an 8-h day of about 10 ppm), the breath concentration of benzene amounts on the average to 0.05 ppm. Concentrations reported by other authors (see Lauwerys 1979) are usually higher. Concentration of benzene in expired air collected 16 h after an integrated exposure to 80 ppm-hours might amount up to 0.12 ppm. Perbellini et al (1988) have estimated that 16 h after the end of the exposure, the alveolar concentration was equal to 6.2% of the average environmental level during the previous workshift.

Because of the nature of the chronic toxic effect of benzene, Berlin et al (1980) have suggested that the concentration of benzene in expired air should return to background value 16 h after the end of exposure. The consumption of alcohol may accelerate benzene elimination in exhaled air and phenol elimination in urine.

Comparing, in coke oven workers, the sensitivity of breath and blood benzene and urinary phenol as biological indicators of uptake (exposure to low concentrations of benzene: about 1 ppm), Drummond et al (1988) concluded that the most useful monitor of low concentrations appears to be breath benzene measured at the end-of-shift. A TWA exposure of 1 ppm would lead to a benzene concentration in exhaled breath (end-of-shift) varying from 0.022 to 0.22 ppm (0.078-0.78 µg/ l, 1-10 nmol/l). Except in smokers, benzole attendants, or tanker drivers, the preshift breath benzene concentrations were below the detection limit.

Money and Gray (1989) determined benzene in exhaled air of coal tar distillation workers 16 h after exposure and concluded that this analysis was not a sensitive method of monitoring benzene exposure (in the order of 1 ppm).

In conclusion, when benzene in air significantly exceeds 1 ppm, benzene in expired air at the end of the shift or prior to the next shift can be used to assess exposure. At exposure levels around or below 1 ppm, the concentration of benzene already present in expired air of some subjects nonoccupationally exposed to benzene (e.g., smokers) does not permit to use a preshift sample. In that case, expired air must be collected during or at the end of the exposure period.

Blood analysis

Benzene. The measurement of benzene in blood has rarely been studied as a method of evaluating exposure. It is likely, however, that benzene in blood follows the same pattern of change as in breath, with which it is in equilibrium. In three volunteers exposed to 25 ppm benzene for 2 h, the benzene concentration in blood at the end of the exposure period was approximately 20 µg/100 ml and decreased to 1 µg/100 ml 15 h after the end of exposure. In a group of workers exposed to approximately 7 ppm benzene, Cavigneaux and Lebbe (1969) found benzene levels in blood between 10 to 30 µg/100 ml 14 to 18 h after the end of exposure. Karacic et al (1987) detected benzene in blood of subjects working in a shoe factory where solvents, glues, and paints containing benzene as a nondeclared constituent were used. The median value of samples taken a half hour after the workshift was 25 µg/100 ml (range: 6-112) (the level in the control group was below the detection limit: 5.1 µg/100 ml). Unfortunately, the authors have not measured benzene in air.

In a study on workers in the coke oven industry, where the exposure level was approximately about 1 ppm benzene, the concentrations of benzene in blood were only just above the detection limit (i.e., 0.05 µmol/l, 4.2 µg/l) (Drummond et al 1988). Provided a sufficiently sensitive method is used, the determination of benzene in blood may be a useful monitoring method for detecting low-level exposure to benzene. Since bone marrow retains 20 times as much benzene as the blood when equilibrium is reached, the finding of low levels of benzene in blood may be of toxicological importance (Braier et al 1981). Further studies are needed, however, to better assess the relationship between low level exposure to benzene (≤1 ppm) and its concentration in blood collected during exposure or before the next shift.

For a TWA of 10 ppm, ACGIH (1991-1992) recommends the following BEI: 50 mg phenol/l urine (end-of-shift), benzene in exhaled air: mixed-exhaled 0.08 ppm and end-exhaled 0.12 ppm (prior to next shift). However, ACGIH is currently considering to lower the TLV to 0.1 ppm.

We propose the following biological values: 45 mg phenol/g creatinine in urine (end-of-shift) for a TWA of 10 ppm, <20 mg phenol/g creatinine in urine when TWA is <10 ppm; 1.4 mg muconic acid/g creatinine in urine for a TWA of 1 ppm; benzene in exhaled air <0.022 ppm during exposure when the TWA is <1 ppm; benzene in blood <5 µg/100 ml during exposure for a TWA of 1 ppm.

REFERENCES

ACGIH. American Conference of Governmental Industrial Hygienists. Threshold Limit Values for Chemical Substances and Physical Agents and Biological Exposure Indices. Cincinnati, 1991-1992.

W. Bechtold, G. Lucier, L. Birnbaum et al. Muconic acid determinations in urine as biological exposure index for workers occupationally exposed to benzene. *Am. Ind. Hyg. Assoc. J.* 52:473, 1991.

W. Bechtold, J. Sun, L. Birnbaum et al. S-phenylcysteine formation in hemoglobin as a biological exposure index to benzene. *Arch. Toxicol.* 66:303, 1992.

M. Berlin, J. Gage, B. Gullberg et al. Breath concentration as an index of the health risk from benzene: studies on the accumulation and clearance of inhaled benzene. *Scand. J. Work Environ. Health* 6:104, 1980.

M. Berlin. Low level benzene exposure in Sweden: effect on blood elements and body burden of benzene. *Am. J. Ind. Med.* 7:365, 1985.

I. Blank and D. McAuliffe. Penetration of benzene through human skin. *J. Inv. Derm.* 85:522, 1985.

L. Braier, A. Levy, K. Dnor and A. Pardo. Benzene in blood and phenol in urine in monitoring benzene exposure in industry. *Am. J. Ind. Med.* 2:119, 1981.

F. Brugnone, L. Perbellini, G. Faccini and F. Pasini. Benzene in the breath and blood of general public. Proceedings of 4th International Conference on Indoor Air Quality and Climate. Vol. 1. Berlin (West), 17-21 Aug., p. 133, 1987.

F. Brugnone, L. Perbellini, G. Faccini et al. Breath and blood levels of benzene, toluene, cumene, styrene in non-occupationally exposed. *Int. Arch. Occup. Environ. Health* 61:303, 1989.

A. Cavigneaux and J. Lebbe. Interpretation des résultats des épreuves biologiques spécifiques de l'exposition au benzène. *Cahiers de Notes Documentaires.* No. 35, INRS, Paris, 1969.

M. De Bortoli, H. Knoppel, E. Pecchio et al. Measurements of indoor air quality and comparison with ambient air: a study on 15 homes in northern Italy. Commission of the European Communities. Luxembourg, 1985.

L. Drummond, R. Luck, A. Afacan and H. Wilson. Biological monitoring of workers exposed to benzene in the coke oven industry. *Br. J. Ind. Med.* 45:256, 1988.

P. Ducos, R. Gaudin, A. Robert et al. Improvement in HPLC analysis of urinary trans,trans-muconic acid, a promising substitute for phenol in the assessment of benzene exposure. *Int. Arch. Occup. Environ. Health* 62:529, 1990.

V. Fiserova-Bergerova and J. Pierce. Biological monitoring. Dermal absorption. *Appl. Ind. Hyg.* 4:F14, 1989.

Ph. Grandjean. Skin penetration. Hazardous chemicals at work. Report prepared for the Commission of the European Communities. Taylor and Francis, Luxembourg. Eur 12599 En, 1990.

B. Holmberg and P. Lundberg. Benzene: Standards, occurrence, and exposure. *Am. J. Ind. Med.* 7:375, 1985.

O. Inoue, K. Seiji, M. Kasahara et al. Quantitative relation of urinary phenol levels to breathzone benzene concentrations: a factory study. *Br. J. Ind. Med.* 43:692, 1986.

O. Inoue, K. Seiji, T. Watanabe et al. Mutual metabolic suppression between benzene and toluene in man. *Int. Arch. Occup. Environ. Health* 60:15, 1988a.

O. Inoue, K. Seiji, M. Kasahara et al. Determination of catechol and quinol in the urine of workers exposed to benzene. *Br. J. Ind. Med.* 45:487, 1988b.

O. Inoue, K. Seiji, H. Nakatsuka et al. Urinary t,t-muconic acid as an indicator of exposure to benzene. *Br. J. Ind. Med.* 46:122, 1989a.

O. Inoue, K. Seiji, H. Nakatsuka et al. Excretion of 1,2,4-benzenetriol in the urine of workers exposed to benzene. *Br. J. Ind. Med.* 46:559, 1989b.

F. Jongeneelen, H. Dirven, C. Leijdekkers et al. S-phenyl-*n*-acetylcysteine in urine of rats and workers after exposure to benzene. *J. Anal. Toxicol.* 11:100, 1987.

V. Karacic, L. Skender and D. Prpic-Majic. Occupational exposure to benzene in the shoe industry. *Am. J. Ind. Med.* 12:531, 1987.

C. Krause, W. Mailhan, R. Nagel et al. Occurrence of volatile organic compounds in the air of 500 homes in the Federal Republic of Germany. Proceedings of 4th International Conference Indoor Air Quality and Climate Vol. 1 Berlin (West), 17-21 Aug., p. 102, 1987.

R. Lauwerys. Human biological monitoring of industrial chemicals: Benzene. Eds.: L. Alessio, A. Berlin, R. Roi, M. Boni. Commission of the European Communities, Luxembourg. Eur 6570 En, 1979.

C. Money and C. Gray. Exhaled breath analysis as a measure of workplace exposure to benzene ppm. *Ann. Occup. Hyg.* 33:257, 1989.

G. Mueller, M. Koelbel, M. Heger and K. Norpoth. Urinary S-phenylmercapturic acid and phenylguanine as indicators of benzene exposure. In: *Biological Monitoring of Exposure to Chemicals Organic Compounds.* Eds.: M. Ho, H. Dillon. Wiley Interscience, New York, p. 91, 1987.

K. Norpoth, W. Stücker, E. Krewet and G. Müller. Biological monitoring of benzene exposure by trace analysis of phenylguanine. *Int. Arch. Occup. Environ. Health* 60:163, 1988.

L. Perbellini, G. Faccini, F. Pasini et al. Environmental and occupational exposure to benzene by analysis of breath and blood. *Br. J. Ind. Med.* 45:345, 1988.

A. Susten, B. Dames, J. Burg and R. Niemeier. Percutaneous penetration of benzene in hairless mice: an estimate of dermal absorption during tire-building operations. *Am. J. Ind. Med.* 7:325, 1985.

L. Wallace and C. Clayton. Volatile organic compounds in 600 US homes: major sources of personal exposure. Proceedings of 4th International Conference on Indoor Air Quality and Climate. Vol. 1. Berlin (West), 17-21 Aug., p. 183, 1987.

L. Wallace, R. Jurgens, L. Sheldon and E. Pellizzari. Volatile organic chemicals in 10 public-access buildings. Proceedings of 4th International Conference on Indoor Air Quality and Climate. Vol. 1. Berlin (West). 7-21 Aug., p. 188, 1987.

R. Wester, H. Maibach, L. Gruenke, J. Craig. Benzene levels in ambient air and breath of smokers and non smokers in urban and pristine environments. *J. Toxicol. Environ. Health* 18:567, 1986.

2.2. Toluene

Toxicokinetics

Absorption of toluene results mainly from inhalation of its vapor. In human studies, uptake of toluene has been estimated by different authors to be 40-60% of the total amount inhaled (WHO 1985; ATSDR 1989). Significant amounts may also be absorbed through the skin if there is contact with the liquid form (Dutkiewicz and Tyras 1968; Guillemin et al 1974). Aitio et al (1984) observed in three human volunteers that the immersion of one hand for 5 min in toluene resulted in highly increased toluene levels in blood taken from the same arm (range: 2-5.4 μmol/l) as compared to the other arm. This difference lasted for 3 h. In his review on the risk of skin penetration of hazardous chemicals at work, Grandjean (1990) concluded that although the penetration process appears to be slow, toluene may penetrate the skin in significant amount and should therefore be regarded as a skin-absorption hazard. Skin absorption of toluene vapor is negligible (Piotrowski 1967). Following absorption, toluene is rapidly distributed, with the highest levels observed in adipose tissue followed by bone marrow, adrenals, kidneys, liver, brain, and blood (WHO 1985). There are large interindividual variations in the uptake and elimination rates from blood. In two workers admitted to hospital because of coma due to an accidental occupational exposure to a mixture of solvents, a half-life between 19 and 21 h was calculated for toluene in blood and alveolar air (Brugnone et al 1983). More recent data obtained from workers exposed to toluene indicate a mean toluene half-life of 3.8 h (2.6-6 h) in the alveolar air and of 4.5 h (3-6.2 h) in blood (Brugnone et al 1986). The authors concluded that the 17-hour interval between two consecutive workshifts was insufficient for the complete elimination of absorbed toluene, resulting in a possible accumulation during the workweek. In exposed subjects, Nise and Orbaek (1988) effectively observed a significant increase in toluene blood levels during the workweek. Furthermore, they found toluene in blood as long as 13 d after the end of exposure. After exposure has ceased, the toluene concentration in venous blood decreased nonlinearly and the elimination curves was composed of at least three exponential components with median estimated half-times of 9 min, 2 h and 90 h. The adipose tissue appeared to have a median half-time of 79 h (Nise et al 1989).

A fraction (approximately 20%) of the absorbed toluene is excreted unchanged in the expired air (Srbova and Teisinger 1952). The remainder is oxidized by transformation of the methyl radical into a carboxyl radical, which is mainly conjugated with glycine to produce hippuric acid (Figure 10). Conjugation of benzoic acid with glucuronic acid seems to occur only when the glycine conjugation reaction is saturated, i.e., after heavy toluene absorption (exposure of 780 ppm at rest or 270 ppm during moderately heavy work) (Riihimaki 1979). Only a very small fraction (less than 1% of the dose) of toluene is hydroxylated to ortho, meta and para-cresol (Bakke and Scheline 1970; Woiwode et al 1979). Hippuric acid is excreted in urine with a biological half-life of about 3 h. Therefore, elimination is practically complete within 18 h after the end of exposure (Ogata et al 1970; von Oettingen et al 1942). During repeated exposure over the week, preexposure levels of hippuric acid of each day returned to the baseline and were not statistically different from those of the referent subjects (De Rosa et al 1986, 1988). At variance with previous findings, Nise (1992) reported a similar half-time for hippuric acid and o-cresol in urine of about 44 h.

Inoue et al (1986) have reported sex and ethnic differences in the urinary excretion of toluene metabolites among Chinese, Turkish, and Japanese workers under similar exposure conditions. Hasegawa et al (1983) and Baelum (1990) also demonstrated sex differences in the metabolism of toluene. Besides sex and genetic differences, many factors influence the pharmacokinetics of toluene (Baelum 1991).

Wallen (1986) described, in occupationally exposed workers, a tendency for an enhanced clearance of toluene from the blood in relation to personal habits such as smoking and/or moderate chronic ethanol intake. This confirms the findings of Waldron et al (1983), who observed in workers exposed to toluene that blood toluene concentrations were the lowest in those who drank alcohol regularly. On the contrary, the biotransformation of toluene was inhibited by ethanol and consequently blood toluene concentrations were increased following the intake of a large dose of alcohol (Waldron et al 1983). This is in agreement with the observations of Dossing et al (1984). Nise (1992) observed that alcohol consumption was associated with a decrease of excretion of hippuric acid, while o-cresol was not influenced.

The study of Wigaeus et al (1988) has demonstrated the faster elimination of toluene in smokers than in nonsmokers. At the same exposure level, smokers have a higher urinary concentration of o-cresol than nonsmokers, but hippuric acid excretion does not seem to be affected by smoking habits (Nise 1992). On the other hand, blood levels of toluene, which is a ubiquitous pollutant, were found to be higher in smokers than in nonsmokers (Hajimiragha 1989).

A survey conducted by Inoue et al (1988) on factory workers showed that the formation of hippuric acid and o-cresol from toluene was reduced by co-exposure to benzene. Wallen et al (1985) reported that coexposure of human volunteers to 55 ppm toluene and 25 ppm p-xylene caused a reduction in the apparent clearance of both solvents without affecting their respective kinetics in blood or in exhaled air. The results of Tardif et al (1991) indicated that simultaneous exposure to relatively low concentrations of toluene (50 ppm) and xylene (40 ppm) affected neither the concentration of either solvent in blood and in end-exhaled air nor the urinary excretion of their respective metabolites. On the other hand, simultaneous exposure to higher concentrations of toluene (95 ppm) and xylene (80 ppm) resulted in increased concentrations of both solvents in blood and in end-exhaled air as measured near the end of a 4-h exposure. Moreover, the urinary excretion of hippuric acid was delayed as a result of mixed exposure.

Biological monitoring

Several biological tests have been proposed for evaluating toluene exposure, namely hippuric acid in urine, benzoic acid in urine, o-cresol in urine, hippuric acid in blood, toluene in blood, and toluene in expired air.

Urine analyses

Hippuric acid. Hippuric acid is a normal constituent of urine, originating mainly from food containing benzoic acid or benzoates. In nonoccupationally exposed workers, the concentration of hippuric acid in spot urine samples rarely exceeds 1.5 g/g creatinine (Buchet and Lauwerys 1973). In 1942, von Oettingen et al reported that exposure to concentration of 50 to 800 ppm of toluene is associated with and followed by an increased excretion of hippuric acid, which is roughly parallel to the intensity of the exposure. Pagnotto and Lieberman (1967) have compared the postexposure hippuric acid/creatinine ratio with toluene exposure in small groups of individuals with nearly similar toluene exposure. According to their data, an exposure to 100 ppm (375 mg/m³) toluene would, on a group basis, produce a urinary hippuric acid content of about 2.8 g/g creatinine in samples collected at the end of the work shift. It should be noted, however, that Pagnotto and Lieberman (1967) have used a nonspecific spectrophotometric technique, which also measures methylhippuric acid and therefore gives higher values than more specific chromatographic techniques when xylene is also present in the work environment. According to Ikeda and Ohtsuji (1969), the average hippuric acid concentration in urine collected at the end of the work shift in a group of workers exposed to an atmospheric toluene concentration of 100 ppm is estimated at 2.0

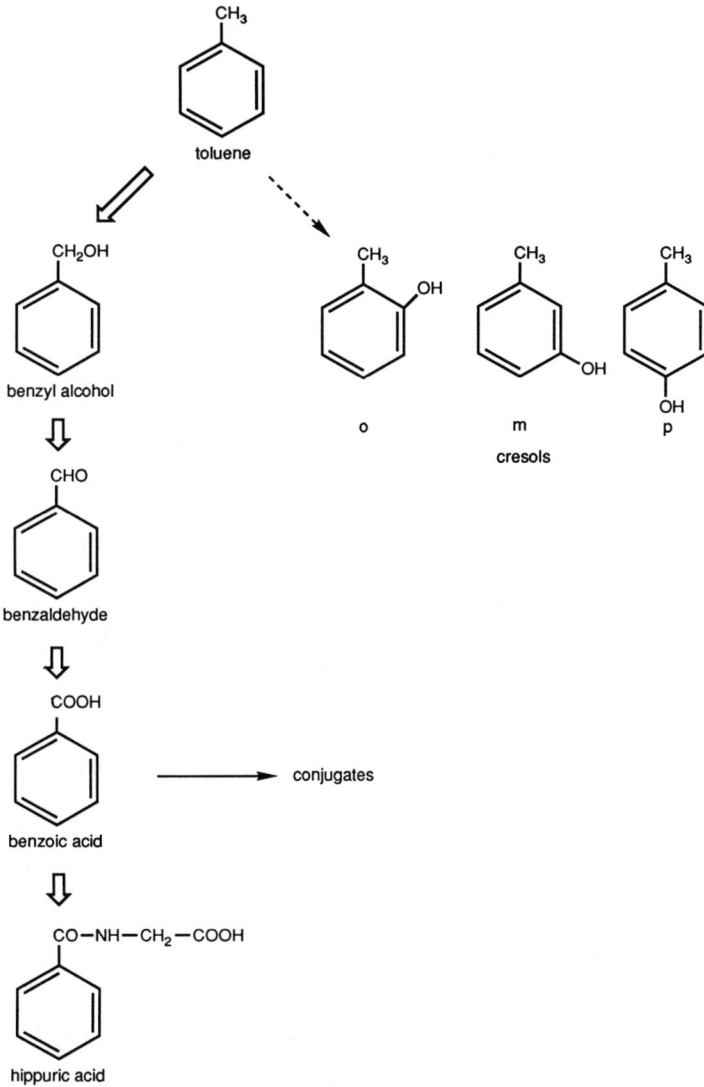

FIGURE 10 Metabolic pathways of toluene.

g/l (specific gravity 1.016) or 2.35 g/g creatinine. Individual values (5th and 95th percentiles) could range from 1.4 to 3.9 g/g creatinine (Imamura and Ikeda 1973). It should be stressed that their results on the toluene concentration in air (photo-gravure printing industry) were obtained with detection tubes and therefore represent semiquantitative spot sample determinations. The authors recognize that fluctuations of environmental toluene concentration in the workshop surveyed are among the possible causes of wide variation in hippuric acid excretion. Tokugana et al (1974) have found that in seven workers exposed for 7 h to 40.2 ppm toluene (range 28.6-53.7) the excess hippuric acid excreted within 24 h after the start of exposure (i.e., after subtracting the amount of hippuric acid normally present when the workers are not exposed) amounts to 482 mg (range: 364-719). Unfortunately, they did not indicate the background level of hippuric acid found in these workers. Angerer (1976) reported that, in 36 workers exposed to a mean toluene concentration of 27 ppm, the mean urinary hippuric acid concentration amounts to 2.04 g/l with a SD of 1.37 g/l.

Ogata et al (1970) exposed 23 male volunteers for 3 h in the morning and 4 h in the afternoon or just for 3 h in the morning to toluene vapor. They collected the urine for about a day after exposure. They demonstrated that, on a group basis, there was an excellent correlation between the total amount of hippuric acid excreted (exposure period plus 18 h after exposure) and the total exposure (concentration × time). They have also studied the correlation between the intensity of exposure (0, 100, and 200 ppm) and the concentration of hippuric acid in urine collected either during the first and second periods of exposure, and during the whole exposure period. In a group of five persons exposed to 100 ppm toluene for 7 h, the average hippuric acid concentration in urine collected during the second period of exposure was about 2.81 µg/l (corrected for a SG of 1.024) or 1.87 g/l (corrected for a SG of 1.016), 90% of the values were included in the range 1 to 2.75 g/l (corrected for a SG of 1.016). Ogata et al (1970) suggested that, for screening, the presence of more hippuric acid in the urine of a man than that corresponding to 2 standard deviations less than the average quantity excreted by subjects exposed to the permissible level should be taken as evidence that the man may have been exposed to a concentration greater than that level. This will be true of about 5% of the men. Hence, for a permissible TWA of 100 ppm, the screening (5th percentile) concentration of hippuric acid in urine collected during the second period of the shift should be 1 g/l (corrected for a SG of 1.016). We have indicated above that the upper normal limit for toluene may exceed 1 g/l. Thus, on an individual basis, separation of the exposed from the nonexposed can hardly be done with a single urine analysis for hippuric acid. Imamura and Ikeda (1973) and Engstrom et al (1976), and more recently De Rosa et al (1985), came to the same conclusion. On a group basis, however, this test appears sufficiently sensitive.

More recent data have confirmed that end-of-shift hippuricuria levels of exposed subjects are significantly correlated with the mean daily environmental concentration. The results of the study of Hasegawa et al (1983) indicate that the mean concentration of hippuric acid in urine after exposure to toluene at the current occupational exposure limit of 100 ppm was about 3 g/l (3.381 g/l for males and 2.496 g/l for females). Even at low exposure concentrations, De Rosa et al (1985) showed that the hippuricuria of a group of workers at the end of a workshift was statistically different from that of controls. For rather low exposures (100 mg/m^3 or 27 ppm) there was no clear distinction between exposed subjects and controls, at least on an individual basis, but, nevertheless, by taking into account the mean hippuric acid concentration in the exposed subjects, which in their survey, always turned out to be significantly different from those of the controls, they were able to find an association between toluene in air and hippuric acid in urine (De Rosa et al 1985). On the basis of their data an exposure to 100, 150, 200 mg/m^3 of toluene corresponds to a hippuric acid concentration of about 1.05, 1.4, and 1.75 g/g creatinine or 1.95, 2.46, and 2.98 g/l corrected for SG 1024 (De Rosa et al 1985, 1987).

The end-of-shift hippuric acid value corresponding to a toluene TWA of 100 ppm (375 mg/m^3) is 2.75 g/g creatinine (De Rosa et al 1987, 1988). This is also in agreement with the survey of Foo et al (1991) and Ogata and Taguchi (1988), who calculated that a mean value of 2.7 g hippuric acid/g creatinine (95% confidence limits 1.3-5.7) and 2.46 g/g creatinine, respectively, corresponds to an average exposure to 100 ppm toluene.

The data reported by Campbell et al (1987) indicate that an exposure limit of 100 ppm toluene corresponds to an equivalent biological value of 1.15 mmol/mmol creatinine (95% confidence limits 0.5-2.6) or 1.9 g/g creatinine (95% confidence limit: 0.8-4.2). Lower values (1.2 g/g creatinine) were obtained by Inoue et al (1988).

Table 3 summarizes some of the relationships reported by several authors. The excretion rate of hippuric acid at the end of the exposure period is much more closely related to the time-weighted toluene load than concentration alone

Table 3

**Correlations Between Airborne Concentration of Toluene (x)
and Hippuric Acid Concentration in Postshift Sample (y)**

Authors	Regression Equation	Concentration Corresponding to a TWA of 100 PPM	Correlation Coefficient
Hasegawa et al (1983)	y (mg/l) = 270.7 + 31.1x (ppm)	3.38 g/l	0.803
	y (mg/g cr.) = 361.3 + 15.8x (ppm)	1.94 g/g creatinine	0.606
	y (mg/l*) = 273.2 + 16.2x (ppm)	1.89 g/l*: SG 1016	0.768
	y (mg/l) = 296.2 + 22x (ppm)	2.49 g/l	0.830
	y (mg/g cr.) = 262.3 + 17.4x (ppm)	2 g/g creatinine	0.794
	y (mg/l*) = 207.9 + 13.6x (ppm)	1.56 g/l*: SG 1016	0.829
De Rosa et al			
(1985)	y (g/g cr.) = 0.2365 + 0.0083x (mg/m^3)	3.349 g/g creatinine	0.80
(1987)	y (g/g cr.) = 0.38 + 0.006x (mg/m^3)	2.63 g/g creatinine	0.88
	y (g/l*) = 0.92 + 0.01x (mg/m^3)	4.67 g/l*: SG 1024	0.84
(1988)	y (g/g cr.) = 0.249 + 0.007x (mg/m^3)	2.874 g/g creatinine	0.86
Campbell et al (1987)	y (mmol/mol cr.) = 0.15 + 0.01x (ppm)	1.15 mmol/mol creatinine 1.82 g/g creatinine	0.71 (r^2)
Droz et al (1987)	y (mol/mol cr.) = 0.540 + 0.00169x x = ppm.h	0.553 mol/mol creatinine 0.87 g/g creatinine	0.714
Inoue et al (1988)	y (mg/l) = 140.9 + 10.47x (ppm)	1.19 g/l	0.751
	y (mg/g cr.) = 159.8 + 10.77x (ppm)	1.24 g/g creatinine	0.827
	y (mg/l*) = 145 + 9.98x (ppm)	1.145 g/l*: SG 1016	0.801
Ogata et al (1988)	y (g/l) = 0.57 + 0.020x (ppm)	2.57 g/l	0.68
	y (g/g cr.) = 0.36 + 0.021x (ppm)	2.46 g/g creatinine	0.86
	y (g/l*) = 0.55 + 0.022x (ppm)	2.75 g/l*: SG 1024	0.80
Foo et al (1991)	y (g/g cr.) = 0.125x$^{0.667}$	2.7 g/g creatinine	0.73

(Veulemans and Masschelein 1979; Veulemans et al 1979; Wilczok and Bieniek 1978). On a group basis, a TWA exposure of 100 ppm corresponds to a hippuric excretion rate of 4 mg/min according to Veulemans et al (1979) and to 2.6 mg/min according to Wilczok and Bieniek (1978). In the study of Baelum et al (1987), the background excretion rate of hippuric acid was 0.97 ± 0.75 mg/min (1.25 ± 1.05 g/g creatinine) and rose to 3.74 ± 1.4 mg/min (3.90 ± 1.85 g/g creatinine) during the last 3 h of a 7-h exposure of 71 volunteers to 100 ppm toluene. Unfortunately, for practical reasons, the collection of a timed urine sample is frequently impossible. The ACGIH (1991-1992) recommends 3 mg/min as a biological exposure indice (last 4 h of the shift).

In summary, exposure to 100 ppm toluene for 8 h at rest should theoretically result in an additional total excretion of about 1 g of hippuric acid. During the performance of light work, a total of 2 to 3 g of hippuric acid is excreted (Cohr and Stokholm 1979). But in practice, a urine specimen is usually collected at the end of the workday, when the excretion of hippuric acid is at its maximum. In a group of workers exposed to 100 ppm toluene for 8 h, the mean hippuric acid concentration in urine samples collected at the end of the exposure period amounts approximately to 2.5 g/g creatinine (1.58 mol/mol creatinine) which is the BEI recommended by the ACGIH (1991-1992). This test can only be used to assess exposure of groups of workers.

Benzoic acid. Bardodej (1968) has suggested that it is theoretically preferable to measure total benzoic acid concentration in urine (free plus conjugated), since benzoic acid produced after toluene exposure is conjugated not only to glycine, but also to glucuronic acid. It has, however, been found by Engström et al (1976) that there was no difference between the direct determination of hippuric acid and the determination of benzoic acid after alkaline hydrolysis of urine.

o-Cresol. Only a minor fraction of inhaled toluene vapor is oxidized on the aromatic ring with the production of cresols. Since *o*-cresol is not a major constituent of normal urine (normal concentration less than 0.3 mg/l), its determination has been proposed as a biological monitoring method (Angerer 1979). Woiwode and Drysch (1981) exposed ten volunteers to approximately 200 ppm toluene for 4 h and, at the end of the exposure period, found an average concentration of 1.6 mg *o*-cresol/l of urine (SG 1017).

A TWA toluene exposure to 100 ppm corresponds to an end-of-shift urinary *o*-cresol concentration of 1 mg/l, according to Pfäffli (1979), and about 1.5 mg/l (1.47 mg/l for males, 1.38 mg/l for females), according to Hasegawa et al (1983). Dossing et al (1983) give a higher estimation (2.2 mg/l). Values higher than 5.3 mg *o*-cresol/l urine might indicate an exposure higher than 200 ppm (Angerer 1985).

On the basis of the data of De Rosa's group an exposure to 100, 150, and 200 mg toluene/m^3 (27, 40.5, and 54 ppm) corresponds to an end-of-shift urinary *o*-cresol excretion of approximately 250, 375, and 475 µg/g creatinine or 488, 662, 835 µg/l corrected for SG 1024 (De Rosa et al 1985, 1987). Extrapolated values at 100 ppm are 785 µg/g creatinine, 1443 µg/l (SG 1024), 990 µg/l (SG 1016) (De

Rosa et al 1987). According to Inoue et al (1988), the same exposure level leads to a mean urinary excretion of 965.6 µg/l, 975.6 µg/g creatinine, or 881.4 µg/l corrected for specific gravity of 1016. The values observed by Baelum are closer to those of Dossing and co-workers. During the last 3 h of a 7-h exposure of volunteers to 100 ppm toluene, the excretion rate (mean ± SD) is 2.04 ± 0.84 µg/min (2.05 ± 1.18 mg/g creatinine), while the background rate is 0.05 ± 0.05 g/min (0.08 ± 0.14 mg/g creatinine) (Baelum et al 1987).

o-Cresol urinary levels turn out to be generally well correlated to hippuric acid urinary levels. However, it should be noted that the correlation between postshift urinary excretion of o-cresol and environmental toluene levels is weaker than the relationship between postshift urinary levels of hippuric acid and toluene concentrations in air.

Nise (1992) has also recently indicated that the impact of smoking and the large interindividual variation make o-cresol excretion an unsuitable means of monitoring low-level exposure in an individual worker.

Furthermore, the small concentration of o-cresol in urine may create analytical difficulties (De Rosa et al 1985, 1987; Hasegawa et al 1983; Angerer 1985). ACGIH (1991-1992) has proposed a BEI of 1 mg/g creatinine (end-of-shift sample), which is in agreement with the proposal of Lauwerys (1990). This value correponds to a toluene TWA of 100 ppm.

In our opinion the main interest of measuring o-cresol in urine is to determine whether the finding of a slight increased urinary concentration of hippuric acid is really due to toluene exposure or may result from dietary intake of benzoic acid. It has also been suggested that interindividual differences in o-cresol production may be of toxicological importance since its precursor is an epoxide (Hasegawa et al 1983).

Table 4 summarizes some of the relationships reported by several authors.

Toluene. Ghittori et al (1987) have proposed the urinary concentration of the solvent itself as a biological indicator of exposure. On the basis of a study conducted on 80 workers, they extrapolated that the 4-h exposure urinary concentration of the solvent corresponding to a TWA of 100 ppm is about 220 µg/l and suggested 194 µg/l as biological equivalent exposure limit (95% lower confidence limit of the regression equations).

Blood analyses

Hippuric acid in serum. Angerer et al (1975) described a gas chromatographic method for hippuric acid determination in serum. They exposed 98 rotogravure printers to an average of 101 mg/m^3 (27 ppm) of toluene. Results showed mean (±SD) levels of 0.43 ± 0.30 mg of hippuric acid/100 ml of serum and 2.2 ± 1.5 g of hippuric acid/l of urine. There was a low (r: 0.37), but significant correlation between hippuric acid concentration in serum and in urine. A control group had 0.19 ± 0.1 mg of hippuric acid per 100 ml of serum.

It seems that this biological marker offers no advantage over hippuric acid in urine for assessing exposure to toluene.

Table 4
Correlations Between Airborne Concentration of Toluene (x) and *o*-Cresol Concentration in Postshift Urine Sample (y)

Authors	Regression Equation	Concentration Corresponding to a TWA of 100 PPM	Correlation Coefficient
Hasegawa et al (1983)	y (μg/l) = 362.1 + 11.1x (ppm)	1.472 mg/l	0.607
	y (μg/g cr.) = 344.2 + 5.1x (ppm)	0.854 mg/g creatinine	0.396
	y (μg/l*) = 290 + 5.2x (ppm)	0.810 mg/l*: SG 1016	0.485
	y (μg/l = 147.3 + 12.3x (ppm)	1.377 mg/l	0.627
	y (μg/g cr.) = 168.5 + 9.3x (ppm)	1.098 mg/g creatinine	0.611
	y (μg/l*) = 118.4 + 7.4x (ppm)	0.858 mg/l*: SG 1016	0.617
De Rosa et al			
(1985)	y (μg/g cr.) = 62.7914 + 2.3225x (mg/m³)	0.934 mg/g creatinine	0.68
(1987)	y (μg/g cr.) = 64.54 + 1.92x (mg/m³)	0.784 mg/g creatinine	0.63
	y (μg/l*) = 140.79 + 3.47x (mg/m³)	1.442 mg/l*: SG 1024	0.62
Droz et al (1987)	y (mmol/mol cr.) = 0.280 + 0.00164x	1.592 mmol/mol creatinine	0.618
	x = ppm.h	1.52 mg/g creatinine	
Inoue et al (1988)	y (mg/l) = 146.6 + 8.19x (ppm)	0.965 mg/l	0.647
	y (mg/g cr.) = 161.6 + 8.14x (ppm)	0.975 mg/g creatinine	0.714
	y (mg/l*) = 144.4 + 7.37x (ppm)	0.881 mg/l*: SG 1016	0.720

Toluene. Engström et al (1976) found a low but statistically significant correlation (r: 0.64; n: 20) between toluene in blood and urinary hippuric acid (blood and urine samples collected at the end of an 8-h working day). On the contrary, Szadkowski et al (1973) found no correlation between toluene concentration in blood and hippuric acid in urine. von Oetingen et al (1942) exposed two volunteers to toluene vapor for 8 h and determined the toluene content of venous blood at the end of the exposure. They found the relationship presented in Table 5 between air and blood toluene concentration.

Szadkowski et al (1973) reported that in a control group the normal upper limit of the toluene concentration in blood was 15 μg toluene/100 ml blood (average: 5.3; n: 30). In an occupationally exposed group, they found a correlation between the toluene concentration in air and in blood. The correlation was, however, very

Table 5
Relationship Between Air and
Blood Toluene Concentration During Exposure

Toluene in Air (ppm)	Toluene in Blood (mg/100 ml)
200	0.41-0.73
300	>0.60-0.73
400	0.87-1.17
600	>0.66-0.95
800	1.82-2.64

From von Oetingen et al 1942.

low (r: 0.29). On the contrary, Apostoli et al (1982) found an excellent correlation (r: 0.88) between environmental toluene concentrations measured with personal samplers and the toluene levels in blood during exposure. In addition, these authors were unable to detect toluene in blood of non-exposed persons. Their data suggest that, in workers exposed to less than 50 ppm toluene (188 mg/m^3), the toluene concentration in blood during exposure is about 3 times higher than in air. A similar ratio was found by Angerer and Behling (see below), but on volunteers at rest the ratio seems closer to 1. In six subjects exposed at rest to 100 ppm toluene (375 mg/m^3), Astrand et al (1972) found a mean value for venous blood toluene concentration during exposure of 0.045 mg/100 ml (SD: 0.015). As expected, this concentration increased during exercise (0.135 ± 0.013 mg/100 ml after an exercise of 50 W for 20 min). For an exposure to 200 ppm, the concentration at rest amounted to 0.061 mg/100 ml (SD: 0.010). Under steady-state conditions, Veulemans and Masschelein (1978) found also a constant relation between uptake rate of toluene and toluene concentration in venous blood. During exposure to 50, 100, and 150 ppm toluene at rest, the average venous blood concentration reaches a plateau value of about 0.02, 0.04, and 0.06 mg/100 ml, respectively. Under nonsteady-state conditions, however, no simple relation exists between uptake rate and venous blood concentration of toluene.

According to Angerer and Behling (1981), an atmospheric concentration of 200 ppm toluene corresponds to a blood concentration higher than 0.26 mg/100 ml. According to Brugnone and Perbellini (1985), an exposure to 100 ppm would lead to a blood concentration at the end of the shift of about 975 µg/l, which is much higher than the results reported by Astrand et al (1972).

On the contrary to alveolar toluene concentrations, the correlation between blood toluene concentrations measured in venous blood samples collected at the end of the workshift and environmental toluene concentrations measured during the 7 h of the entire workshift was better (r: 0.84) than the correlation between blood and instantaneous environmental toluene concentrations measured at the end of the workshift (Brugnone et al 1986).

According to Nise and Orbaek (1988), an air level of 100 mg/m^3 (27 ppm) corresponds to an average blood level of 2.9 µmol/l (0.027 mg/100 ml) and an air level of 300 mg/m^3 (81 ppm) to 8.2 µmol/l (0.075 mg/100 ml) (samples taken at the end of the work).

Brugnone et al (1986) and Nise and Orbaek (1988) found a correlation between preshift blood toluene and the exposure during the previous day. However, the correlation was relatively weak (r: 0.36 and r: 0.40, respectively), the amount of toluene accumulated in body fat being probably of importance. It has indeed been suggested that the preshift level at the end of the week is an approximate index of the accumulated weekly exposure (Nise and Orbaek 1988). Nise et al (1989) have also suggested that blood toluene concentrations on Monday morning might be used as an index of the exposure intensity during the previous week.

From the data of Campbell et al (1987), exposure to 100 ppm toluene would be expected to produce a blood concentration of 20 µmol/l (0.184 mg/100 ml) at the end of the shift (95% limit confidence 8-55).

Foo et al (1988, 1991) established that the expected toluene concentration in the finger-prick blood was 140 µg/100 ml after 8 h of exposure to 100 ppm toluene (95%, 0.7-5.1).

The ACGIH (1991-1992) recommends 100 µg/100 ml as BEI for a toluene TWA of 100 ppm.

Breath analysis

Toluene.   By exposing volunters to toluene, Astrand et al (1972) have shown that the content of solvent in alveolar air samples collected during exposure is related to the intensity of exposure. At rest, the mean value for alveolar air concentration found during exposure to 100 ppm of toluene is 68 mg/m^3, or 18.1 ppm (SE: 1.41 ppm; n: 15). The corresponding value for an exposure to approximately 200 ppm at rest is 37.5 ppm. The correlation between environmental exposure to toluene and its alveolar concentration measured in workers during the workshift seems to agree with these estimates (Brugnone et al 1980). Astrand et al (1972) are of the opinion that "neither alveolar air samples (nor venous blood samples) taken at given intervals after the conclusion of a period of exposure provide sufficiently accurate information on the average amount of solvent in inspired air at a working place or on the magnitude of an individual's uptake".

According to Brugnone et al (1985), an exposure to 100 ppm toluene would lead to an alveolar concentration of about 125 mg/m^3 (during exposure). In another study (Brugnone et al 1986), these authors found a better correlation between the alveolar and environmental toluene concentrations measured by instantaneous samples collected simultaneously at various times during the workshift (r: 0.94) than between alveolar toluene concentrations measured at the end of the workshift and the TWA exposure to toluene during the entire workshift (r: 0.52).

Following Campbell et al (1987) after exposure to 100 ppm toluene, the breath concentration is about 580 nmol/l (14.25 ppm) when the sample is taken at the end of the shift and 54 nmol/l (1.3 ppm) when the breath sample is collected the next morning.

Foo et al (1991) extrapolated from their study that, at the end of the shift, the mean end-expired breath toluene concentration is 16 ppm (58 µg/l) after an 8-h exposure to 100 ppm.

The ACGIH (1990-1991) recommends 20 ppm in end-exhaled air (during shift) as BEI for a toluene TWA of 100 ppm.

Not enough investigations have been performed to evaluate whether or not analysis of expired air (or blood) collected 16 h after the end of exposure (i.e., before the next shift) can provide, as for benzene, an estimation of the magnitude of the intake of toluene of the previous day. Following Campbell et al (1987), with prolonged exposure an obese person will deposit more solvent in the adipose tissue, so this should be reflected in the breath concentration 16 h postexposure. Droz et al (1987) have also mentioned that the sample of the next morning might be better correlated with long-term exposure.

Analysis of expired air and/or blood during exposure reflects current intake. During exposure to 100 ppm toluene at rest, the average concentration of toluene in venous blood and in air will amount approximately to 40 µg/100 ml and 70 mg/ m^3, respectively. During light physical exercise, the blood concentration may amount to 110 µg/100 ml. Determination of the average hippuric acid concentration in urine collected at the end of the workshift is still the most practical method to evaluate whether or not the overall hygiene conditions are satisfactory. A group average below 2 g/l (SG 1.016) or 2 g/g creatinine suggests that the atmosphere probably contained less than 100 ppm toluene. On an individual basis, one must recognize that this threshold probably has a good specificity, but a very low sensitivity. Therefore, it is certainly not a suitable individual index of toluene exposure at low air concentrations.

There is no general agreement concerning the validity of the different biological indicators. For instance, in a 4-year-period study aimed to validate blood, breath, and urinary monitoring for toluene exposure, Campbell et al (1987) concluded urinary hippuric acid to be an unreliable method for monitoring workers, particularly at low exposure; blood toluene to be a sensitive, specific, and suitable indicator and breath sample to be a suitable indicator and a suitable alternative to blood sample.

Foo et al (1991) do agree on the fact that the use of hippuric acid is inappropriate and that the measurement of toluene in blood is a specific exposure indicator; however, as it is an invasive procedure, they state toluene concentration in expired air to be the most reliable indicator of exposure. On the contrary, for Droz et al (1987) (who did not measure blood toluene) hippuric acid in urine presents the highest validity and is the best indicator followed by toluene in expired air.

Currently, the determination of *o*-cresol in urine does not seem to offer advantages over the analysis of hippuric acid. However, it might be useful in case of doubt to confirm exposure to toluene.

Practically, the following biological values can be proposed corresponding to a TWA exposure to 100 ppm toluene: 2.5 g hippuric acid/g creatinine (end-shift); 1 mg *o*-cresol/g creatinine (end-shift); 20 ppm toluene in expired air (during the shift); 0.1 mg toluene/100 ml blood (during the shift).

REFERENCES

ACGIH. American Conference of Governmental Industrial Hygienists. Threshold Limit Values for Chemical Substances and Physical Agents and Biological Exposure Indices. Cincinnati, 1991-1992.

A. Aitio, K. Pekari and J. Järvisalo. Skin absorption as a source of error in biological monitoring. *Scand. J. Work Environ. Health* 10:317, 1984.

J. Angerer, V. Kassebart, D. Szadkowski and G. Lehnert. Chronische Lösungsmittelbelastung am Arbeitsplatz. III. Eine gaschromatographische Method zur Bestimmung von Hippursäure im Serum. *Int. Arch. Arbeitsmed.* 34:199, 1975.

J. Angerer. Chronische Lösungsmittelbelastung am Arbeitsplatz. IV. Eine dünn-schicktchromatographisch densitometrische Method zur Bestimmung von Hippursäure im Harn. *Int. Arch. Occup. Environ. Health* 36:287, 1976.

J. Angerer. Occupational chronic exposure to organic solvents. VII. Metabolism of toluene in man. *Int. Arch. Occup. Environ. Health* 43:63, 1979.

J. Angerer and K. Behling. Chronische Lösungsmittelbelastung am Arbeitsplatz. *Int. Arch. Occup. Environ. Health* 48:137, 1981.

J. Angerer. Occupational chronic exposure to organic solvents. XII. O-cresol excretion after toluene exposure. *Int. Arch. Occup. Environ. Health* 56:323, 1985.

P. Apostoli, F. Brugnone, L. Perbellini et al. Biomonitoring of occupational toluene exposure. *Int. Arch. Occup. Environ. Health* 50:153, 1982.

I. Astrand, H. Ehrner-Samuel, A. Kilbom and P. Ovrum. Toluene exposure. I. Concentration in alveolar air and blood at rest and during exercise. *Scand. J. Work Environ. Health* 9:119, 1972.

ATSDR. Agency for Toxic Substances and Disease Registry. Toxicological profile for toluene. US Public Health Service. ATSDR/TP 89/23. 1989.

J. Baelum, M. Dossing, S. Hansen et al. Toluene metabolism during exposure to varying concentrations combined with exercise. *Int. Arch. Occup. Environ. Health* 59:281, 1987.

J. Baelum. Toluene in alveolar air during controlled exposure to constant and to varying concentrations. *Int. Arch. Occup. Environ. Health* 62:59, 1990.

J. Baelum. Human solvent exposure. Factors influencing the pharmacokinetics and acute toxicity. *Pharmacol. Toxicol.* 68 Suppl. 1, 1991.

O. Bakke and R. Scheline. Hydroxylation of aromatic hydrocarbons in the rat. *Toxicol. Appl. Pharmacol.* 16:691, 1970.

Z. Bardodej. Beurteilung der Gefahrdung durch Toluol in der Industrie mittels der Hippursäurebestimmung im Harn. *Arbeitsmed. Socialmed. Arbeitshyg.* 3:254, 1968.

F. Brugnone, L. Perbellini, E. Gaffuri and P. Apostoli. Biomonitoring of industrial solvent exposures in workers' alveolar air. *Int. Arch. Occup. Environ. Health* 47:245, 1980.

F. Brugnone, L. Perbellini, P. Apostoli et al. Decline of blood and alveolar toluene concentration following two accidental human poisonings. *Int. Arch. Occup. Environ. Health* 53:157, 1983.

F. Brugnone and L. Perbellini. Biological monitoring of occupational exposure to solvents by analysis of alveolar air and blood. In: Organic solvents and the Central Nervous System. Environmental Health Series No. 5. Report on a joint WHO/Nordic Council of Ministers Working Group. Copenhagen, 10-14 June, p 56, 1985.

F. Brugnone, E. De Rosa, L. Perbellini and G.B. Bartolucci. Toluene concentrations in the blood and alveolar air of workers during the workshift and the morning after. *Br. J. Ind. Med.* 43:56, 1986.

J.-P. Buchet and R. Lauwerys. Measurement of urinary hippuric and m-methylhippuric acids by gas chromatography. *Br. J. Ind. Med.* 30:125, 1973.

L. Campbell, D. Marsh and H. Wilson. Towards a biological monitoring strategy for toluene. *Ann. Occup. Hyg.* 31:121, 1987.

K. Cohr and J. Stokholm. Toluene, a toxicologic review. *Scand. J. Work Environ. Health* 5:71, 1979.

E. De Rosa, F. Brugnone, G. Bartolucci et al. The validity of urinary metabolites as indicators of low exposures to toluene. *Int. Arch. Occup. Environ. Health* 56:135, 1985.

E. De Rosa, G.B. Bartolucci and M. Sigon. Environmental and biological monitoring of workers exposed to low levels of toluene. *Appl. Ind. Hyg.* 1:132, 1986.

E. De Rosa, G. Bartolucci, M. Sigon et al. Hippuric acid and ortho-cresol as biological indicators of occupational exposure to toluene. *Am. J. Ind. Med.* 11:529, 1987.

E. De Rosa, G. Bartolucci, L. Perbellini et al. Environmental and biological monitoring of exposure to toluene, styrene and *n*-hexane. *Appl. Ind. Hyg.* 3:332, 1988.

DFG. Deutsche Forschungsgemeinschaft. *Maximum Concentrations at the Workplace and Biological Tolerance Values for Working Materials* Report No. XXVII. Commission for the Investigation of Health Hazards of Chemical Compounds in the Work Area. VCH, Weinheim, 1991.

M. Dossing, J. Baelum, S. Hansen et al. Urinary hippuric acid and ortho-cresol excretion in man during experimental exposure to toluene. *Br. J. Ind. Med.* 40:470, 1983.

M. Dossing, J. Baelum, S. Hansen and S. Lundqvist. Effect of ethanol, cimetidine and propanol on toluene metabolism in man. *Int. Arch. Occup. Environ. Health* 54:309, 1984.

P. Droz, M. Berode, M. Boillat and M. Lob. Biological monitoring and health surveillance of rotogravure printing workers exposed to toluene. In: *Biological Monitoring of Exposure to Chemicals. Organic Compounds.* Eds.: M. Ho, H.Dillon. Wiley Interscience, New York, p. 111, 1987.

T. Dutkiewicz and H. Tyras. The quantitative estimation of toluene skin absorption in man. *Arch. Gewerbepathol. Gewerbehyg.* 24:253, 1968.

K. Engström, K. Husman and J. Rantanen. Measurements of toluene and xylene metabolites by gas chromatography. *Int. Arch. Occup. Environ. Health* 36:153, 1976.

S. Foo, W. Phoon and N. Khoo. Toluene in blood after exposure to toluene. *Am. Ind. Hyg. Assoc. J.* 49:255, 1988.

S. Foo, J. Jayaratnam, C. Ong et al. Biological monitoring for occupational exposure to toluene. *Am. Ind. Hyg. Assoc. J.* 52:212, 1991.

S. Ghittori, M. Imbriani, G. Pezzagno and E. Capodaglio. The urinary concentration of solvents as a biological indicator of exposure: proposal for the biological equivalent exposure limit for nine solvents. *Am. Ind. Hyg. Assoc. J.* 48:786, 1987.

Ph. Grandjean. *Skin penetration: hazardous chemicals at work.* Report prepared for the Commission of the European Communities. Taylor and Francis, Luxembourg, Eur 12599, 1990.

M. Guillemin, J.-C. Murset, M. Lob and J. Riquez. Simple method to determine the efficiency of a cream used for skin protection against solvents. *Br. J. Ind. Med.* 31:310, 1974.

H. Hajimiragha. Levels of benzene and other volatile aromatic compounds in the blood of non-smokers and smokers. *Int. Arch. Occup. Environ. Health* 61:513, 1989.

K. Hasegawa, Sh. Shiojima, A. Koizumi and M. Ikeda. Hippuric acid and *o*-cresol in the urine of workers exposed to toluene. *Int. Arch. Occup. Environ. Health* 52:197, 1983.

M. Ikeda and H. Ohtsuji. Significance of urinary hippuric acid determination as an index of toluene exposure. *Br. J. Ind. Med.* 26:44, 1969.

T. Imamura and M. Ikeda. Lower fiducial limit of urinary metabolite level as an index of excessive exposure to industrial chemicals. *Br. J. Ind.* 30:289, 1973.

O. Inoue, K. Seiji, T. Watanabe et al. Possible ethnic differences in toluene metabolism: a comparative study among Chinese, Turkish, and Japanese solvent workers. *Toxicol. Lett.* 34:67, 1986.

O. Inoue, K. Seiji, T. Watanabe et al. Mutual metabolic suppression between benzene and toluene in man. *Int. Arch. Occup. Environ. Health* 60:15, 1988.

L. Lowry. Review of biological monitoring tests for toluene. In: *Biological Monitoring of Exposure to Chemicals. Organic Compounds.* Eds.: M. Ho, H. Dillon. Wiley Interscience, New York, p. 99, 1987.

G. Nise and P. Orbaek. Toluene in venous blood during and after work in rotogravure printing. *Int. Arch. Occup. Environ. Health* 60:31, 1988.

G. Nise, R. Attewell, S. Skerfving and P. Orbaek. Elimination of toluene from venous blood and adipose tissue after occupational exposure. *Br. J. Ind. Med.* 46:407, 1989.

G. Nise. Urinary excretion of O-cresol and hippuric acid after toluene exposure in rotogravure printing. *Int. Arch. Occup. Environ. Health* 63:377, 1992.

M. Ogata, K. Tomokuni and Y. Takatsuka. Urinary excretion of hippuric acid and m- or p-methylhippuric acid in the urine of persons exposed to vapors of toluene and m- or p-xylene as a test of exposure. *Br. J. Ind. Med.* 27:43, 1970.

M. Ogata and T. Taguchi. Simultaneous determination of urinary creatinine and metabolites of toluene, xylene, styrene, ethylbenzene and phenol by automated HPLC. *Int. Arch. Occup. Environ. Health* 61:131, 1988.

L. Pagnotto and L. Lieberman. Urinary hippuric acid excretion as an index of toluene exposure. *Am. Ind. Hyg. Assoc. J.* 28:129, 1967.

O. Pfäffli, H. Savolainen, P.L. Kalliomake and P. Kalliokoski. Urinary *o*-cresol in toluene exposure. *Scand. J. Work Environ. Health* 5:286, 1979.

J. Piotrowski. Quantitative estimate of the absorption of toluene in people. *Med. Pracy* 18:213, 1967 (in Polish).

V. Riihimaki. Conjugation and urinary excretion of toluene and m-xylene metabolites in man. *Scand. J. Work Environ. Health* 5:135, 1979.

J. Srbova and J. Teisinger. Absorption and elimination of toluene in man. *Prac. Lek.* 4:41, 1952.

D. Szadkowski, R. Pett, J. Angerer et al. Chronische Lösungsmittelbelastung am Arbeitsplatz. II. Schadstoffspiegel in Blut und Metabolitenelimination im Harn in ihrer Bedeutung als Überwachungskriterien bei toluolexponierten Tiefdruckern. *Int. Arch. Arbeitsmed.* 31:265, 1973.

R. Tardif, S. Lapare, G. Plaa and J. Brodeur. Effect on simultaneous exposure to toluene and xylene on their respective biological exposure indices in humans. *Int. Arch. Occup. Environ. Health* 63:279, 1991.

R. Tokunaga, S. Takahata, M. Onoda et al. Evaluation of the exposure to organic solvent mixture. *Int. Arch. Arbeitsmed.* 33:257, 1974.

H. Veulemans and R. Masschelein. Experimental human exposure to toluene. II. Toluene in venous blood during and after exposure. *Int. Arch. Occup. Environ. Health* 42:105, 1978.

H. Veulemans and R. Masschelein. Experimental human exposure to toluene. III. Urinary hippuric acid excretion as a measure of individual solvent uptake. *Int. Arch. Occup. Environ. Health* 43:53, 1979.

H. Veulemans, E. Van Vlem, H. Janssens and R. Masschelein. Exposure to toluene and urinary hippuric acid excretion in a group of heliorotagravure printing workers. *Int. Arch. Occup. Environ. Health* 44:99, 1979.

W. von Oetingen, P. Neal and D. Danahue. The toxicity and potential dangers of toluene. *J. Am. Med. Assoc.* 118:579, 1942.

H. Waldron, N. Cherry and J. Johnston. The effects of ethanol on blood toluene concentrations. *Int. Arch. Occup. Environ. Health* 51:365, 1983.

M. Wallen, S. Holm and M. Nordqvist. Co-exposure to toluene and p-xylene in man: uptake and elimination. *Br. J. Ind. Med.* 42:111, 1985.

M. Wallen. Toxicokinetics of toluene in occupationally exposed volunteers. *Scand. J. Work Environ. Health* 12:588, 1986.

E. Wigaeus Hjelm, Ph. Näslund and M. Wallen. Influence of cigarette smoking on the toxicokinetics of toluene in man. *J. Toxicol. Environ. Health* 25:155, 1988.

T. Wilczok and G. Bieniek. Urinary hippuric acid concentration after occupational exposure to toluene. *Br. J. Ind. Med.* 35:330, 1978.
W. Woiwode, R. Wordarz, K. Drysch and H. Weichardt. Metabolism of toluene in man:
gas chromatographic determination of *o*-, m- and p-cresol in urine. *Arch. Toxicol.*
43:93, 1979.
W. Woiwode and K. Drysch. Experimental exposure to toluene: further consideration of
cresol formation in man. *Br. J. Ind. Med.* 38:194, 1981.
WHO. IPCS. International Programme on Chemical Safety. Environmental health criteria
52: Toluene. World Health Organization. Geneva, 1985.

2.3. Xylene

Toxicokinetics

As a result of its widespread use, xylene is ubiquitously present in indoor and
outdoor environments. There are three isomers of xylene: ortho, meta, and para. The
pulmonary retention of xylene vapors in man at rest amounts to about 55-65% of
the quantity inhaled (Sedivec and Flek 1967, 1976a; Wallen et al 1985; Liira et al
1988; Norström et al 1989). The proportional retention does not change with the
intensity and the duration of exposure, but with the ventilatory activity. There appear
to be a first short phase of absorption which occurs within 15 min of initiation of
exposure, and a second longer phase representing the establishment of an equilibrium between the inhaled xylene and blood (ATSDR 1990). In eight volunteers
experimentally exposed to *p*-xylene (4-h inhalation), the blood solvent concentration reached its maximum at about 150 min after the onset of exposure and
decreased during the latter part of the exposure irrespective of the exposure conditions (high or low) (Wallen et al 1985). Experiments on volunteers have demonstrated that liquid xylene is absorbed through the skin (Engström et al 1977;
Lauwerys et al 1978). Extensive skin exposure could result in systemic toxicity and
it should therefore be regarded as a skin-absorption hazard (Grandjean 1990). After
its absorption, xylene distributes to all the tissues, primarily the adipose tissue. It has
been calculated that, in man, more than 70% of the xylene absorbed is metabolized
and less than 10% is excreted unchanged with the expired air (Astrand et al 1978;
Sedivec and Flek 1967, 1976a; Riihimaki et al 1979a). The estimated pulmonary
elimination of xylene in the study of Liira et al (1988), conducted on eight volunteers
exposed to 100 ppm xylene, was 8.4 to 10.5% of the total uptake, the amount of
methylhippuric acid excreted during a 24-h period represented 71.4 to 80.3%.The
biotransformation undergone by the xylene isomers can be summarized as follows
(Figure 11) (Ogata et al 1970; Flek and Sedivec 1975). The main metabolic pathway
is oxidation to the corresponding toluic acids (methylbenzoic acids). In man, the
toluic acids are mainly conjugated with glycine to form *o*-, *m*-, and *p*-methylhippuric
acids (toluric acids), which are excreted in urine. In man, it seems that no
toluylglucuronic acid is formed (Sedivec and Flek 1976a). Hydroxylation of the
aromatic ring with formation of xylenols also occurs *in vivo*, but it has been
estimated that in man less than 3% of absorbed xylene is excreted as xylenols (Flek
and Sedivec 1975; Riihimaki et al 1979a; Sedivec and Flek 1976a). For example,

in the study by Engström et al (1984), which involved four volunteers exposed to 150 ppm *m*-xylene , the profile of urinary metabolites during the 24-h period from onset of exposure consisted in 97.4% *m*-methylhippuric acid, 2.5% 2,4-dimethylphenol, 0.05% 3-methylbenzylalcohol. In six volunteers exposed to *o*-xylene, the cumulative excretion of *o*-methylhippuric acid was 53% of the total uptake 3 hours postexposure (Norström et al 1989). As for any industrial solvent, the metabolites of xylene are rapidly excreted. There seems to be at least two distinct phases of elimination: a relatively rapid one (elimination from the muscles) and a slower one (elimination from the adipose tissue), the excretion of methylhippuric acid being complete within a day or two after the end of exposure (ATSDR 1990). In a subject exposed to 17 mg/m^3 of *m*-xylene for 95 min, the peak concentration of *m*-methylhippuric acid in urine was observed 4 hours after the end of exposure. Levels returned to baseline within 11 h after exposure (Lowry et al 1987).

In practice, it is usual to find that the amount of methylhippuric acid excreted reaches a maximum at the end of the exposure period (Sedivec and Flek 1976a). Sedivec and Flek (1976a) have found that, when volunteers are exposed to a mixture of xylene isomers, the excretion curve of total toluic acids is the same as that obtained when exposure is to pure isomers.

Several authors have examined whether coexposure to other substances may influence methylhippuric acid production from xylene. Experiments on volunteers who were given a moderate dose of ethanol (0.8 g/kg) prior to 4-h inhalation of xylene (145 and 280 ppm) have shown that urinary methylhippuric acid declined by about 50% while blood xylene level rose about 1.5- to 2.0-fold (Riihimaki et al 1982). This suggests that ethanol decreases the metabolic clearance of xylene by about one-half during xylene inhalation.

Coexposure to xylene and methylethylketone resulted in increased concentration of xylene in blood and decreased urinary excretion of methylhippuric acid (Liira et al 1988). However, no significant differences in the urinary excretion of methylbenzoic acid were observed after exposure to *m*-xylene (45 to 74 ppm) and exposure to *m*-xylene together with *n*-hexane, toluene (slight decrease 10-15%, but not significant), methylisobutylketone, or *n*-butyl alcohol (Jakubowski and Kostrzewski 1989). Engström et al (1984) also observed a mutual inhibition of solvent metabolism in case of coexposure to ethylbenzene and xylene.

The ingestion of aspirin (1500 mg over 5 h) by volunteers during a 4-h inhalation of *m*-xylene (100 ppm) produced a reduction of the excretion of the major metabolites of both *m*-xylene and aspirin of about 50% (Campbell et al 1988).

Biological monitoring

The biological tests that have been considered for evaluating xylene exposure are methylhippuric acid in urine, xylene in blood, and xylene in expired air.

Urine analysis

Methylhippuric acid. Methylhippuric acid is not normally present in urine. Several authors have attempted to quantitatively correlate the urinary excretion of

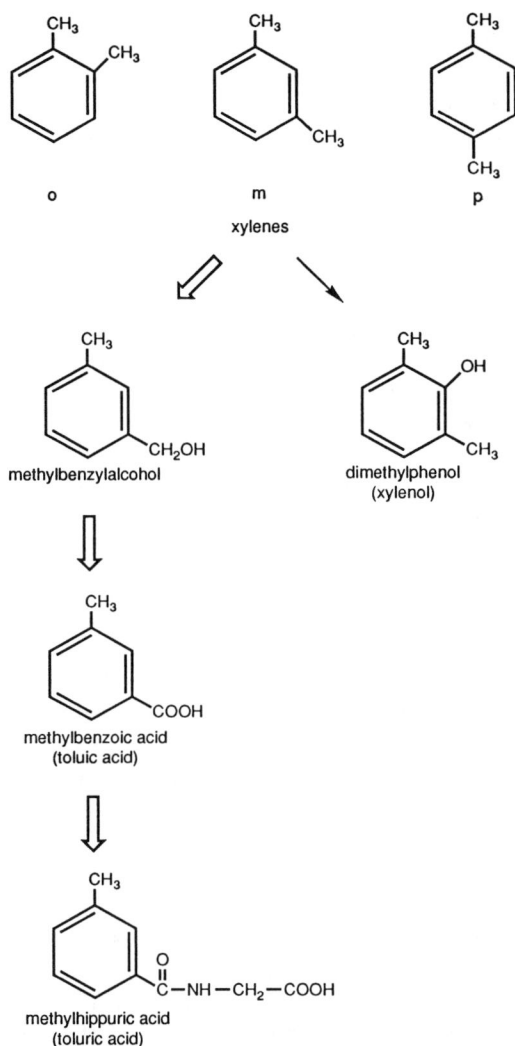

FIGURE 11 Main metabolic pathways of xylene.

xylene metabolites with exposure level. Of course, the rapid excretion of methylhippuric acid limits its utility to the assessment of very recent exposure (ATSDR 1990). Ogata et al (1970) exposed four or five volunteers for 7 h to 100 ppm *m*- or *p*-xylene or to 200 ppm *m*-xylene. They found that total *p*- or *m*-methylhippuric acid excretion during and for 18 h after exposure was linearly related to exposure levels. An integrated exposure (700 ppm-h) to *m*- or *p*-xylene corresponded to a total urinary excretion of about 1.5 g of *m*- or *p*-methylhippuric acid. For the same integrated exposure, the average *m*-methylhippuric acid concentration in urine collected during the second period of exposure amounted to about 2.63 g/l (corrected for a SG of 1.024) or 1.75 g/l (corrected for a SG of

1.016), 90% of the values were included in the range 0.75-2.75 g/l (corrected for a SG of 1.016). Data for p-methylhippuric acid were rather similar. The authors confirmed the usual observation that the ranges of concentration were considerably decreased by correcting results to a constant urine density. As expected, the standard deviations of rates of excretion were still smaller. Ogata et al (1970) have also suggested that, for screening purposes, the presence of more methylhippuric acid in the urine of a man than that corresponding to 2 standard deviations less than the average quantity excreted by subjects exposed to the permissible level should be taken as evidence that the man may have been exposed to a concentration greater than this level. This will be true of about 5% of the subjects tested. According to their data, the screening (5th percentile) concentration of m- and p-methylhippuric acids in urine collected during the second period of the shift should be 0.73 and 0.89 g/l, respectively (corrected for a SG of 1.016), for a TLV of 100 ppm.

Sedivec and Flek (1976b) also exposed volunteers for 8 h to xylene and followed the excretion of methylhippuric acid (measured as toluic acid). They found that the amount of metabolite excreted reached a maximum at the end of exposure and then decreased exponentially. There was a linear relationship between the mean vapor concentration in the air and the amount of metabolite excreted. In persons exposed for 8 h to a constant xylene concentration of about 46 ppm (ventilation 9 l/min), the concentration of methylhippuric acid found in the urine collected during the last 2 h of exposure ranged from 0.88 to 1.65 g/g creatinine. When these concentrations were calculated for the urine excreted during the whole exposure period (8 h), the values found were 0.67 to 1.45 g/g creatinine. The authors studied the correlation between exposure levels and the amount of methylhippuric acid excreted during the shift (8 h) by using different methods of expressing the quantity of metabolite excreted. The variability of the results decreased in the following order: mg/l > mg/l (corrected for SG) > mg/g creatinine = mg/unit time > mg/kg actual body weight = mg/kg ideal body weight > mg/l of air ventilated. For an average exposure of 92 ppm (400 mg/m^3) for 8 h, the average urinary concentration of methylhippuric acid in the total-shift urine sample amounted to about 2.0 g/g creatinine or 1.53 g/l (SG 1.016) with a range of approximately 1.28 to 2.71 g/g creatinine or 0.73 to 2.27 g/l (SG 1.016). The same authors have shown that, if the concentration of vapors in the atmosphere was constant, the short-term (last 2 h of exposure), all-shift, or all-day samples of urine could be used with equal success for estimation of the level of exposure.

Mikulski et al (1972) have studied the relationship between urinary metabolite excretion and xylene concentration in air among 51 painters using as solvent a mixture of toluene (10-20%) and xylene (80-90%). Unfortunately, they used a method that does not distinguish between hippuric and methylhippuric acid and it is therefore difficult to draw precise conclusions from their data. It can, however, be estimated that, in a group of workers exposed for 8 h to an average xylene concentration of 93 ppm and to only 7 ppm toluene, the increment in hippuric acid excretion (in fact, mainly methylhippuric acid) over control values was about 0.68 g/l (SG 1.016; urine collected during the second part of the shift). This result is lower than those reported by Ogata et al (1970) and Sedivec and Flek (1976b). The discrepancy is probably due to the nonspecificity of the technique used by Mikulski et al (1972). Indeed, only the analytical techniques involving a chromatographic step are sufficiently specific to measure methylhippuric acid in urine with accuracy. In painters exposed to xylene, Engström et al (1978) found a good relationship between the TWA exposure and urinary methylhippuric acid concentration at the end of the workday. Amounts of 665 and 1280 mg methylhippuric acid per g of creatinine corresponded to 50 and 100 ppm xylene, respectively. The authors recognize, however, that these values may be too low because their urine analyses ignore the absorbed ortho-isomer, which represented 10 to 15% of the technical xylene used by the workers. In volunteers exposed to about 90 ppm m-xylene for 6 h, Riihimaki et al (1979b) found in urine collected at the end of the exposure period a mean concentration of methylhippuric acid of 0.85 g/g creatinine at rest and 1.27 g/g creatinine under light work. Ogata and Taguchi (1986) extrapolated the urinary concentration of methylhippuric acids corresponding to a TWA exposure of 100 ppm xylene to be 2.05 g/g creatinine.

However, in eight volunteers exposed to 100 ppm xylene for 4 h, Liira et al (1988) reported a lower mean urinary excretion of 4.94 mmol methylhippuric acid/24 h (954 mg/24 h).

From observations on painters exposed to various concentrations of xylene and other solvents, Lundberg and Sollenberg (1986) estimated the correlation between the 8-h TWA xylene exposure and different methods of expressing methylhippuric acid excretion: the amount of methylhippuric acid excreted in approximately 24 h showed the highest correlation to exposure (r: 0.84) but not much higher than the excretion in the afternoon expressed as rate of excretion (r: 0.81); when expressed as amount per liter the correlation was 0.76. The body weight adjustment improved the correlation coefficients. According to these results, an 8-h exposure to 100 ppm would yield a methylhippuric acid excretion of around 14.2 μmol/kg body weight/h (2.7 mg/kg body weight/h) during the last few h of a workshift. This is in agreement with the average estimate, of Riihimaki (1984): 14 μmol/kg/h (1000 μmol/h or about 3.2 mg/min) or 1300 mmol/mol creatinine (or about 2.2 g/g creatinine).

Kawai et al (1991) studied the urinary excretion of methylhippuric acid isomers in the second half of a working week in 121 workers exposed predominantly to the three xylene isomers: a mean exposure (geometric mean) to 3.8 ppm xylenes (0.8 ppm o-xylene, 2.1 ppm m-xylene, 0.9 ppm p-xylene) lead to a mean methylhippuric acid excretion (GM) of 57.8 mg/l (14.0 mg/l o-MHA, 24.6 mg/l m-MHA, 16.5 mg/l p-HA). This study suggests that for a TWA exposure of 100 ppm the urinary concentration of methylhippuric acid isomers at the end of a shift would be 1.8 g/l or 1.7 g/g creatinine.

In conclusion, the available data suggest that, in a group of workers exposed to 100 ppm xylene, the mean methylhippuric acid concentration in urine collected in the second part of a 7-h exposure period would amount to about 1.5 to 2 g/g creatinine with a range between 1.0 to 3.0 g/g creatinine. The BEI recommended by ACGIH (1991-1992) are 1.5 g/g creatinine (end-of-shift sample) and 2 mg/min (last 4 h of the shift) corresponding to a xylene TWA of 100 ppm (434 mg/m^3). The DFG (1991) recommends 2 g/l (end-of-shift).

Blood analysis

Xylene. Measurement of blood level of xylene is limited by the rapid metabolism of xylene (ATSDR 1990). During exposure, xylene concentration in blood is proportional to recent uptake (Riihimaki et al 1979b). Angerer and Lehnert (1979) have reported that, in workers doing light physical work, exposure to 100 ppm xylene produced concentrations in blood ranging from 0.36 to 0.44 mg per 100 ml (*m*- and *p*-xylene) and 0.22 to 0.32 mg per 100 ml (*o*-xylene). On volunteers exposed to 200 ppm xylene and doing light physical work (50 W) or exposed to 100 ppm xylene and doing heavy physical work (150 W), Astrand et al (1978) have found mean concentrations of xylene in venous blood of 0.6 and 0.52 mg/ 100 ml, respectively.

In volunteers exposed to 90 ppm xylene, Riihimaki et al (1979b) measured at the end of exposure a mean concentration of xylene in blood of 0.13 mg/100 ml at rest and 0.21 mg/100 ml under light work. The corresponding values 18 h after the end of exposure were 0.006 and 0.016 mg/100 ml, respectively. As indicated above, they also found that ethanol administration prior to xylene exposure increases the blood xylene concentration (Riihimaki et al 1982). The DFG (1991) recommends 1.5 mg/l (end-of-shift).

Breath analysis

Xylene. Exhaled air sampling during the workday may be used for the estimation of momentary exposure. In painters, Engström et al (1978) found that the xylene concentration in the exhaled air amounts to approximately 8% of the mean level in the ambient air sample collected during the preceding 0.5-h period. The same authors were unable to estimate the average amount of xylene in ambient air from exhaled air samples or venous blood taken at several intervals after the termination of exposure. The ratio between expired air concentrations and environmental concentrations reported by Engström et al (1978) on painters is lower than that (24-36%) found by Astrand et al (1978) on volunteers.

In conclusion, the following biological values correspond to a TWA of 100 ppm xylene: 0.3 mg xylene/100 ml blood during the exposure and 1.5 g methylhippuric acid/g creatinine at the end of the shift.

REFERENCES

ACGIH. American Conference of Governmental Industrial Hygienists. Threshold Limit Values for Chemical Substances and Physical Agents and Biological Exposure Indices. Cincinnati, 1991-1992.
ATSDR. Agency for Toxic Substances and Disease Registry, U.S. Public Health Service. Toxicological profile for total xylenes. ATSDR/TP-90/30, 1990.

J. Angerer and G. Lehnert. Occupational chronic exposure to organic solvents. VIII. Phenolic compounds — metabolites of alkylbenzenes in man. Simultaneous exposure to ethylbenzene and xylenes. *Int. Arch. Occup. Environ. Health* 43:145, 1979.

I. Astrand, J. Engström and P. Ovrum. Exposure to xylene and ethylbenzene. I. Uptake, distribution and elimination in man. *Scand. J. Work Environ. Health* 4:185, 1978.

L. Campbell, H. Wilson, A. Samuel and D. Gompertz. Interactions of *m*-xylene and aspirin metabolism in man. *Br. J. Ind. Med.* 45:127, 1988.

DFG. Deutsche Forschungsgemeinschaft. *Maximum Concentrations at the Workplace and Biological Tolerance Values for Working Materials* Report No. XXVII. Commission for the Investigation of Health Hazards of Chemical Compounds in the Work Area. VCH, Weinheim, 1991.

K. Engström, K. Husman and V. Riihimaki. Percutaneous absorption of *m*-xylene in man. *Int. Arch. Occup. Environ. Health* 39:181, 1977.

K. Engström, K. Husman, P. Pfaffli and V. Riihimaki. Evaluation of occupational exposure to xylene by blood, exhaled air and urine analysis. *Scand. J. Work Environ. Health* 4:114, 1978.

K. Engström, V. Riihimaki and A. Laine. Urinary disposition of ethylbenzene and *m*-xylene in man following separate and combined exposure. *Int. Arch. Occup. Environ. Health* 54:355, 1984.

J. Flek and V. Sedivec. Metabolism of isomeric xylene in man. *Prac. Lek.* 27:9, 1975.

Ph. Grandjean. *Skin penetration: Hazardous chemicals at work.* Report prepared for the Commission of the European Communities. Taylor and Francis, Luxembourg, Eur 12599 En, 1990.

M. Jakubowski and P. Kostrzewski. Excretion of methylbenzoic acid in urine as a result of single and combined exposure to *m*-xylene. (Polish). *J. Occup. Med.* 2:238, 1989.

T. Kawai, Mizunuma, T. Yasugi et al. Urinary methylhippuric acid isomer levels after occupational exposure to an exposure mixture. *Int. Arch. Occup. Environ. Health* 63:69, 1991.

K. Lauwerys, T. Dath, J. Lachapelle et al. The influence of two barrier creams on the percutaneous absorption of *m*-xylene in man. *J. Occup. Med.* 20:17, 1978.

J. Liira, V. Riihimaki, K. Engström and P. Pfäffli. Coexposure of man to *m*-xylene and methyl ethyl ketone: kinetics and metabolism. *Scand. J. Work Environ. Health* 14:322, 1988.

L. Lowry, T. Thoburn, F. Phipps et al. Xylene exposure in a histology laboratory investigated by environmental and biological monitoring. In: *Biological Monitoring of Exposure to Chemicals. Organic Compounds.* Eds.: M. Ho, H. Dillon. Wiley Interscience, New York, p. 143, 1987.

I. Lundberg, J. Sollenberg. Correlation of xylene exposure and methyl hippuric acid excretion in urine among paint industry workers. *Scand. J. Work Environ. Health* 12:149, 1986.

P. Mikulski, R. Wiglusz, A. Bublewski and J. Uselis. Investigation of exposure of ships painters to organic solvents. *Br. J. Ind. Med.* 29:450, 1972.

A. Norstroem, B. Andersson, J. Levin et al. Biological monitoring of *o*-xylene experimental exposure in man: determination of urinary excretion products. *Chemosphere* 18:1513, 1989.

M. Ogata, K. Tomokuni and Y. Takatsuka. Urinary excretion of hippuric acid and *m*- or *p*-methylhippuric acid in the urine of persons exposed to vapours of toluene and *m*- or *p*-xylene as a test of exposure. *Br. J. Ind. Med.* 27:43, 1970.

M. Ogata and T. Taguchi. Quantitative analysis of urinary glycine conjugates by high performance liquid chromatography: excretion of hippuric acid and methylhippuric acids in the urine of subjects exposed to vapours of toluene and xylenes. *Int. Arch. Occup. Environ. Health* 58:121, 1986.

V. Riihimaki. Xylene. In: *Biological Monitoring and Surveillance of Workers Exposed to Chemicals*. Eds.: A. Aïtio, V. Riimäki, H. Vainio. Hemisphere Publishing, Washington D.C., p. 83, 1984.

V. Riihimaki, P. Pfäffli, K. Savolainen and K. Pekari. Kinetics of *m*-xylene in man. General features of absorption, distribution, biotransformation and excretion in repetitive inhalation exposure. *Scand. J. Work Environ. Health* 5:217, 1979a.

V. Riihimaki, P. Pfäffli and K. Savolainen. Kinetics of *m*-xylene in man. Influence of intermittent physical exercise and changing environmental concentrations on kinetics. *Scand. J. Work Environ. Health* 5:232, 1979b.

V. Riihimaki, K. Savolainen, P. Pfäffli et al. Metabolic interaction between *m*-xylene and ethanol. *Arch. Toxicol.* 49:253, 1982.

V. Sedivec and J. Flek. Absorption and excretion of xylene in man. *Prac. Lek.* 26:243, 1967 (English summary only).

V. Sedivec and J. Flek. The absorption, metabolism and excretion of xylenes in man. *Int. Arch. Occup. Environ. Health* 37:205, 1976a.

V. Sedivec and J. Flek. Exposure test for xylenes. *Int. Arch. Occup. Environ. Health* 37:219, 1976b.

M. Wallen, S. Holm and M. Byfalt Nordqvist. Coexposure to toluene and *p*-xylene in man: uptake and elimination. *Br. J. Ind. Med.* 42:111, 1985.

2.4. Ethylbenzene

Toxicokinetics

Inhalation is the most important route of absorption. Liquid ethylbenzene may be absorbed through the skin, but not the vapors (Gromiec and Piotrowski 1984). Ethylbenzene accumulates primarily in the intestine, liver, kidney, and fat (ATSDR 1990). Ethylbenzene follows approximately the same metabolic pathway as styrene (Figure 12).

Bardodej and Bardodejova (1970) exposed human volunteers to concentrations of ethylbenzene ranging from 23 to 85 ppm for 8 h. They found that 64% of the vapors were retained in the respiratory tract; 64, 25, and 5% of the retained dose was eliminated as mandelic acid, phenylglyoxylic acid, and methylphenylcarbinol, respectively. No elevation of hippuric acid excretion was demonstrated. Engström et al (1984) studied the biotransformation of ethylbenzene in four volunteers exposed to 150 ppm (655 mg/m^3) of the solvent; ring oxidation accounted for 4% (4-ethylphenol, *p*- and *m*-hydroxyacetophenones); mandelic and phenylglyoxylic acids amounted to 90% of the ethylbenzene metabolites (71.5 and 19.1%, respectively) and phenylethanol to 4%. The combined exposure to ethylbenzene and *m*-xylene decreased the amounts of metabolites found as compared to separate exposure and delayed the excretion of most metabolites (Engström et al 1984).

Gromiec and Piotrowski (1984) found, in six volunteers exposed to concentrations of 18, 34, 80, and 200 mg/m^3, a pulmonary retention of 49 ± 5% (mean ± SD). Total excreted mandelic acid accounted for 55 ± 2% of retained ethylbenzene.

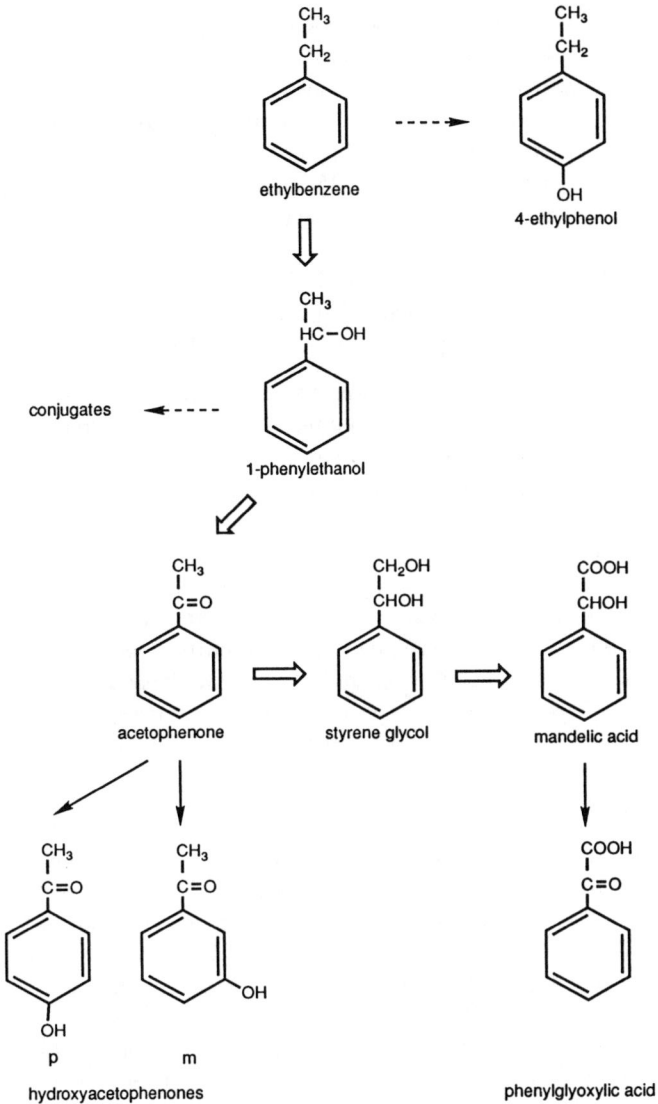

FIGURE 12 Main metabolic pathways of ethylbenzene.

The elimination of this metabolite was biphasic with biological half-lives of 3.1 and 24.5 h.

Biological monitoring

Urine analyses

Mandelic acid. From their experiments on volunteers, Bardodej and Bardodejova (1970) have calculated that the concentration of mandelic acid collected during the last 2 h of an 8-h exposure to 100 ppm amounts to 2 g/l (SG 1.024) or 1.5 g/g creatinine. According to Gromiec and Piotrowski (1984), an exposure to 100 mg/m^3 (23 ppm) ethylbenzene corresponds to a urine concentration of 220 mg mandelic acid/g creatinine (urine collected during the last 2 h of the exposure period). By extrapolating this result to higher exposure levels, it can be estimated that a TWA exposure of 100 ppm (435 mg/m^3) is associated with a urinary concentration of mandelic acid of about 1 g/g creatinine (921 mg/g creatinine)(end-of-shift sample).

The ACGIH (1991-1992) recommends a concentration of mandelic acid in urine of 2 g/l or 1.5 g/g creatinine as BEI (sample collected the last 4 h of the last shift of the workweek) corresponding to a TWA of 100 ppm (434 mg/m^3).

Ethylphenol. Small amounts of ethylphenol were detected in urine of workers exposed to ethylbenzene. Angerer and Lehnert (1979) have estimated that, in case of exposure to 100 ppm ethylbenzene, an excretion of ethylphenol between 12 and 15 mg should be expected in the course of 24 h. Angerer and Wulf (1985) have investigated the levels of alkylbenzenes in blood and their metabolites in urine in 35 spraymen exposed to a mixture of *o-, p-, m*-xylene, and ethylbenzene (average concentration 2.1, 7.9, 2.8, and 4 ppm, respectively). The average (±SD) concentrations of ethylbenzene in blood and 1-phenylethanol in urine were 61.4 mg/l (±62.3) and 1.9 mg/l (±2.3), respectively (end-of-shift samples). The levels of 1-phenylethanol in urine and the concentration of ethylbenzene in air were well correlated (r: 0.65; $p < 0.001$): an exposure to 100 ppm corresponding to a urinary excretion of about 39.4 mg/l (end-of-shift).

Blood analysis

Ethylbenzene. Angerer and Lehnert (1979) have also indicated that an exposure to 100 ppm ethylbenzene produces concentrations of ethylbenzene in blood ranging from 0.15 to 0.2 mg per 100 ml. In another field study, this same exposure level was estimated to result in a blood concentration of ethylbenzene (end-of-shift) of about 0.0945 mg/100 ml (r: 0.51; $p < 0.001$) (Angerer and Wulf 1985).

In practice, measurement of mandelic acid in urine appears to be the best indicator of recent exposure to ethylbenzene. On a group basis, a urinary concentration of mandelic acid of 1 to 1.5 g/g creatinine at the end of the workshift corresponds to a TWA exposure of 100 ppm ethylbenzene.

As mandelic acid is also a metabolite of styrene (styrene glycol, styrene oxide, phenylglyoxylic acid, α-phenyl-aminoacetic acid), it may sometimes be relevant to confirm exposure to ethylbenzene by determining its concentrations in blood and expired air. Samples collected during or at the end of exposure reflect very recent exposure and samples collected 16 h after the shift reflect exposure during the previous shift (Lauwerys 1989); 0.15 mg ethylbenzene/100 ml blood (during exposure) can be suggested as maximum permissible concentration.

The ACGIH (1991-1992) also recommends the determination of ethylbenzene in end-exhaled air (collected prior to the shift) as a confirmatory test for ethylbenzene exposure. This Committee estimates that the concentrations in end-exhaled air collected 16 h after the fourth exposure of the workweek should be about 2 ppm (87 mg/m^3) for a TWA of 100 ppm (434 mg/m^3).

REFERENCES

ACGIH. American Conference of Governmental Industrial Hygienists. Threshold Limit Values for Chemical Substances and Physical Agents and Biological Exposure Indices. Cincinnati, 1991-1992.

Agency for toxic substances and disease registry, U.S. Public Health Service. Toxicological profile for ethylbenzene. ATSDR/TP 90/15, 1990.

J. Angerer and G. Lehnert. Occupational chronic exposure to organic solvents. VIII. Phenolic compounds — metabolites of alkylbenzenes in man. Simultaneous exposure to ethylbenzene and xylenes. *Int. Arch. Occup. Environ. Health* 43:145, 1979.

J. Angerer and H. Wulf. Occupational chronic exposure to organic solvents. XI. Alkylbenzene exposure of varnish workers: effects on hematopoietic system. *Int. Arch. Occup. Environ. Health* 56:307, 1985.

Z. Bardodej and E. Bardodejova. Biotransformation of ethylbenzene, styrene and alpha-methylstyrene in man. *Am. Ind. Hyg. Assoc. J.* 31:106, 1970.

K. Engström, V. Riimäki and A. Laine. Urinary disposition of ethyl-benzene and *m*-xylene in man following separate and combined exposure. *Int. Arch. Occup. Environ. Health* 54:355, 1984.

J. Gromiec and J. Piotrowski. Urinary man, delic acid as an exposure test for ethylbenzene. *Int. Arch. Occup. Environ. Health* 55: 61, 1984.

R. Lauwerys. Biological indicators for the assessment of human exposure to industrial chemicals: Ethylbenzene, methylstyrene, isopropylbenzene. Eds.: L. Alessio, A. Berlin, M. Boni, R. Roi. Commission of the European Communities, Luxembourg, Eur 12174 En, 1989.

2.5. α-Methylstyrene

Atrolactic acid has been found in urine of men inhaling α-methylstyrene (Bardodej and Bardodejova 1970). Clinical studies are needed to evaluate whether or not the determination of this metabolite can be used for monitoring exposure to α-methylstyrene.

A fraction of the absorbed α-methylstyrene is probably also eliminated unchanged with the expired air (Lauwerys 1990) (Figure 13).

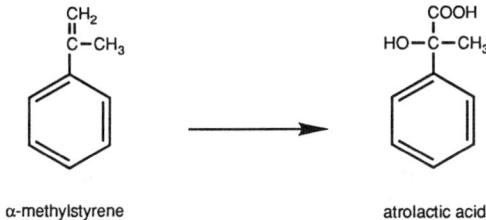

α-methylstyrene   atrolactic acid

FIGURE 13 Main metabolic pathway of α-methylstyrene.

REFERENCES

Z. Bardodej and E. Bardodejova. Biotransformation of ethylbenzene, styrene and alpha-methylstyrene in man. *Am. Ind. Hyg. Assoc. J.* 31:106, 1970.
R. Lauwerys. *Toxicologie Industrielle et Intoxications Professionnelles.* 3rd ed. Masson, Paris, 1990.

2.6. Isopropylbenzene (Cumene)

Toxicokinetics

Cumene seems, as many other organic solvents, a ubiquitous environmental pollutant. Brugnone et al (1989) have detected cumene in alveolar air and blood of persons nonoccupationally exposed to this substance.

Cumene can enter the organism by inhalation of its vapor and probably also by skin contact. It has been shown that, in rabbits, cumene is converted into dimethylphenylcarbinol (2-phenyl-2-propanol), methylbenzylcarbinol, and 2-phenylpropionic acid (Robinson et al 1955) (Figure 14).

Biological monitoring

Urine analysis

Dimethylphenylcarbinol. Senczuk and Litewka (1976) have investigated the possibility of evaluating the intensity of cumene exposure by the determination of dimethylphenylcarbinol in urine. They exposed volunteers for 8 h to cumene

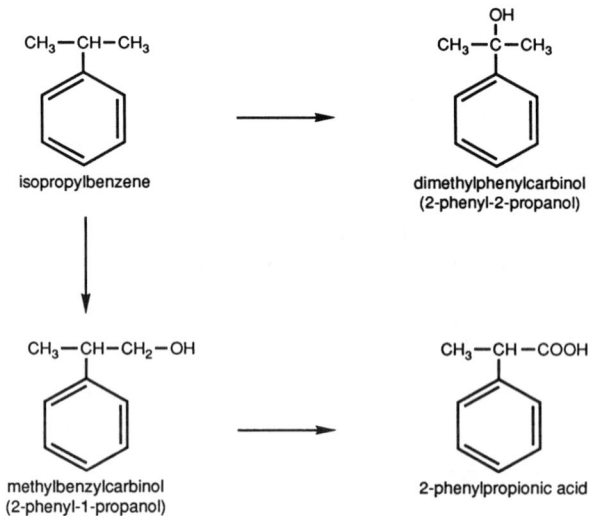

FIGURE 14 Main metabolic pathways of isopropylbenzene.

vapors ranging from 240 to 720 mg/m^3. During exposure, there was a rapid increase in the rate of excretion of the metabolite, the maximum value being in the urine fraction collected during the last 2 h of exposure. No metabolite was found in urine 48 h after the end of exposure. There was an excellent correlation between the amount of absorbed cumene and the amount of dimethylphenylcarbinol excreted during 24 h or the excretion rate of dimethylphenylcarbinol during the last 2 h of exposure, but correlation between the absorbed dose of cumene and the concentration of dimethylphenylcarbinol in urine collected at the end of exposure was not satisfactory. The authors have suggested evaluation of the intensity of exposure by measuring the excretion rate of dimethylphenylcarbinol during the last 2 h of the shift. According to their data, in man and woman an exposure to 50 ppm cumene vapor (in the absence of skin absorption) could be associated with an excretion rate of about 15 and 10 mg dimethylphenylcarbinol per hour, respectively.

REFERENCES

F. Brugnone, L. Perbellini, G. Faccini et al. Breath and blood levels of benzene, toluene, cumene, styrene in non occupational exposure. *Int. Arch. Occup. Environ. Health* 61:303, 1989.

D. Robinson, J. Smith and R. Williams. Studies in detoxication; metabolism of alkylbenzenes. Isopropylbenzene (cumene) and derivatives of hydratropic acid. *Biochem. J.* 59:153, 1955.

W. Senczuk and B. Litewka. Absorption of cumene through the respiratory tract and excretion of dimethylphenylcarbinol in urine. *Br. J. Ind. Med.* 33:100, 1976.

2.7. Styrene

Toxicokinetics

In industry, absorption of styrene occurs through inhalation of vapors and by skin contact with the liquid form. Skin absorption of styrene in its vapor phase is negligible (Wieczorek 1985). Dutkiewicz and Tyras (1970) have studied the cutaneous absorption of styrene in man. They have calculated that, when applied undiluted on the skin of the forearm, the speed of absorption is about 12 mg/cm^2/h (range: 9-12). In aqueous solutions, the amount absorbed is a function of its concentration. For aqueous solutions containing 70 to 250 mg styrene per liter, the rate of absorption ranges from 0.04 to 0.18 mg/cm^2 × hour. Berode et al (1985) have estimated that at an absorption rate of about 1 µg/cm^2 × min, contact of one hand (500 cm^2) for 30 min with liquid styrene corresponds to an absorption of 15 mg of styrene. The pulmonary absorption has also been investigated in volunteers exposed to known concentrations of styrene vapors. The absorption ranges from 42 to 90% (Bardodej and Bardodejova 1970; Fernandez and Caperos 1977; Fiserova-Bergerova and Teisinger 1965; Oltramare et al 1974; Stewart et al 1968; Wigaeus et al 1983, 1984; Pezzagno et al 1985).

A small fraction (less than 5%) of the absorbed styrene is eliminated un-changed with the expired air (Fernandez and Caperos 1977; Oltramare et al 1974; Stewart et al 1968). An even smaller amount (<2%) is eliminated unchanged in urine. The major fraction is oxidized to styrene oxide, which is hydrolyzed to styrene glycol. This metabolite is then oxidized to mandelic acid. In man, mandelic acid is either excreted as such in urine or further oxidized to phenylglyoxylic acid, which is also excreted by the kidneys (Figure 15) (Bardodej and Bardodejova 1970; Ohtsuji and Ikeda 1970). When exposure is high (i.e., atmospheric concen-tration exceeding 100 ppm), the *in vivo* concentration of mandelic acid is suffi-cient to significantly stimulate another pathway, i.e., the decarboxylation of mandelic acid into benzyl alcohol, the precursor of hippuric acid (Ikeda et al 1974). Man, however, seems to have a relatively poor ability to convert mandelic acid to benzyl alcohol (Bardodej and Bardodejova 1970; Stewart et al 1968). Therefore, in workers moderately exposed to styrene (TWA less than 100 ppm) only an increased urinary excretion of mandelic and phenylglyoxylic acids is found. A very small percentage of styrene (less than 1%) is also oxidized on the aromatic ring with production of styrene-7-8-oxide, which is converted into 4-vinylphenol (Figure 15). The latter metabolite has been isolated in the urine of workers exposed to styrene (Pfäffli et al 1981).

After the end of exposure, the pattern of urinary excretion of the two main metabolites is different. The biological half-life of phenylglyoxylic acid in urine is greater than that of mandelic acid, and is a function of the intensity of exposure (Philippe et al 1974).

Wigaeus et al (1983, 1984) have determined mean elimination half-times of 41 min for the rapid elimination phase of styrene in blood, 72 min for styrene glycol in blood, 3.6 h and 8.8 h, respectively, for mandelic acid (MA) and phenylglyoxylic acid in urine. According to Perbellini et al (1990), the elimination kinetics of MA and PGA are at least biphasic due to continuous formation from styrene released from fatty tissues. In the experiment of Wigaeus et al (1983), the elimination rate of MA seemed effectively to decrease 15-20 h after exposure, but because of the low concentrations and the few individual values after this time, they were unable to estimate the slower half-time. Perbellini et al (1990) also observed a slower elimination for PGA than MA; they nevertheless stressed out that the two metabo-lites have the same renal clearance, the difference coming from the different rates of formation.

In controlled human studies, Wieczorek and Piotrowski (1988) found that MA and PGA appeared in urine as early as in the first 2 h, the maximum excretion rates being observed during the last 2 h of exposure. But after cessation of exposure, they described a biphasic elimination with half-times fairly close for both metabo-lites: 2.5 h for the first phase and 30 h for the second phase. There was no indication of substantial styrene accumulation in the body under repeated indus-trial exposure. However, the physiologic model developed by Perbellini et al (1988) indicates that there is a tendency to styrene accumulation in fat tissues, the half-life in fat tissue ranging from between 32 and 46 h and the complete disappearance from 7 to 10 d.

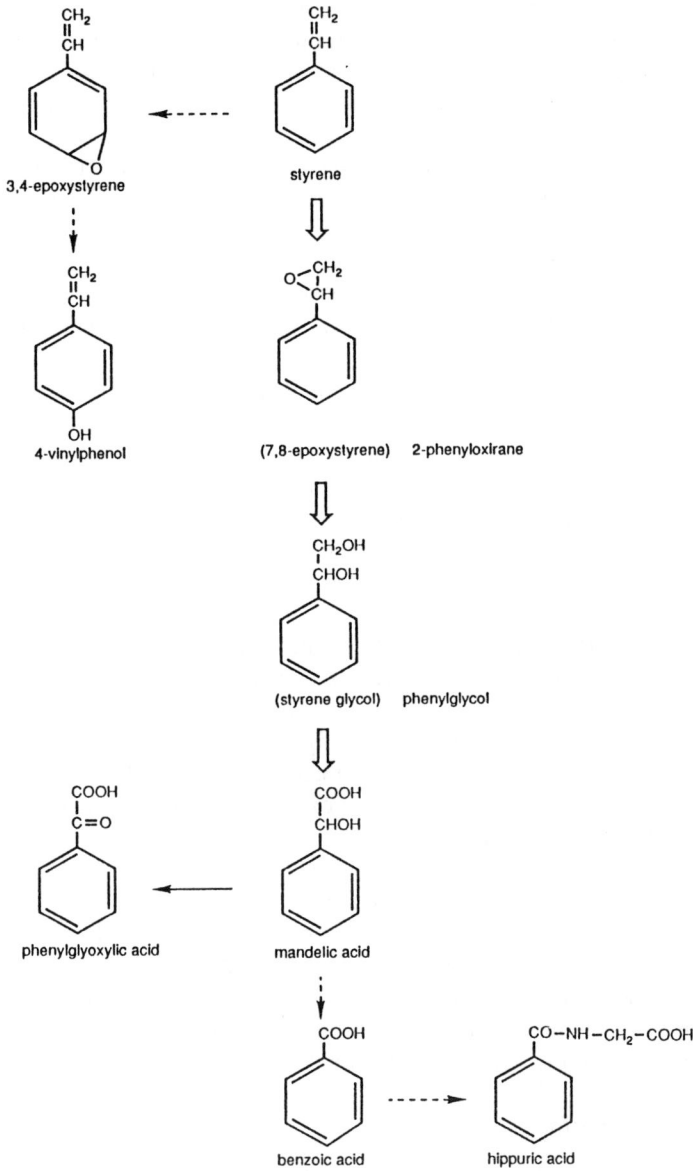

FIGURE 15 Main metabolic pathways of styrene.

The study of Imbriani et al (1990) on 64 subjects occupationally exposed to styrene also suggests an accumulation of styrene during the workday and week. According to Perbellini et al (1990), urine styrene concentration in the morning represents 14% of that found at the end of the workshift of the previous day. Blood concentration before the workshift amounts to 11% of that at the end of the previous shift. Mandelic acid concentration in the urine collected before the workshift is approximately 25% of that found at the end of the previous day.

It seems that following long-term exposure, styrene may induce its metabolism. Both styrene and nonconjugated styrene glycol were effectively eliminated faster from occupationally exposed workers than from volunteers with no previous exposure to solvents (Löf et al 1986b). However, the authors could not exclude the potential influence of other components in the work environment on the higher clearance of the solvent.

No significant interactions of acetone on the kinetics of styrene could be demonstrated on volunteers acutely exposed to both solvents (Wigaeus et al 1984). However, Dolara et al (1983) have shown enzyme induction in humans simultaneously exposed to styrene and acetone.

A midday intake of alcohol (dose producing blood alcohol concentration of about 60 mg/100 ml) inhibited the excretion of mandelic acid, so that the peak excretion was delayed from the end of the exposure period until 3 h later. Mandelic acid concentration was 56% of the level found during the alcohol-free control exposure 1 h after alcohol ingestion, and styrene glycol was increased by 15-fold. After ethanol intake the mandelic acid concentration in the end of shift samples was less than 30% of the value obtained in the "nonalcoholic shift" (Wilson et al 1983).

Berode et al (1986) have also demonstrated that ethanol blocks the oxidation of styrene glycol to MA and PGA. When the presence of ethanol is suspected, the ratio of mandelic acid and styrene glycol in blood might be used to assess whether or not the metabolic interference may be significant. The normal ratio is 85 ± 33 and if this ratio is smaller than 30, the ethanol influence is likely to be too high to allow any interpretation of the urinary concentration of the metabolites (Berode et al 1986; Guillemin and Berode 1988).

Droz and Guillemin (1986) have assessed the effect of physical workload and body build on styrene concentration in breath. The physical workload has a large impact on styrene concentration in breath measured at the end of the shift, and the body build mainly influences prior to shift biological levels. Pezzagno et al (1988) have also demonstrated the importance of the work load on the kinetics of styrene: the urinary concentration after 4 h exposure during a physical exercise involving a pulmonary ventilation of 30 l/min is more than three times the value at rest.

Biological monitoring

The following biological tests have been investigated for evaluating exposure to styrene: styrene concentration in expired air, in blood, in urine; mandelic acid in urine; phenylglyoxylic acid in urine; hippuric acid in urine.

Breath analysis

Styrene. Stewart et al (1968) exposed volunteers to 99.4 ppm styrene vapor for 7 h. Styrene was readily detected in the expired breath samples of all the subjects during the exposure period. The mean breath concentration was 27.4 ppm 30 min before the end of exposure with a range from 22 to 40 ppm. On a group basis, Gotell et al (1972) found that the concentration of styrene in expired air at different time intervals after the end of exposure was related to the intensity of exposure.

Short-term exposure tests (30-min exposure) to styrene were performed by Astrand et al (1974). The volunteers were exposed to approximately 50 and 150 ppm of styrene during rest and light physical exercise on a bicycle ergometer. Their results are summarized in Table 6. Alveolar concentration increased only slightly as exercise intensity and ventilation increased. If Pezzagno et al (1985) do agree with this statement, other authors tend, on the contrary, to show the importance of the workload on styrene concentration in expired air (Droz and Guillemin 1986). After only a minute or so, alveolar concentration reached a plateau value that persisted through the exposure period. The authors were not able to draw any reliable conclusions about the extent of exposure on the basis of the decline of alveolar air concentrations after exposure (Astrand et al 1974).

Oltramare et al (1974) exposed four volunteers to styrene vapors. They found a significant correlation between inspired and alveolar air concentration of styrene during exposure. According to their results for 50 and 100 ppm styrene in ambient air, the alveolar air concentrations during exposure at rest would be 5.7 and 10 ppm, respectively. Oltramare et al (1974) also found that the concentration rapidly leveled off after the beginning of exposure. They are of the opinion that large individual differences limit the usefulness of determining the pulmonary clearance of styrene after the end of exposure for estimating the intensity of the latter.

Brugnone et al (1980) have estimated the relationship between environmental exposure to styrene and its alveolar concentration in workers during exposure. They stated that the alveolar concentration during the workshift corresponds on the average to 16% of the inspired air concentration, but their published data suggest a much higher ratio.

In subjects exposed at rest to styrene air concentration ranging between 28.4 and 172.2 mg/m^3 (273-1654 μmol/m^3) the mean alveolar concentration was 112

Table 6
Relationship Between Exposure to Styrene
and Its Concentration in Expired Air

Exposure (30 min)	Alveolar Concentration of Styrene During Exposure (ppm); mean ± SE (n = 14)
50 ppm	
Rest	8.8 ± 0.7
50 watts	9.5 ± 1.3
150 ppm	
Rest	23.0 ± 1.1
50 watts	29.9 ± 2.3
100 watts	32.5 ± 3.2
150 watts	35.4 ± 3.0

From Astrand et al 1974.

$\mu mol/m^3$. During the first 10 min following the end of the exposure the alveolar concentration values at 0.5, 2.5, and 10 min decreased of about 83, 86, and 90% of the average end-exposure value. The estimated alveolar air concentration corresponding to an average air concentration of 50 ppm is 27.5 mg/m^3 (6.4 ppm) (Pezzagno et al 1985).

In a study conducted on workers exposed to fluctuating concentrations of styrene, Perbellini et al (1990) obtained a good correlation (r: 0.9113; $p < 0.001$) between the mean environmental and expired air styrene concentrations: a TWA exposure at 50 ppm, leading to an average concentration of 37.7 mg/m^3 (8.76 ppm) at the end of the shift.

Blood analysis

Styrene. In the course of their experiments on volunteers exposed for 7 h to 99.4 ppm styrene, Stewart et al (1968) found that venous blood samples obtained 10 min before the end of the exposure had a mean styrene concentration of 0.1 mg/100 ml with a range from 0.09 to 0.13 mg/100 ml. During their short-term experiments on volunteers, Astrand et al (1974) have also determined the concentration of styrene in venous blood. Their results are summarized in Table 7.

The blood concentration increased not only as exposure increased, but also as exercise intensity increased. Venous blood concentrations also rose sharply during the 30-min exposure without any sign of leveling off. During exposure, there was thus no good correlation between alveolar air and blood concentration of styrene.

In eight male subjects exposed for 2 h to about 2.88 $mmol/m^3$ of styrene vapor (300 mg/m^3, 70 ppm) during light physical exercise, the arterial blood concentration of styrene reached a relatively stable level of about 20 µmol/l (0.2 mg/100 ml) after 75 min of exposure (Wigaeus et al 1983, 1984). On the basis of a field study, Apostoli et al (1983) have calculated that a mean styrenemia of 81 µg/100 ml corresponded to 215 mg/m^3 environmental exposure (50 ppm).

Table 7
Relationship Between Exposure to Styrene
and Its Concentration in Venous Blood

Exposure (30 min)	Concentration of Styrene in Venous Blood (mg/100 ml) at the End of Exposure (mean ± SE)
50 ppm	
Rest (n = 5)	0.03 ± 0
50 watts (n = 5)	0.14 ± 0.01
150 ppm	
Rest (n = 13)	0.1 ± 0.01
50 watts (n = 10)	0.47 ± 0.05
100 watts (n = 14)	0.79 ± 0.11
150 watts (n = 14)	1.24 ± 0.08

From Astrand et al 1974.

In 17 workers handling polyester resins, Bartolucci et al (1987) found also that the concentration of styrene in blood taken at the end of the workshift was significantly correlated with exposure. The correlation was even better when only the last 2 h of styrene concentration in air was taken into account.

According to the physiologic model of Perbellini et al (1988), the end-of-shift venous blood concentration of styrene resulting from a mean exposure to 215 mg styrene/m^3, for 8 h, and for 5 d, would range between 71.5 and 113 µg/100 ml on Monday and between 75 and 126 µg/100 ml on Friday. Table 8 summarizes some correlations between airborne concentration of styrene and its concentration in venous blood.

Other biological indicators. Some authors have stressed the potential interest of measuring styrene glycol and styrene 7,8-oxide in blood (Wigaeus et al 1983, 1984; Löf et al 1986a,b), but the available data are still too limited to propose these measurements for assessing styrene uptake.

An increased level of styrene hemoglobin adducts was detected in workers exposed to less than 50 ppm styrene (Brenner et al 1991).

Urine analyses

Mandelic acid. During normal working conditions, the concentration of mandelic acid in urine increases progressively to reach a maximum at the end of the exposure period (Engström et al 1976; Philippe et al 1974; Wilson et al 1983). Its biological half-life in urine after the end of exposure amounts to a few h (Engström et al 1976; Ikeda et al 1974). In the urine of workers exposed to approximately 100 ppm styrene, Korn et al (1984) observed a L/D mandelic acid enantiomer ratio of nearly 1.5 (1.1-2.1), subject to individual variations.

Bardodej and Bardodejova (1966), who exposed volunteers to styrene vapors for 8 h, have calculated that the concentration of mandelic acid collected during the last 2 h of an 8-h exposure to 50 ppm styrene amounts to 1.5 g/l (SG 1.024) or 1g/g creatinine. The range of individual value was not reported by the authors.

Table 8
Correlations Between Airborne Concentration of Styrene (x) and Its Concentration in End-of-shift Venous Blood Sample (y)

Authors	Regression Equation	Concentration Corresponding to a TWA of 50 PPM	Correlation Coefficient
Apostoli et al (1983)	y (µg/l) = −246 + 4.92x (µg/l)	0.812 µg/l	0.9215
Bartolucci et al	y (µg/l) = 9.90 + 4.32x (mg/m^3)	0.939 µg/l*	0.74
(1987)	y (µg/l) = 68.14 + 4.23x (mg/m^3)	0.978 µg/l**	0.9
Imbriani et al (1990)	y (µg/l) = 153.9 + 1.646x (mg/m^3)	0.508 µg/l	0.72
Perbellini et al (1990)	y (µg/l) = 297 + 6.5x (µg/l)	1.694 µg/l	0.8764
Ghittori et al (1990)	y (µg/l) = 170.2 + 2.07x (mg/m^3)	0.615 µg/l	0.77

*8-h styrene TWA exposure.
**Last 2 h of styrene TWA exposure.

The ratio mandelic acid/phenylglyoxylic acid varies with environmental styrene concentration, being greater at higher styrene concentrations (Bardodej and Bardodejova 1966; Ohtsuji and Ikeda 1970).

Caperos et al (1979) have exposed volunteers to 103 or 206 ppm styrene vapor for 8 h during five consecutive days. They found that the amount of mandelic and phenylglyoxylic acids excreted in urine correlates with the intake of styrene, but not with its environmental concentration. This demonstrates again that individual differences in ventilatory parameters may play a major role in the uptake of a solvent and that environmental concentration is only a crude estimate of the true internal dose. Engström et al (1976) studied in workers the correlation between the amount of urinary mandelic acid sampled at the end of the workshift and the styrene concentration in the ambient air. A strong linear correlation (r: 0.93) was found between both parameters. An excretion of about 2.3, 1.2, and 0.7 g mandelic acid/g creatinine at the end of the shift corresponds to a TWA exposure of about 100, 50, and 25 ppm, respectively.

According to Gotell et al (1972), 1 g of mandelic acid per liter in a urine sample (SG 1.024) taken after an 8-h workshift indicates a time-weighted exposure of 50 ppm. They also reported that when styrene exposure exceeds 150 ppm the urinary levels of phenylglyoxylic and mandelic acids are no longer proportional to the solvent concentration in air, but this observation is not in agreement with the experience of other authors (Engström et al 1976). Apostoli et al (1983) found that a mean airborne concentration of styrene of 215 mg/m³ entails a mean mandelic acid concentration of 495 mg/l in urine collected at the end of the shift.

On the basis of a mathematical model, Droz and Guillemin (1983) have estimated that a styrene TWA of 50 ppm corresponds to a mandelic acid concentration of 800 mg/g creatinine and 150 mg/g creatinine in urine collected at the end of the shift and the next morning, respectively.

Bowman et al (1990), who monitored urinary mandelic acid concentrations in 324 workers from the fiberglass-reinforced plastic industry, recommend for the monitoring of workers exposed to 50 ppm styrene, 8 h a day a biological exposure indice of 1 g/l (corrected for specific gravity 1.024).

Lindstrom et al (1976), who studied 98 male workers exposed to styrene in the manufacture of polyester plastic products, found that high mandelic acid concentration in urine excreted at the end of the shift (more than 1.7 g/l) was related to visuomotor inaccuracy and to poor psychomotor performance and lowered vigilance. Abnormalities in the electroencephalogram (mainly in the occipital region) have been found in workers whose mandelic acid concentration in urine collected at the end of the workshift exceeded 0.7 g/l (Seppalainen and Harkonen 1976). These changes were not related to the duration of exposure. In the same workers, no significant electroneuromyographic changes were found. Guillemin et al (1978) found during a study on volunteers and on workers from polyester factories that the parameter best correlated with exposure was the sum of mandelic and phenylglyoxylic acid concentration in a urine sample collected 14 h after the end of the exposure (next morning). On the average, a TWA exposure to 50 ppm styrene for 8 h produces a total concentration of mandelic and phenylglyoxylic acids of 300 mg/g creatinine in the next morning sample. They have made a similar observation on workers from fiberglass-reinforced polyester plants (see below). The results obtained by Ohtsuji and Ikeda (1970) are reported under phenylglyoxylic acid in urine.

Table 9 summarizes some correlations between airborne concentration of styrene and mandelic acid concentration in urine.

Phenylglyoxylic acid. The excretion of phenylglyoxylic acid is also a very sensitive index of exposure (Caperos et al 1979; Ohtsuji and Ikeda 1970; Philippe et al 1974). Ohtsuji and Ikeda (1970) have determined phenylglyoxylic and mandelic acids in urine of workers exposed to styrene during the production of plastic containers. The urine samples were collected during the second period of the workshift. It should, however, be pointed out that the estimation of styrene concentration in air was very approximate, since it was performed with detector tubes. Their results are summarized in Table 10.

On the basis of a limited survey among 43 workers exposed to styrene vapors, Philippe et al (1974) concluded that when phenylglyoxylic acid concentration in urine collected during the second period of the workshift does not exceed 350 mg/g creatinine the exposure is probably below 100 ppm (430 mg/m^3).

In workers employed in a fiberglass-reinforced plastic factory, the mean phenylglyoxylic acid urinary concentration (end-of-shift samples) corresponding

Table 9
Correlations Between Airborne Concentration of Styrene (x) and Mandelic Acid (MDA) Concentration in Urine (y)

Authors	Regression Equation	Concentration Corresponding to a TWA of 50 PPM	Correlation Coefficient
Guillemin et al (1982)	y (mg/g cr.) = 72.39 + 2.174x x = ppm.8h	942 mg/g creatinine*	0.749
	y (mg/g cr.) = 66.62 + 0.352x x = ppm.8h	207.4 mg/g creatinine**	0.620
Ikeda et al (1982)	y (mg/l) = 118 + 10.1x (ppm)	623 mg/l*	0.84
	y (mg/g cr.) = 55 + 15.4x (ppm)	825 mg/g creatinine*	0.86
	y (mg/l) = 181 + 12.5x (ppm)	806 mg/l SG 1016*	0.70
Apostoli et al (1983)	y (mg/l) = 65 + 2x (µg/l)	495 mg/l*	0.6684
Bartolucci et al (1987)	y (mg/g cr.) = 18.60 + 4.14x x = mg/m^3	908.7 mg/g creatinine*	0.74
	y (mg/g cr.) = 15.06 +1.28x x = mg/m^3	290.3 mg/g creatinine**	0.82
Imbriani et al (1990)	y (mg/l) = –86.05 + 4.94x x = mg/m^3	976 mg/l*	0.76
Takahashi et al (1989)	y (mg/l) = 34.95 + 7.1x (ppm)	890 mg/l (SG 1024)*	0.72
Perbellini et al (1990)	y (mg/l) = 169 + 2.7x (mg/m^3)	749.5 mg/l*	0.4677

*End-of-shift sample.
**Next morning sample.

Table 10

Relationship Between Styrene Exposure and Metabolite Concentration in Urine (Collected During the Second Period of the Shift)

No. of subjects	Styrene Exposure (ppm)	Metabolites in Urine (mg/l)		
		Phenylglyoxylic Acid (PGA)	Mandelic Acid (MA)	Ratio MA/PGA
6	10-30	381	875	2.3
		(298-487)	(505-1515)	
4	7-20	287	473	1.7
		(183-449)	(319-702)	
4	1-20	201	310	1.5
		(153-263)	(189-518)	
7	<1	98	137	1.4
		(46-213)	(49-385)	
9	0	19	92	4.8
		(11-34)	(47-178)	

From Ohtsuji and Ikeda 1970.

to a mean environmental exposure to 215 mg/m³ (50 ppm) in styrene was found to be 260 mg/l (Apostoli et al 1983).

Ikeda et al (1982) have found that a TWA exposure to 100 ppm corresponds on a group basis to a concentration of 438 mg phenylglyoxylic acid/g creatinine in urine collected at the end of the shift (with a 95% confidence interval ranging from 356 to 520 mg/g creatinine). The same researchers have also concluded that the sum of mandelic acid and phenylglyoxylic acid concentration after correction for creatinine concentration is the best styrene exposure indicator when the urine samples are collected at the end of the shift. According to their data, an exposure to 100 ppm styrene for 8 h leads at the end of the shift to a mean urinary concentration of the sum of both metabolites (phenylglyoxylic acid expressed in mandelic acid equivalents) of 2033 mg/g creatinine (range 1779 to 2354). These values should be divided by 2 when the TWA is 50 ppm. We have already indicated that Guillemin et al (1978) had made the same observation, but had suggested analysis of the urine collected 16 h after the end of exposure. In workers from polyester plants, they found the following urinary metabolite concentrations corresponding to an 8-h exposure, end-of-shift sample: mandelic acid, 1640; phenylglyoxylic acid, 510; sum of both metabolites, 2150 mg/g creatinine, respectively; next-morning sample: mandelic acid, 330; phenylglyoxylic acid, 330; sum of both metabolites, 660 mg/g creatinine, respectively (Guillemin et al 1982).

Table 11 summarizes some correlations between airborne concentration of styrene and phenylglyoxylic acid concentration in urine.

Comparing the correlation between the TWA exposure to styrene and various biological indicators (styrene in blood, mandelic acid in urine (MA), phenylglyoxylic acid in urine (PGA), sum of both metabolites in urine end-of-shift, next morning), Bartolucci et al (1986, 1987) and De Rosa et al (1988) also obtained the best association with the sum of metabolites in urine (MA + PGA)

Table 11
Correlations Between Airborne Concentration of Styrene (y) and Phenylglyoxylic Acid (PGA) Concentration in Urine (x)

Authors	Regression Equation	Concentration Corresponding to a TWA of 50 ppm	Correlation Coefficient
Guillemin et al (1982)	y (mg/g cr.) = 77.1 + 0.611x x = ppm.8h	321.5 mg/g creatinine*	0.706
	y (mg/g cr.) = 75.33 + 0.308x x = ppm.8h	198.53 mg/g creatinine**	0.721
Ikeda et al (1982)	y (mg/l) = 52 + 3x (ppm)	202 mg/l*	0.65
	y (mg/l) = 78 + 4x (ppm)	278 mg/l SG 1016*	0.54
	y (mg/g cr.) = 38 + 4x (ppm)	238 mg/g creatinine*	0.82
Apostoli et al (1983)	y (mg/l) = 157 + 0.478x x = mg/m^3	259.8 mg/l*	0.5321
Bartolucci et al (1987)	y (mg/g cr.) = 10.37 + 0.82x x = mg/m^3	186.7 mg/g creatinine*	0.69
	y (mg/g cr.) = 3.57 + 0.55x x = mg/m^3	121.8 mg/g creatinine**	0.81
Takahashi et al (1989)	y (mg/l) = 16.19 + 8.38x x = ppm	435.2 mg/l (SG 1024)*	0.80

*End-of-shift sample.
**Next-morning sample.

in the next morning sample. Considering subjects exposed to widely fluctuating concentrations of styrene during the workday, the correlation between TWA exposure and urinary metabolites (MA + PGA) was highly significant for urine collected the next morning and not when the urines were collected at the end of the shift. Instead, in subjects exposed to constant styrene air concentrations, the correlation between environmental data and total urinary metabolites was excellent, both at the end of the shift and the next morning. The results of these authors indicate that urinary (MA + PGA) concentrations of about 1075 mg/g creatinine (end-of-shift) and about 415 mg/g creatinine (next morning) can be expected when styrene TWA exposure amounts to 50 ppm (215 mg/m^3).

Based on a mathematical model, Droz and Guillemin (1983) proposed the following biological limits (corresponding to a TWA exposure to 215 mg/m^3 styrene) for the sum of mandelic acid and phenylglyoxylic acid: 1000 mg/g creatinine at the end of the shift, 300 mg/g creatinine the morning after. Takahashi et al (1989) measured the urinary metabolites of styrene in operators handling styrene (to determine the absolute daily amount of excretion of the metabolites, all the urine samples were collected during the 3-d survey). From the regression equations it emerges that the urinary concentrations of MA and PGA corresponding to 50 ppm TWA are 390 and 435 mg/l, respectively. These authors state that the concentration in urine of these metabolites should be kept to less than 60% of the aforementioned levels: i.e., 248 mg MA/l and 268 mg PGA/l, but it should be noted that their recommendations are based on a 24-h urine collection.

Hippuric acid. As indicated above, man has a poor ability to transform styrene into hippuric acid. Thus, moderate exposure to styrene (less than 60 ppm) is not associated with a significant change in hippuric acid excretion (Ikeda et al 1974). This was confirmed by Takahashi et al (1989) who did not find a significant correlation between the urinary hippuric acid concentration and exposure to low levels of styrene (≤20 ppm). It can be concluded that urinary hippuric acid is an insensitive indicator of exposure to styrene.

Styrene. Several Italian authors have assessed the interest of measuring styrene concentration in urine. Dolora et al (1984) were the first to monitor styrene exposure in workers manufacturing polystyrene plastics, by following urinary excretion of styrene itself. For concentration of styrene in the work environment ranging from 16 to 61 mg/m^3, urinary concentrations of styrene ranged from 0.7 to 4.1 µg/l.

In experimentally and occupationally exposed subjects, Pezzagno et al (1985) also found a linear relationship between urinary styrene concentration and styrene TWA. In occupationally exposed subjects performing moderate work, the urinary styrene concentration corresponding to a TWA of 215 mg/m^3 (50 ppm, 2065 µmol/m^3) was 815 nmol/l or (85 µg/l) with a 95% lower confidence limit of 740 nmol/l (77 µg/l). Further investigations by Ghittori et al (1987) and Imbriani et al (1990) provided similar results: 90 µg/l corresponding to a TWA exposure of 215 mg/m^3. Ghittori et al (1987) have proposed 80 µg/l as a biological exposure limit.

Table 12 summarizes some correlations between airborne concentration of styrene and its concentration in urine (end-of-shift sample).

Droz et al (1989a,b, 1990) have developed a physiological model simulating a repeated constant exposure at 50 ppm (215 mg/m^3) with a constant light physical workload (50 watts) and have calculated the biological indicator levels in samples taken in the end of the shift on Thursday and before the shift on Friday. End-of-shift on Thursday: styrene in alveolar air: 106 µmol/m^3 (11 mg/m^3, 2.56 ppm), in venous blood: 12 µmol/l (0.125 mg/100 ml), in urine: 1.3 µmol/l (135.4 µg/l); phenylglyoxylic acid in blood: 25 µmol/l (0.375 mg/100 ml), in urine: 141 mmol/mol creatinine

Table 12

**Correlations Between Airborne Concentration of Styrene (y) and
Its Concentration in End-of-shift Urine Sample (x)**

Authors	Regression Equation	Concentration Corresponding to a TWA of 50 ppm	Correlation Coefficient
Pezzagno et al (1985)	y (nmol/l) = 154.2 + 0.32x (mmol/m^3)	815 nmol/l (85 µg/l)	0.88
Ghittori et al (1987)	y (µg/l) = 18.9 + 0.328x (mg/m^3)	90 µg/l	0.88
(1990)	y (µg/l) = 16.7 + 0.293x (mg/m^3)	79.7 µg/l	0.83
Imbriani et al (1990)	y (µg/l) = 22.1 + 0.26x (mg/m^3)	78 µg/l	0.79
Perbellini et al (1990)	y (µg/l) = 29 + 0.24x (mg/m^3)	80.6 µg/l	0.7181

(187.13 mg/g creatinine); mandelic acid in blood: 150 μmol/l (2.28 mg/100 ml), in urine: 927 mmol/mol creatinine (1.25 g/g creatinine). Before shift on Friday: styrene in alveolar air: 29 μmol/m³ (6.74 mg/m³), in venous blood: 1.7 μmol/l (17.7 μg/100 ml), in urine: 0.15 μmol/l (15 μg/l); phenylglyoxylic acid in blood: 15 μmol/l (0.225 mg/100 ml), in urine: 113 mmol/mol creatinine (150 mg/g creatinine); mandelic acid in blood: 34 μmol/l (0.52 mg/100 ml), in urine: 284 mmol/mol creatinine (382 mg/g creatinine). The concentration of styrene in blood samples taken 16 h after the end of exposure estimated by Droz et al (1989) is about 10 times higher than that observed by other authors (see, e.g., Ramsey et al 1980).

Perbellini et al (1988) have also used a physico-mathematical model to assess the BEI of styrene which reflect an exposure to a TWA of 215 mg/m³ (50 ppm). The ranges of styrene concentration at the end of the various workshifts are 15 to 25 μg/l in the alveolar air and 71.5 to 126.3 μg/100 ml in the venous blood. The urinary concentrations of mandelic and phenylglyoxylic acid at the end of the shift range from 800 to 1732 mg/l and from 235 to 617 mg/l, respectively (ranges explained by the alveolar ventilation of 6 and 12 l/min). These authors have proposed, as biological limit values, 45 μg/l and 105.3 μg/100 ml for styrene in alveolar air and blood (end-of-shift) (Brugnone and Perbellini 1985). They recently underlined the constant equilibrium between urinary, alveolar, and blood styrene concentrations and concluded that these markers are equivalent for the biological monitoring of styrene exposure (Perbellini et al 1990).

According to Guillemin et al (1988), the most appropriate biological exposure limit values corresponding to the TWA of 50 ppm styrene are 850 mg mandelic acid/g creatinine or 1100 mg/l urine (SG 1018) in the end-of-shift sample and 330 mg mandelic acid + phenylglyoxylic acid/g creatinine or 430 mg/l urine (SG 1018) in the next morning sample.

The biological tolerance (BAT) values recommended by the DFG (1991) are 2 g mandelic acid/l urine and 2.5 g mandelic acid + phenylglyoxylic acid/l urine (end-of-shift sample). The biological exposure indices (BEI) recommended by the ACGIH (1991-1992) are for styrene in exhaled air: 40 ppb prior to shift or 18 ppm during the shift; styrene in blood: 0.55 mg/l end-of-shift or 0.02 mg/l prior to next shift; mandelic acid in urine: 1 g/l or 0.8 g/g creatinine (end-of-shift); phenylglyoxylic acid in urine: 250 mg/l or 240 mg/g creatinine (end-of-shift). These latter values correspond to a styrene TWA of 50 ppm (215 mg/m³).

In summary, among the biological tests proposed for evaluating the intensity of exposure to styrene, mandelic acid excretion in urine is probably the most practical. Phenylglyoxylic acid urinary concentration a minor metabolite is also a useful tool for monitoring styrene exposure. Both metabolites are sensitive indicators of the average exposure level to styrene.

Styrene concentrations in blood and alveolar air are good indicators of the very recent level of exposure and, being specific, can be used to confirm exposure to styrene whenever there is a doubt about the origin of the non-specific metabolites. Samples taken in the morning reflect the level of exposure the day before.

The current knowledge suggests that for a 8 h TWA of 50 ppm, a mean urinary mandelic acid concentration of just below 1 g/g creatinine, and a mean urinary phenylglyoxylic acid concentration probably below 300 mg/g creatinine can be expected. The corresponding mean styrene concentrations in alveolar air and blood are estimated to be just below 10 ppm and about 0.1 mg/100 ml, respectively. This exposure level will also probably lead to a urinary styrene concentration of about 85 μg/g creatinine.

REFERENCES

ACGIH. American Conference of Governmental Industrial Hygienists. Threshold Limit Values for Chemical Substances and Physical Agents and Biological Exposure Indices. Cincinnati, 1991-1992.

P. Apostoli, F. Brugone, L. Perbellini et al. Occupational styrene exposure: environmental and biological monitoring. *Am. J. Ind. Med.* 4:741, 1983.

I. Astrand, A. Kilbom, I. Per Ovrum Walhberg and O. Vesterberg. Exposure to styrene. Concentration in alveolar air and blood at rest and during exercise and metabolism. *Work Environ. Health* 2:64, 1974.

Z. Bardodej and E. Bardodejova. The metabolism of ethylbenzene, styrene and alpha-methylstyrene. Proceedings 15th International Congress on Occupational Health. Vienna, p. 457, 1966.

Z. Bardodej and E. Bardodejova. Biotransformation of ethylbenzene, styrene and alphamethylstyrene in man. *Am. Ind. Hyg. Assoc. J.* 31:206, 1970.

G. Bartolucci, E. De Rosa, P. Gori et al. Biomonitoring of occupational to low styrene levels. *Appl. Ind. Hyg.* 1:125, 1986.

G. Bartolucci, P. Corona, E. De Rosa et al. Environmental and biological monitoring of workers exposed to styrene. In: *Biological Monitoring of Exposure to Chemicals. Organic Compounds.* Eds.: M. Ho, H. Dillon. Wiley Interscience, New York, p. 155, 1987.

M. Berode, P. Droz and M. Guillemin. Human exposure to styrene. VI. Percutaneous absorption in human volunteers. *Int. Arch. Occup. Environ. Health* 55:331, 1985.

M. Berode, P. Droz, M. Boillat and M. Guillemin. Effect of alcohol on the kinetics of styrene and its metabolites in volunteers and in workers. *Appl. Ind. Hyg.* 1:25, 1986.

J. Bowman, J. Held and D. Factor. A field evaluation of mandelic acid in urine as a compliance monitor for styrene exposure. *Appl. Occup. Environ. Hyg.* 5:526, 1990.

D. Brenner, A. Jeffrey, L. Latriano et al. Biomarkers in styrene-exposed boatbuilders. *Mutat. Res.* 261:225, 1991.

F. Brugnone and L. Perbellini. Biological monotoring of occupational exposure to solvents by analysis of alveolar air and blood. In: *Organic Solvents and the Central Nervous System.* Environmental Health 5. WHO, Copenhagen, 1985.

F. Brugnone, L. Perbellini, E. Gaffuri and P. Apostoli. Biomonitoring of industrial solvent exposures in workers' alveolar air. *Int. Arch. Occup. Environ. Health* 47:245, 1980.

J. Caperos, B. Humbert and P. Droz. Exposition au styrène. II. Bilan de l'absorption de l'excrétion et du métabolisme sur des sujets humains. *Int. Arch. Occup. Environ. Health* 42:223, 1979.

E. De Rosa, G. Bartolucci, L. Perbellini et al. Environmental and biological monitoring of exposure to toluene, styrene, n-hexane. *Appl. Ind. Hyg.* 3:332, 1988.

DFG. Deutsche Forschungsgemeinschaft. *Maximum Concentrations at the Workplace and Biological Tolerance Values for Working Materials* Report No. XXVII. Commission for the Investigation of Health Hazards of Chemical Compounds in the Work Area. VCH, Weinheim, 1991.

P. Dolara, M. Lodovici, M. Salvadori et al. Enzyme induction in humans exposed to styrene. *Ann. Occup. Hyg.* 27:183, 1983.

P. Dolara, G. Caderini, M. Lodovici and G. Santoni. Determination of styrene in the urine of workers manufacturing polystyrene plastics. *Ann. Occup. Hyg.* 28:195, 1984.

P. Droz and M. Guillemin. Human styrene exposure. V. Development of a model for biological monitoring. *Int. Arch. Occup. Environ. Health* 53:19, 1983.

P. Droz and M. Guillemin. Occupational exposure monitoring using breath analysis. *J. Occup. Med.* 28:593, 1986.

P. Droz, M. Wu, W. Cumberland and M. Berode. Variability in biological of solvent exposure. I. Development of a population physiological model. *Br. J. Ind. Med.* 46:447, 1989a.

P. Droz, M. Wu and W. Cumberland. Variability in biological monitoring of organic solvent exposure. II. Application of a population physiological model. *Br. J. Ind. Med.* 46:547, 1989b.

P. Droz. Pharmacokinetic modelling of styrene exposure. In: *Esposizione a Stirene. Rischi, Tossicita e Criteri per il Monitoraggio Biologico.* Eds.: E. Capodaglio, L. Manzo. Convegno Internazionale di Pavia, Fundazione Clinica Del Lavoro, p. 19, 18 Maggio 1990.

I. Dutkiewicz and H. Tyras. Recherche sur l'absorption percutanée du styrène chez l'homme. Note établie par l'Institut National de Recherche et de Sécurité (France). Cahiers de Notes Documentaires, No. 59, 2nd trimestre 1970.

K. Engström, H. Harkonen, P. Kalliokoski and J. Rantanen. Urinary mandelic acid concentration after occupational exposure to styrene and its use as a biological exposure test. *Scand. J. Work Environ. Health* 2:21, 1976.

J. Fernandez and J. Caperos. Exposition au styrène. I. Etude expérimentale de l'absorption et de l'excrétion pulmonaires sur des sujets humains. *Int. Arch. Occup. Environ. Health* 40:1, 1977.

V. Fiserova-Bergerova and J. Teisinger. Pulmonary styrene retention. *Ind. Med. Surg.* 34:620, 1965.

S. Ghittori, M. Imbriani, G. Pezzagno and E. Capodaglio. The urinary concentration of solvents as a biological indicator of exposure: proposal for the biological equivalent exposure limit for nine solvents. *Am. Ind. Hyg. Assoc. J.* 48:786, 1987.

P. Gotell, O. Axelson and B. Lindelof. Field studies on human styrene exposure. *Work Environ. Health* 9:76, 1972.

M. Guillemin and M. Berode. Biological monitoring of styrene: a review. *Am. Ind. Hyg. Assoc. J.* 49:497, 1988.

M. Guillemin, D. Bauer, P. Hotz et al. Monitoring of styrene exposure in the polyester industry. *Scand. J. Work Environ. Health* 4:14, 1978.

M. Guillemin, D. Bauer, M. Martin and A. Marazzi. Human exposure to styrene. IV. Industrial hygiene investigations and biological monitoring in the polyester industry. *Int. Arch. Occup. Environ. Health* 51:139, 1982.

M. Ikeda, T. Imamura, M. Hayashi et al. Evaluation of hippuric, phenylglyoxylic and mandelic acids in urine as indices of styrene exposure. *Int. Arch. Occup. Environ. Health* 32:93, 1974.

M. Ikeda, A. Koizumi, M. Miyasaka and T. Watanabe. Styrene exposure and biological monitoring in FRP boat production plants. *Int. Arch. Occup. Environ. Health* 49:325, 1982.

M. Imbriani, F. Gobba, S. Ghittori et al. A biological monitoring of occupational exposure to styrene. Comparison between urinary mandelic acid concentration and styrene concentration in urine and blood. *Appl. Occup. Environ. Hyg.* 4:223, 1990.

M. Korn, R. Wodoz, W. Schoknecht et al. Styrene metabolism in man: gas chromatographic separation of mandelic acid enantiomers in urine of exposed persons. *Arch. Toxicol.* 55:59, 1984.

R. Lindstrom, H. Harkonen and S. Hernberg. Disturbances in psychological functions of workers occupationally exposed to styrene. *Scand. J. Work Environ. Health* 2:129, 1976.

A. Löf, E. Lundgren, E. Nydahl and M. Nordqvist. Biological monitoring of styrene metabolites in blood. *Scand. J. Work Environ. Health* 12:70, 1986a.

A. Löf, E. Lundgren and M. Nordqvist. Kinetics of styrene in workers from plastic industry after controlled exposure: a comparison with subjects not previously exposed. *Br. J. Ind. Med.* 43:537, 1986b.

H. Ohtsuji and M. Ikeda. A rapid colorimetric method for the determination of phenylglyoxylic and mandelic acids. Its application to the urinalysis of workers exposed to styrene vapour. *Br. J. Ind. Med.* 27:150, 1970.

M. Oltramare, E. Desbaumes, C. Imhoff and W. Michiels. Toxicologie du styrène monomère. Recherches expérimentales et cliniques chez l'homme. *Ed. Méd. Hyg.*, Genève, 1974.

M. Perbellini, P. Mozzo, P. Turri et al. Biological exposure of styrene suggested by a physiological mathematical model. *Int. Arch. Occup. Environ. Health* 60:187, 1988.

M. Perbellini, L. Romeo, G. Maranelli et al. Biological monitoring of fluctuating occupational exposures to styrene. In: *Esposizione a Stirene. Rischi, Tossicita e Criteri per il Monitoraggio Biologico*. Eds.: E. Capodaglio, L. Manzo. Convegno Internazionale di Pavia, Fondazione Clinica del Lavoro, p. 133, 18 Maggio 1990.

G. Pezzagno, S. Ghittori and M. Imbriani. Urinary elimination of styrene in experimental and occupational exposure. *Scand. J. Work Environ. Health* 11:371, 1985.

G. Pezzagno, M. Imbriani, S. Ghittori and E. Capodaglio. Urinary concentration environmental concentration and respiratory uptake of same solvents: effect of the work load. *Am. Ind. Hyg. Assoc. J.* 49:546, 1988.

P. Pfäffli, A. Hesso, H. Vainio and M. Hyvonen: 4-Vinylphenol excretion suggestive of arene oxide formation in workers occupationally exposed to styrene. *Toxicol. Appl. Pharmacol.* 60:85, 1981.

R. Philippe, R. Lauwerys, J.P. Buchet and H. Roels. Evaluation de l'exposition des travailleurs au styrène par le dosage de ses métabolites urinaires: les acides mandélique et phénylglyoxylique. 2. Application aux travailleurs fabriquant des polyesters. *Arch. Mal. Prof.* 35:631, 1974.

J. Ramsey, J. Young, R. Karbowski et al. Pharmacokinetics of inhaled styrene in human volunteers. *Toxicol. Appl. Pharmacol.* 53:54, 1980.

S. Seppalainen and H. Harkonen. Neurophysiological findings among workers occupationally exposed to styrene. *Scand. J. Work Environ. Health* 2:140, 1976.

R. Stewart, H. Dodd, E. Baretta and A. Schaffer. Human exposure to styrene vapor. *Arch. Environ. Health* 16:656, 1968.

T. Takahashi, R. Hayashi, A. Ohara and Y. Fujiki. Biological monitoring by urinary metabolites of styrene. *Hojinkai Igaku Shi* 28:91, 1989.

H. Wieczorek. Evaluation of low exposure to styrene. II. Dermal absorption of styrene vapours in humans under experimental conditions. *Int. Arch. Occup. Environ. Health* 57:71, 1985.

H. Wieczorek and J. Piotrowski. Kinetic interpretation of the exposure text for styrene. *Int. Arch. Occup. Environ. Health* 61:107, 1988.

E. Wigaeus Heljm, A. Löf, R. Bjurström and M. Nordqvist. Exposure to styrene. Uptake, distribution, metabolism and elimination in man. *Scand. J. Work Environ. Health* 9:479, 1983.

E. Wigaeus Heljm, A. Löf and M. Nordqvist. Uptake, distribution, metabolism and elimination of styrene in man. A comparison between single exposure and co-exposure with acetone. *Br. J. Ind.* Med 41(4):539, 1984.

M. Wilson, S. Robertson, H. Waldron and D. Gompertz. Effect of alcohol on the kinetics of mandelic acid excretion in volunteers exposed to styrene vapour. *Br. J. Ind. Med.* 40:75, 1983.

2.8. Biphenyl

Toxicokinetics

Biphenyl can be absorbed through the skin and lungs. The half-life of absorbed biphenyl seems to be relatively short, on the order of hours.

Biological monitoring

The 2- and 4-hydroxybiphenyl metabolites have been detected in the urine of workers exposed to biphenyl. A TWA of 1.5 mg/m^3 would result in a mean concentration of both metabolites in urine of about 1.9 mg/g creatinine at the end of the shift. Animal experiments suggest that 4-hydroxybiphenyl is the main urinary metabolite of biphenyl (Meyer 1977). According to Bardodej et al (1980), after an 8-h exposure to 0.45 mg of biphenyl/m^3 (0.09 ppm), the urine of workers contains 0.7-0.8 mg/l of 4-hydroxybiphenyl.

REFERENCES

Z. Bardodej, F. Hladik, V. Rejlkova et al. The value and use of exposure tests. XIX. Exposure test for biphenyl. *Cesk. Hyg.* 25:241, 1980 (in Czech).

F. Dorgelo, G. Verver, G. Wieling et al. Urinary hydroxyphenyl excretion of workers occupationally exposed to a mixture of diphenyl and diphenylether (Dowtherm A). *Int. Arch. Occup. Environ. Health* 56:129, 1985.

T. Meyer. The metabolism of biphenyl. IV. Phenolic metabolites in the guinea pig and the rabbit. *Acta Pharm. Toxicol.* 40:193, 1977.

2.9. Polycyclic Aromatic Hydrocarbons

Toxicokinetics

The polycyclic aromatic hydrocarbons (PAH) are generated when natural or synthetic organic materials are burnt in the presence of suboptimal oxygen supply.

They are present in ambient air mainly due to their release from motor vehicles, domestic coal- or oil- fired heating systems, and various industrial sources. They are also found in high concentration in some workplaces (e.g., coke ovens, aluminium-reduction plant, steel industry, asphalt industry, creosote impregnating plant, gas and petroleum industries, etc.).

Intake may occur through the lungs and gastrointestinal tract. Percutaneous absorption is also possible (Jongeleenen et al 1988a). About 500 PAH have been detected in the air, but in practice, only a few of them are routinely measured.

Biological monitoring

The assessment of occupational exposure is very difficult, since workers are exposed to a mixture of compounds with similar physical and chemical properties. These substances include several potential carcinogens, and occupationally PAH-exposed workers often serve as model populations for assessing or validating biological monitoring methods aiming to detect exposure to genotoxic substances (see Mutagenic and Carcinogenic Substances, page 276).

A practical approach to the problem of exposure assessment to PAH requires the use of a marker, such as a major metabolite, of a PAH representative of this class of compounds.

Urine analysis

1-Hydroxypyrene. 1-Hydroxypyrene seems a specific and sensitive index for the assessment of human uptake of pyrene, a component commonly present in PAH-containing substances (Jongeleenen et al 1985, 1986, 1988a,b, 1989, 1990, 1992; Clonfero et al 1989; Zhen-Hua Zhao et al 1990; Buchet et al 1992; Burgaz et al 1992). Exposures to PAH through diet can sometimes be substantial and can result in high urinary excretion of 1-hydroxypyrene (Buckley and Lioy 1992). In occupationally exposed subjects, the smoking habits appear to be a minor confounder (Jongeleenen et al 1986) or at least not a strong determinant (Burgaz et al 1991). After an oral dose of PAH, urinary 1-hydroxypyrene was eliminated with a half-time of 4.5 h and a t_{max} of 6-7 h (Buckley and Lioy 1992). In workers exposed for 8 h to a mixture of PAH, Buchet et al (1992) have calculated that the half-life for the urinary excretion of the metabolite 1-hydroxypyrene amounts on average to 18 h, indicating a possible accumulation of pyrene in the body during the work week. According to Jongeleenen et al (1992), the half-life of 1-hydroxypyrene in urine might range from 6 to 35 h.

Burgaz et al (1991) found a mean (±SD) 1-hydroxypyrene excretion of 0.61 ± 0.38 µmol/mol creatinine in workers exposed to bitumen fumes during road-paving operations, while the mean 1-hydroxypyrene excretion in the nonexposed group was 0.28 ± 0.17 µmol/mol creatinine.

In workers from a coke oven plant and a graphite electrode-producing factory, Buchet et al (1992) have found good correlation coefficients between pyrene (r: 0.67) or total airborne concentration of 13 PAH (r: 0.72) (collected by personal sampling) and the hydroxypyrene concentration in urine collected at the end of the workshift. Hydroxypyrene excretion doubled when the exposure to pyrene in air increased tenfold.

Jongeleenen et al (1992) have estimated that in coke oven workers, a urinary concentration of 1-hydroxypyrene of 2.3 μmol/mol creatinine after a 3-d working period corresponds to the ACGIH current airborne threshold limit value of coal tar pitch volatiles (0.2 mg/m^3).

REFERENCES

J.-P. Buchet, J.-Ph. Gennart, F. Mercado-Calderon et al. Evaluation of exposure to poly-cyclic aromatic hydrocarbons in a coke production and a graphite electrode manufac-turing plant and assessment of the urinary excretion of 1-hydroxypyrene as a biologi-cal indicator of exposure. *Br. J. Ind. Med.* 49:761, 1992.

T. Buckley and P. Lioy. An examination of the time course from human dietary exposure to polycyclic aromatic hydrocarbons to urinary elimination of 1-hydroxypyrene. *Br. J. Ind. Med.* 49:113, 1992.

S. Burgaz, P. Borm and F. Jongeleenen. Biological monitoring of exposure to bitumen fumes during road paving operations. The 1991 Eurotox Congress. Book of abstracts. Maastricht, The Netherlands, Sept 1991.

S. Burgaz, P. Borm and F. Jongeleenen. Evaluation of urinary excretion of 1-hydroxypyrene and thioethers in workers exposed to bitumen fumes. *Int. Arch. Occup. Environ. Health* 63:397, 1992.

E. Clonfero, M. Zordan, P.Venier et al. Biological monitoring of human exposure to coal tar urinary excretion of total polycyclic aromatic hydrocarbons. 1-Hydroxypyrene and mutagens in psoriatic patients. *Int. Arch. Occup. Environ. Health* 61:363, 1989.

F. Jongeleenen, R. Anzion, Ch. Leijdekkers et al. 1-Hydroxypyrene in human urine after exposure to coal tar and a coal tar derived product. *Int. Arch. Occup. Environ. Health* 57:47, 1985.

F. Jongeleenen, R. Bos, R. Anzion et al. Biological monitoring of polycyclic aromatic hydrocarbons: metabolites in urine. *Scand. J. Work Environ. Health* 12:137, 1986.

F. Jongeleenen, P. Scheepers, P. Groenendijk et al. Airborne concentrations, skin contami-nation and urinary metabolite excretion of past among paving workers exposed to coal tar derived road tars. *Am. Ind. Hyg. Assoc. J.* 19:600, 1988a.

F. Jongeleenen, R. Anzion, P. Scheepers et al. 1-Hydroxypyrene in urine as a biological indicator of exposure to polycyclic aromatic hydrocarbons in several work environ-ments. *Ann. Occup. Hyg.* 32:35, 1988b.

F. Jongeleenen, R. Anzion, J. Theuws and R. Bos. Urinary 1-hydroxypyrene in workers handling petroleum coke. *J. Toxicol. Environ. Health* 26:133, 1989.

F. Jongeleenen, F. Van Leeuwen, S. Oosterink et al. Ambient and biological monitoring of coke oven workers: determinants of the internal dose of polycyclic aromatic hydro-carbons. *Br. J. Ind. Med.* 47:454, 1990.

F. Jongeleenen. Biological exposure limit for occupational exposure to coal tar pitch volatiles at coke ovens. *Int. Arch. Occup. Environ. Health* 63:511, 1992.

Zhao Zhen-Hua, Quand Wen-Li and Haitian De. Urinary 1-hydroxypyrene as an indicator of human exposure to ambient polycyclic aromatic hydrocarbons in a coal-burning environment. *Sci. Total Environ.* 92:145, 1990.

3. HALOGENATED HYDROCARBONS

Except for some halogenated aromatic hydrocarbons (e.g., polychlorinated or polybrominated biphenyls), most aliphatic halogenated hydrocarbons that are widely used as solvents or plastic monomers are non-cumulative chemicals. For several representatives of this group of chemicals, measurement of the substance itself or its metabolites in blood, urine or expired air has been proposed for evaluating recent exposure.

3.1. Monochloromethane (Methyl Chloride)

Toxicokinetics

Methyl chloride, a gas at normal temperature, is readily absorbed from the lungs following inhalation exposure (ATSDR 1990).

In six volunteers exposed to 10 or 50 ppm for 6 h, Nolan et al (1985) observed that blood and alveolar air levels reached an equilibrium during the first hour of exposure. Elimination from blood and excretion in expired air occurred in a biphasic manner when exposure ceased. The half-life for the second phase was 50-90 min, suggesting the absence of accumulation in case of repeated exposure. Approximately 30% of absorbed methyl chloride was exhaled during the first hour after exposure (Morgan et al 1970). The remainder was metabolized mainly by conjugation with glutathione, leading to the production of *S*-methylcysteine, excreted in urine (Van Doorn et al 1980).

Preliminary results obtained on workers exposed to methyl chloride suggest that the biological half-life of *S*-methylcysteine in urine is greater than 16 h. Furthermore, it is possible that some persons are less able to produce *S*-methylcysteine following exposure to methyl chloride. It has been speculated that they are more susceptible to the toxic effects of this substance (Van Doorn et al 1980). Nolan et al (1985) also identified two distinct populations of slow and fast eliminators. As in the study of Van Doorn et al, the urinary excretion of *S*-methylcysteine in the volunteers exposed to chloromethane was quite variable. It did not correlate with exposure and pre- and postexposure levels were not significantly different.

No biological threshold limit values can yet be proposed for methyl chloride in expired air and in blood or of *S*-methylcysteine in urine.

REFERENCES

ATSDR. U.S. Department of Health and Human Services, Public Health Service, Toxicological profile for choloromethane. Agency for Toxic Substances and Disease Registry, ATSDR/TP-90/07, 1990.

A. Morgan, A. Black and D. Belcher. The excretion in breath of some aliphatic halogenated hydrocarbons following administration by inhalation. *Ann. Occup. Hyg.* 13:219, 1970.

R. Nolan, D. Rick and T. Landry. Pharmacokinetics of inhaled methylchloride (CH3Cl) in male volunteers. *Fund. Appl. Toxicol.* 5:361, 1985.

R. Van Doorn, P. Borm, C. Leijdekkers et al. Detection and identification of *S*-methylcysteine in urine of workers exposed to methyl chloride. *Int. Arch. Occup. Environ. Health* 46:99, 1980.

3.2. Monochlorobenzene

Toxicokinetics

In industry, workers are exposed to monochlorobenzene (MCB) mainly through inhalation and possibly through skin contact (e.g., in case of accidental projection) (Fiserova-Bergerova and Pierce 1989). A simulation study indicates a pulmonary retention of about 60% (ACGIH 1991-1992).

Chlorobenzene is partly eliminated unchanged in the expired air. Urinary excretion of unchanged monochlorobenzene is negligible. 4-Chlorocatechol and chlorophenols (free and conjugated) are the main metabolites identified in man, 4-chlorophenol accounting for the major portion of the total chlorophenols (Figure 16). Another minor metabolite is *p*-chlorophenyl mercapturic acid (Ogata and Shimada 1983; Yoshida et al 1986). On the average, the amount of 4-chlorocatechol excreted in urine is at least three times more important than that of 4-chlorophenol (Kusters and Lauwerys 1990; Ogata et al 1991). The follow-up of 21 workers during several consecutive days showed that more than 80% of these two metabolites are eliminated within 16 h after the end of the exposure and no tendency for an increased concentration of the metabolites in preshift urine samples during the workweek was observed. Ogata et al (1991) have calculated elimination half-lives of 2.2 h (phase I) and 17.3 h (phase II) for chlorocatechol; 3 h (phase I) and 12.2 h (phase II) for 4-chlorophenol.

Biological monitoring

Urine analysis

The measurement of monochlorobenzene metabolites in urine of workers exposed to monochlorobenzene has been proposed as a biological monitoring method. Both metabolites, 4-chlorocatechol and 4-chlorophenol, are not normally present in the urine, but can also be found in urine of subjects exposed to dichlorobenzene or 4-chlorophenol at work or after the use of some household products. In a field study, Yoshida et al (1986) observed a mean urinary concentration of about 52 mg 4-chlorocatechol/g creatinine (range: 24-114) in seven workers exposed to a mean air monochlorobenzene concentration of about 3.15 ppm (1.72-5.8 ppm; area sampling) and about 69.7 mg 4-chlorocatechol/g creatinine (range: 51.2-94.9) in four workers exposed in another plant to a mean air MCB concentration of 3.14 ppm (2.7-3.7 ppm).

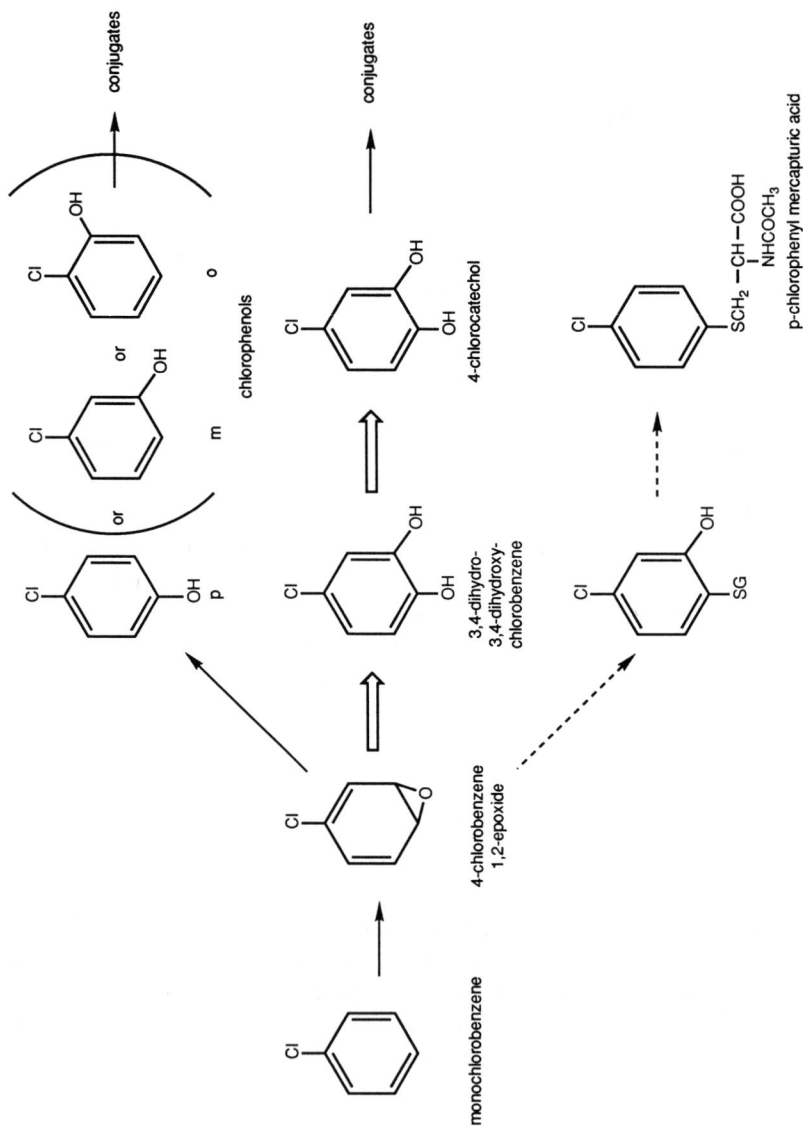

FIGURE 16 Main metabolic pathways of monochlorobenzene.

Kusters and Lauwerys (1990) have assessed the exposure to MCB of 44 workers performing maintenance work in a diphenylmethane-4-4'diisocyanate-producing plant. A statistically significant correlation was found between monochlorobenzene TWA exposure and the concentration of both metabolites in urine collected at the end of the workshift. A mean air concentration of 1.25 ppm (range: <0.05-106) (personal sampling, n: 251) led to mean postshift urinary concentrations of 1.8 mg 4-chlorophenol/g creatinine (<0.2-21.1) and 3.8 mg 4-chlorocatechol/g creatinine (<0.2-57.9).

On the basis of a study conducted on five volunteers (11.8 to 60.2 ppm; 3 h in the morning, 4 h in the afternoon, 1-h break between), Ogata et al (1991) have estimated that the levels of 4-chlorocatechol, 4-chlorophenol, and total chlorophenols in urine taken during the last 2 h of exposure to 1 ppm monochlorobenzene are 6.56, 1.13, and 2.83 mg/g creatinine.

The values of 150 mg 4-chlorocatechol/g creatinine and 25 mg 4-chlorophenol/ g creatinine (urine collected at the end of the shift) are proposed as biological exposure indice by ACGIH (1991-1992) (TLV 10 ppm or 46 mg/m^3). However, since there are large discrepancies in the results obtained by different authors, more studies are necessary to establish reliable biological threshold values.

REFERENCES

V. Fiserova-Bergerova and J. Pierce. Biological monitoring. V. Dermal absorption. *Appl. Ind. Hyg.* 4:F14, 1989.

E. Kusters and R. Lauwerys. Biological monitoring of exposure to monochlorobenzene. *Int. Arch. Occup. Environ. Health* 62:329, 1990.

M. Ogata, T. Taguchi, N. Hirota et al. Quantitation of urinary chlorobenzene metabolites by HPLC: concentrations of 4-chlorocatechol and chlorophenols in urine and of chlorobenzene in biological specimens of subjects exposed to chlorobenzene. *Int. Arch. Occup. Environ. Health* 63:121, 1991.

M. Ogata and Y. Shimada. Differences in urinary monochlorobenzene metabolites between rats and humans. *Int. Arch. Occup. Environ. Health* 53:51, 1983.

M. Yoshida, M. Sunaga and I. Hara. Urinary metabolite levels in workers exposed to chlorobenzene. *Ind. Health* 24:255, 1986.

3.3. Monobromomethane (Methyl Bromide)

Toxicokinetics

It is likely that this chemical, which is gaseous at normal temperature, and can be absorbed through the lung and the skin, follows the same metabolic pathway as methyl chloride. Workers exposed to methyl bromide probably excrete *S*-methylcysteine in urine, but this hypothesis needs confirmation. A fraction of the absorbed methyl bromide is also excreted unchanged in expired air (Tanaka et al 1991).

Biological monitoring

Increased bromine levels can be detected in the blood and urine of persons exposed to this compound (Rathus and Landy 1961; Tanaka et al 1991).

Blood analysis

Bromine. Verberk et al (1979) have reported that, in workers with blood bromine levels higher than 1.2 mg/100 ml, there is a 3.5 times higher risk of slight electroencephalogram disturbances than in those with blood bromide levels below 1.2 mg/100 ml. These data are not yet conclusive (Zielhuis 1978). According to Yamano et al (1987), the normal concentration of bromine in plasma is 0.37 mg/100 ml (SD: 0.15). It has been suggested that values of bromide ion in blood of 5 mg/100 ml and above are indicative of potentially hazardous exposure (World Health Organization 1975).

S-Methylcysteine adduct in hemoglobin. Methyl bromide reacts with cysteine on hemoglobin to form a *S*-methylcysteine adduct (Iwasaki 1988a,b). An increase concentration of this adduct has been found in workers exposed to methyl bromide (Iwasaki et al 1989).

Urine analysis

Bromine. Tanaka et al (1991) found that workers involved in fumigating with methyl bromide excreted more bromine in urine than control subjects. Despite the fact that the workers wore a gas mask, there was a linear relationship between ambient concentration and bromine levels in urine collected at the end of the workshift. In control subjects the mean urinary bromine concentration amounted to 6.3 mg/l (SD: 2.5) with a 95% confidence limit of 10 mg/l.

REFERENCES

K. Iwasaki. Determination of *S*-methylcysteine in mouse hemoglobin following exposure to methylbromide. *Ind. Health* 26:187, 1988a.

K. Iwasaki. Individual differences in the formation of hemoglobin adducts following exposure to methylbromide. *Ind. Health* 26:257, 1988b.

K. Iwasaki, I. Ito and J. Kagawa. Biological exposure monitoring of methylbromide workers by determination of hemoglobin adducts. *Ind. Health* 27:181, 1989.

E. Rathus and P. Landy. Methylbromide poisoning. *Br. J. Ind. Med.* 18:53, 1961.

S. Tanaka, S. Abuku, Y. Seki and S. Imamiya. Evaluation of methylbromide exposure on the plant quarantine fumigators by environmental and biological monitoring. *Ind. Health* 29:11, 1991.

M. Verberk, T. Rooyakkers-Beemster, M. De Vlieger and A. VanVliet. Bromine in blood, EEG and transaminases in methyl bromide workers. *Br. J. Ind. Med.* 36:59, 1979.

WHO. World Health Organization. Data sheet on pesticides. No. 5, Methylbromide, Geneva, 1975.

Y. Yamano, I. Ito, N. Nagao and S. Ishizu. A simple determination method of bromide ion in plasma of methylbromide workers by lead space chromatography. *Sangyo Igaku* 29:196, 1987.

R. Zielhuis. Biological monitoring. *Scand. J. Work Environ. Health* 4:1, 1978.

3.4. Dichloromethane (Methylene Chloride)

Toxicokinetics

Methylene chloride is a volatile solvent that is easily absorbed by the lung (retention: 55-70%) (Astrand et al 1975; Di Vincenzo and Kaplan 1981a) and probably also by direct skin contact with the liquid form (Stewart and Dodd 1964). It is partly eliminated in expired air, the remaining being metabolized. Two main pathways have been identified. One involves an oxidation by a P450 cytochrome-dependent system leading to the production of carbon monoxide; the other involves conjugation with GSH with production of formaldehyde (Figure 17) (Astrand et al 1975; Di Vincenzo and Kaplan 1981a; Ratney et al 1974; Stewart et al

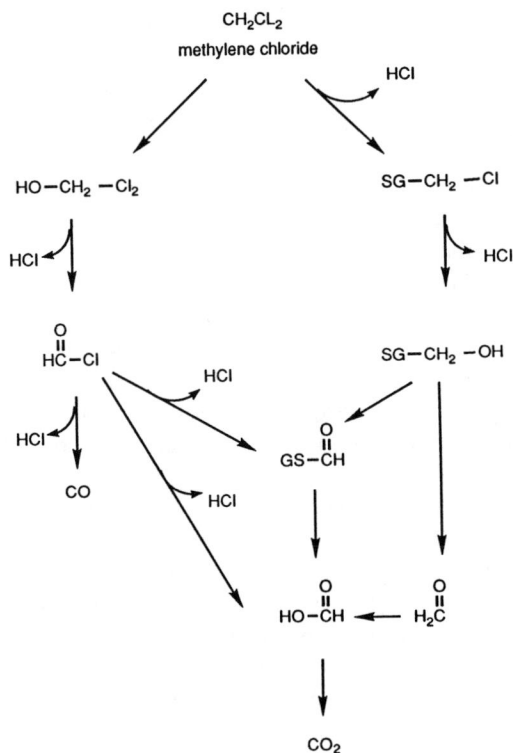

FIGURE 17 Main metabolic pathways of dichloromethane.

1972a,b; Ott et al 1983; Kubic and Anders 1975; Kubic et al 1974; Ahmed and Anders 1978).

Biological monitoring

Breath and blood analyses

Measurement of methylene chloride in blood or in expired air, and carboxyhemoglobin in blood can be used to monitor the magnitude of exposure. A detailed study of the relationship between these parameters has been undertaken by Di Vincenzo and Kaplan (1981a,b). At the end of exposure of nonsmoking, sedentary volunteers for 7.5 h to methylene chloride vapor concentrations ranging from 50 to 200 ppm, the concentration of the solvent in alveolar air (ppm) and in blood (mg/100 ml) and the percent carboxyhemoglobin saturation, respectively, were on the average as follows (DiVincenzo and Kaplan 1981a):

> for 50 ppm :   15, 0.03 and 1.9
> for 100 ppm: 35, 0.08 and 3.25
> for 150 ppm: 55, 0.12 and 5.3
> for 200 ppm: 80, 0.18 and 6.8

The carboxyhemoglobin level exceeds 8% after an exposure to 250 ppm methylene chloride for 7.5 h (Hake et al 1975). For a 200-ppm exposure, the mean postexposure and tidal air concentration for methylene chloride decreases to 1 ppm at 16 h (DiVincenzo and Kaplan 1981a). According to Stewart et al (1976), 16-20 h after exposure to 50 or 100 ppm methylene chloride for 7.5 h, the level in expired air is less than 0.1 ppm.

Di Vincenzo and Kaplan (1981a) have estimated that an 8-h exposure to about 150 ppm of methylene chloride vapor is equivalent to an 8-h exposure to 35 ppm of CO, inasmuch as either exposure under sedentary conditions will increase blood carboxyhemoglobin levels to about 5% of saturation at the end of the exposure. An exposure to 1000 ppm methylene chloride increases the carboxyhemoglobin level up to 10-15% (Rodkey and Collison 1977; Stewart et al 1972a,b).

As expected, physical exercise performed during exposure to methylene chloride vapor will produce higher blood carboxyhemoglobin saturations than those found in sedentary workers (Astrand et al 1975; Di Vincenzo and Kaplan 1981b). Hence under moderate workload an exposure to 100 ppm methylene chloride for 7.5 h may cause a carboxyhemoglobin saturation of about 5% at the end of the exposure period (Di Vincenzo and Kaplan 1981b). The combined effect of smoking and exposure to methylene chloride produces an additive increase in blood carboxyhemoglobin values (Di Vincenzo and Kaplan 1981b).

Regression analysis of data collected in 14 workers engaged in furniture stripping with methylene chloride showed that the methylene chloride TWA exposure concentration was quadratically correlated to postshift alveolar breath (r: 0.98) and blood concentrations (r: 0.87) of methylene chloride (McCammon et al

1991; Glaser et al 1991). These workers were divided into three subgroups depending on their job and hence their TWA exposure levels: 191 ppm, 145 ppm, and 31 ppm. The mean alveolar breath (and blood) postshift concentrations were 60 ppm (4.2 ppm) 48 ppm (2.2 ppm), and 7.1 ppm (1.3 ppm), respectively. For the nonsmoking workers, environmental concentrations of methylene chloride were correlated with postshift percent carboxyhemoglobin levels (r: 0.85), but less with alveolar CO levels (r: 0.55). When the smokers were included the correlation coefficients decreased (r: 0.32 and 0.40, respectively).

The DFG (1991) recommends 5% carboxyhemoglobin and 1 mg dichloromethane/l blood (end-of-shift) as biological tolerance values.

In conclusion it is tentatively proposed that workers whose concentration of the solvent in expired air (ppm) and in blood (mg/100 ml) at the end of the shift does not exceed 15 ppm and 0.05 mg/100 ml have not been exposed to a TWA exceeding 50 ppm. In nonsmokers this level of exposure would not increase the carboxyhemoglobin level above 2%.

REFERENCES

A. Ahmed and M. Anders. Metabolism of dihalomethanes to formaldehyde and inorganic halide. II. Studies on the mechanism of the reaction. *Biochem. Pharmacol.* 27:2021, 1978.

I. Astrand, P. Ovrum and A. Carlsson. Exposure to methylene chloride. I. Its concentration in alveolar air and blood during rest and exercise and its metabolism. *Scand. J. Work Environ. Health* 1:78, 1975.

DFG. Deutsche Forschungsgemeinschaft. *Maximum Concentrations at the Workplace and Biological Tolerance Values for Working Materials.* Report No. XXVII. Commission for the Investigation of Health Hazards of Chemical Compounds in the Work Area. VCH, Weinheim, 1991.

G. Di Vincenzo and C. Kaplan. Uptake, metabolism and elimination of methylene chloride vapor by humans. *Toxicol. Appl. Pharmacol.* 59:130, 1981a.

G. Di Vincenzo and C. Kaplan. Effect of exercise or smoking on the uptake, metabolism and excretion of methylene chloride vapor. *Toxicol. Appl. Pharmacol.* 59:141, 1981b.

R. Glaser, J. Arnold, Ch. McCammon and F. Phipps. Evaluation of a new solid sorbent sampler for alveolar methylene chloride used in tandem with a bag for sampling alveolar carbon monoxide. *Appl. Occup. Environ. Hyg.* 6:380, 1991.

C. Hake, R. Stewart, A. Wu and S. Graff. Carboxyhemoglobin levels of humans exposed to methylene chloride. Presented at the 14th annual meeting of the Society of Toxicology. Williamsburg, VA, 1975.

V. Kubic, M. Anders, R. Engels et al. Metabolism of dihalomethanes to carbon monoxide. I. In vivo studies. *Drug. Metab. Disp.* 2:53, 1974.

V. Kubic and M. Anders. Metabolism of dihalomethanes to carbon monoxide. III. In vitro studies. *Drug. Metab. Disp.* 3:104, 1975.

Ch. McCammon, R. Glaser, V. Wells et al. Exposure of workers engaged in furniture stripping to methylene chloride as determined by environmental and biological monitoring. *Appl. Occup. Environ. Hyg.* 6:371, 1991.

M. Ott, L. Skory, B. Holder et al. Health evaluation of employees occupationally exposed to methylene chloride. *Scand. J. Work Environ. Health* 9:31, 1983.

R. Ratney, D. Wegman and H. Elkins. The in vivo conversion of methylene chloride to carbon monoxide. *Arch. Environ. Health* 28:223, 1974.

F. Rodkey and H. Collison. Biological oxidation of (^{14}C)-methylene chloride to carbon monoxide and carbon dioxide by rat. *Toxicol. Appl. Pharmacol.* 40:33, 1977.

R. Stewart and H. Dodd. Absorption of carbon tetrachloride, trichloroethylene, tetrachloroethylene, methylene chloride and 1,1,1-trichloroethane through the human skin. *Am. Ind. Hyg. Assoc. J.* 25:439, 1964.

R. Stewart, T. Fisher, M. Hosko et al. Carboxyhemoglobin elevation after exposure to dichloromethane. *Science* 176:295, 1972a.

R. Stewart, T. Fisher, M. Hosko et al. Experimental human exposure to methylene chloride. *Arch. Environ. Health* 25:342, 1972b.

R. Stewart, C. Hake and A. Wu. Use of breath analysis to monitor methylene chloride exposure. *Scand. J. Work Environ. Health* 2:57, 1976.

3.5. Trichloromethane (Chloroform) and Tetrachloromethane (Carbon Tetrachloride)

Both solvents can be detected in blood and in expired air of exposed persons (Morgan et al 1970; Stewart et al 1961; Stewart and Dodd 1964). The paucity of human data precludes the evaluation of the relationship between environmental exposure to these solvents and their levels in biological media. Hence, no biological threshold limit values can yet be proposed.

REFERENCES

A. Morgan, A. Black and D. Belcher. The excretion in breath of some aliphatic halogenated hydrocarbons following administration by inhalation. *Ann. Occup. Hyg.* 13:219, 1970.

R. Stewart, A. Arbor, H. Gay et al. Human exposure to carbon tetrachloride vapor; relationship of expired air concentration to exposure and toxicity. *J. Occup. Med.* 3:586, 1961.

R. Stewart and H. Dodd. Absorption of tetrachloride, trichloroethylene, tetrachloroethylene, methylene chloride and 1,1,1-trichloroethane through the human skin. *Am. Ind. Hyg. Assoc. J.* 25:439, 1964.

3.6. 1,1,1-Trichloroethane (Methylchloroform)

Toxicokinetics

Because of its extensive use in industry and in domestic products, 1,1,1-trichloroethane is a ubiquitous pollutant which can be detected in the blood and breath of individuals nonoccupationally exposed to volatile halogenated hydrocarbons (Hajimiragha et al 1986; Wallace et al 1985, 1986).

Trichloroethane enters the organism through inhalation of vapor and to a small extent by skin contact with the liquid form. The concentration of 1,1,1-trichloroethane was determined in blood collected from both forearms of two subjects after one of their hands was soaked for 5 min in trichloroethane as a stimulation of the washing of hands with solvent after work. The trichloroethane concentration on the "soaked side" was 35-fold higher than in the contralateral side. Intraindividual differences were marked, but it was not until after 5 h that the difference between the two arms vanished (Aitio et al 1984). From physicochemical data, skin penetration of vapors rate was calculated to be 1.64 mg trichloroethane/cm^2/h (Fiserova-Bergerova and Pierce 1989).

The alveolar retention decreases during the course of exposure and increases with ventilation (Humbert and Fernandez 1977; Monster et al 1979). Pezzagno et al (1988) have noted that the uptake of trichloroethane is higher during light exercise than during heavy exercise; the explanation being the proportionally lesser increase of the pulmonary ventilation in comparison with the corresponding decrease of its retention value. Trichloroethane is mainly eliminated unchanged with the expired air. Only a very small fraction (less than 10%) is metabolized *in vivo* with the production of trichloroacetic acid and trichloroethanol, which are excreted in urine. The latter metabolite can also be detected in expired air (Monster et al 1979). The elimination of 1,1,1-trichloroethane from blood into exhaled air is exponential. Three phases have been identified with half-times of 1-9 h, 6-20 h, and greater than 26 h.

Trichloroethanol and trichloroacetic acid reach their maximum concentrations in blood almost directly after exposure and up to 20 to 40 h, respectively. Their half-lives in blood have been estimated to be 10-27 h for trichloroethanol and 70-100 h for trichloroacetic acid. This can result in a transient accumulation of these metabolites during the workweek (Monster et al 1979; Nolan et al 1984; Monster 1987).

Biological monitoring

Breath analysis

1,1,1-Trichloroethane. The concentration of 1,1,1-trichloroethane in exhaled air of volunteers exposed at rest to about 500 ppm of the solvent for 6.5-8 h amounts to 50-70 ppm 0.5 h after the end of the exposure and to 10-20 ppm 16 h later (Stewart et al 1969; Rowe et al 1963). According to Astrand et al (1973), the alveolar air concentrations of 1,1,1-trichloroethane during exposure to 350 ppm amount on the average to 179 ppm at rest and 239 ppm under a workload of 50 watts. These authors have proposed that during exposure the alveolar air content should not exceed 50 ppm, i.e., the mean value minus one standard deviation obtained after a 10-min exposure to 350 ppm, if impairment of psychological functions is to be avoided with any degree of certainty. They have also recommended the analysis of alveolar air during ongoing exposure and not in the aftermath of exposure because the "washing out" curves vary from individual to individual. Monster (1986, 1987) investigated exposure to trichloroethane in four workshops. He estimated that for a TWA exposure of 100 mg/m³ (18 ppm), 8 h a day, 5 d a week, the concentrations of trichloroethane in exhaled air 5 to 15 min after cessation of 5 d of exposure or 16 h after exposure or 64 h after exposure were 60, 12, or 2% of the average exposure concentration, respectively. On this basis, one can expect an alveolar air concentration of 42 ppm 16 h after an average exposure to 350 ppm or 6 ppm 16 h after an average exposure to 50 ppm. When extrapolating the data obtained from human volunteers, the trichloroethane mean alveolar air concentrations at the end of the work week after daily exposure to 50 ppm were for the same time period of sampling 210 mg/m³ (39ppm), 13 mg/m³ (2.4 ppm) and 8 mg/m³ (1.5 ppm). An explanation for this difference being probably a higher accumulation in adipose tissue with the longer occupational exposure time in comparison with the experimental studies.

The BEI recommended by the ACGIH (1991-1992) is 40 ppm, prior to the last shift of the week (TLV 350 ppm). The BAT (biological tolerance value) recommended by the DFG (1991) is 20 ml/m³ (prior to the shift after several days of exposure) (MAK 200 ppm).

Trichloroethanol. The mean concentrations of trichloroethanol in alveolar air were 0.014 mg/m³ and 0.007 mg/m³, respectively, 5-15 min and 16 h after the exposure of human volunteers to a methylchloroform TWA of 50 ppm (Monster 1987).

Blood analyses

Trichloroethane, trichloroethanol, trichloroacetic acid. In volunteers exposed at rest to 140 ppm 1,1,1-trichloroethane for 4 h, the concentrations (in mg/100 ml) of the solvent and its metabolites in blood at the end of exposure were: 0.19 (trichloroethane), 0.019 (trichloroethanol), and 0.010 (trichloroacetic acid). The corresponding blood levels 16 h after the end of exposure were 0.004 (trichloroethane), 0.005 (trichloroethanol), and 0.025 (trichloroacetic acid) mg/ 100 ml (Monster et al 1979). On the basis of his study on workers Monster (1986, 1987) extrapolated that a TWA exposure to 50 ppm (8 h/d, 5 d/week) would lead to mean blood concentrations of 0.09 mg/100 ml trichloroethane, 0.016 mg/100 ml trichloroethanol and 0.23 mg/100 ml trichloroacetic acid 5-15 min after exposure. The DFG (1991) recommends 550 μg trichloroethane/l blood (prior to the shift after several days of exposure) as BAT (TLV 200 ppm) and the ACGIH (1991-1992) 1 mg trichloroethanol /l blood (end-of-shift, end of the week) as BEI (TLV 350 ppm).

Urine analyses

1,1,1-Trichloroethane. Both in experimentally and occupationally exposed subjects the urinary concentration (end of 4 h exposure) of 1,1,1-trichloroethane showed a linear relationship to the environmental TWA concentration. The urinary concentration corresponding to a TWA of 1900 mg/m³ (350 ppm), derived from the regression line established in occupationally exposed workers was found to be 860 µg/l. The proposed biological exposure limit, based on 4 h exposure, being 805 µg/l (Imbriani et al 1988; Ghittori et al 1987).

Trichloroethanol, trichloroacetic acid. In workers from a printing plant exposed to moderate concentrations (less than 60 ppm) of trichloroethane, Seki et al (1975) found a linear relationship between the concentration of the solvent in air and that of its metabolite in urine. In the urine collected at the end of the exposure period from workers exposed to an average concentration of 53 ppm 1,1,1-trichloroethane, the mean (range) concentrations (in mg/l) of trichloroethanol and trichloroacetic acid were 9.9 (6.8-14.5) and 3.6 (2.4-5.5), respectively.

It is evident that the amount of trichloro-derivatives excreted in urine after 1,1,1-trichloroethane exposure is much lower than after similar exposure to tri-chloroethylene. Stewart and Rowe (1967) reported that, in humans exposed to 500 ppm of trichloroethane for 7 h a day for 5 d, the amount of total trichloro-derivatives (trichloroethanol plus trichloroacetic acid) found in urine never ex-ceeded one-tenth of the amount found after exposure to 200 ppm trichloroethyl-ene. It is therefore likely that, at the end of an 8-h exposure to 350 ppm trichloroethane, the urinary concentration of total trichloro-compounds will be less than 50 mg/g creatinine. This estimate is in agreement with that made by Caperos et al (1982). The latter authors have calculated that during a week of steady exposure to 350 ppm trichloroethane the average urinary concentration of trichloroethanol at the end of the shift approximates 30 mg/l. They have proposed, however, that the most suitable method to estimate the exposure is by determina-tion of urinary trichloroethanol both before and after a workshift. Monster (1986) estimated the mean trichloroethanol and trichloroacetic acid urinary concentra-tions of subjects exposed to a TWA of 18 ppm (100 mg/m^3) (8 h/d, 5 d/week) to be 1.6 and 0.7 mg/g creatinine respectively, directly after exposure. According to estimations based on experimental studies, a TWA exposure of 50 ppm (275 mg/m^3) (8 h/day, 5 d/week) would lead to mean urinary concentrations of 2.5 and 4.9 mg/g creatinine at the end of the exposure, 1.8 and 2.5 mg/g creatinine 16 h after exposure, and 1.5 and 0.9 mg/g creatinine 64 h after exposure for trichloroethanol and trichloroacetic acid, respectively (Monster 1986).

The BEI recommended by the ACGIH (1991-1992) are 10 mg/l and 30 mg/l for trichloroacetic acid and trichloroethanol, respectively (end-of-shift, end-of-workweek) corresponding to a TWA of 350 ppm (1910 mg/m^3).

Droz et al (1989a,b) have developed a physiological model taking into account variability in exposure, physical workload, body build, and liver and renal func-tions. The biological indicators corresponding to a normal work cycle (8 h/d, 5 d/week, 4 weeks to reach a steady-state situation) after exposure at a TWA of 350 ppm are the following. End-of-work on Thursday: trichloroethane in alveolar air: 4340 µmol/m^3 (106 ppm), in venous blood: 35 µmol/l (467 µg/100 ml), in urine: 10 µmol/l (1.3 mg/l); trichloroacetic acid in blood: 87 µmol/l (14 mg/l), in urine: 8.4 mmol/mol creatinine (12 mg/g creatinine); trichloroethanol in blood: 6.6 µmol/l (98.6 µg/100 ml), in urine: 38 mmol/mol creatinine (50 mg/g creatinine).

On Friday before the shift: trichloroethane in alveolar air: 1270 μmol/m³ (31 ppm), in venous blood: 5.1 μmol/l (68 μg/100 ml), in urine: 1.2 μmol/l (160 μg/l); trichloroacetic acid in blood: 89 μmol/l (1.45 mg/100 ml), in urine: 8.7 mmol/mol creatinine (12.6 mg/g creatinine); trichloroethanol in blood: 3.4 μmol/l (55.5 μg/100 ml), in urine: 24.1 mmol/mol creatinine (32.1 mg/g creatinine).

Trichloroethane concentration in alveolar air before the workshift is an indicator of the mean exposure of the previous day(s). A TWA exposure to 350 ppm will probably lead to a mean alveolar air concentration below 40 ppm 16 h after the exposure. The corresponding mean blood and urinary trichloroethane concentration are approximately 100 μg/100 ml blood and 1 mg/l urine at the end of the shift (end-of-workweek).

Due to its long half-life, trichloroacetic acid is an indicator of the integral exposure of the week; a mean urinary concentration below 14 mg/l in the end of the shift (end-of-workweek) seems indicative of a mean exposure concentration below 350 ppm.

Trichloroethanol is a measure of very recent exposure, and a TWA exposure to 350 ppm corresponds approximately to mean blood and urinary concentrations of 0.1 mg/100 ml and 30 mg/g creatinine at the end of the shift (end-of-workweek).

REFERENCES

ACGIH. American Conference of Governmental Industrial Hygienists. Threshold Limit Values for Chemical Substances and Physical Agents and Biological Exposure Indices. Cincinnati, 1991-1992.

A. Aitio, K. Pecari and J. Järvisalo. Skin absorption as a source of error in biological monitoring. *Scand. J. Work Environ. Health* 10:317, 1984.

I. Astrand, A. Kilbom, I. Wahlberg and P. Ovrum. Methylchloroform exposure. I. Concentration in alveolar air and blood at rest and during exercise. *Work Env. Health* 10:69, 1973.

J. Caperos, P. Droz, C. Hake et al. 1,1,1-trichloroethane exposure, biologic monitoring by breath and urine analyses. *Int. Arch. Occup. Env. Health* 49:293, 1982.

DFG. Deutsche Forschungsgemeinschaft. *Maximum Concentrations at the Workplace and Biological Tolerance Values for Working Materials* Report No. XXVII. Commission for the Investigation of Health Hazards of Chemical Compounds in the Work Area. VCH, Weinheim, 1991.

P. Droz, M. Wu, W. Cumberland and M. Berode. Variability in biological monitoring of solvent exposure. I. Development of a population physiological model. *Br. J. Ind. Med.* 46:447, 1989a.

P. Droz, M. Wu, W. Cumberland and M. Berode. Variability in biological monitoring of solvent exposure. II. Application of a population physiological model. *Br. J. Ind. Med.* 46:547, 1989b.

V. Fiserova-Bergerova and J. Pierce. Biological monitoring. V. Dermal absorption. *Appl. Ind. Hyg.* 4:F14, 1989.

S. Ghittori, M. Imbriani, G. Pezzagno and E. Capodaglio. The urinary concentration of solvents as a biological indicator of exposure: proposal for the biological equivalent exposure limit for nine solvents. *Am. Ind. Hyg. J.* 48:786, 1987.

H. Hajimiragha, V. Ewers, R. Jansen-Roseck and A. Brockhaus. Human exposure to volatile halogenated hydrocarbons from the general environment. *Int. Arch. Occup. Environ. Health* 58:141, 1986.

B. Humbert and J. Fernandez. Exposition au 1,1,1-trichloroéthane: contribution à l'étude de l'absorption de l'excrétion et du métabolisme chez des sujets humains. *Arch. Mal. Prof.* 38:415, 1977.

M. Imbriani, S. Ghittori, G. Pezzagno et al. 1,1,1-trichloroethane (methyl chloroform) in urine as biological index of exposure. *Am. J. Ind. Med.* 13:211, 1988.

A. Monster, G. Boersma and H. Steenweg. Kinetics of 1,1,1-trichloroethane in volunteers; influence of exposure concentration and workload. *Int. Arch. Occup. Env. Health* 42:293, 1979.

A. Monster. Biological monitoring of chlorinated hydrocarbon solvents. *J. Occup. Med.* 8:583, 1986.

A. Monster. Biological monitoring of exposure to 1,1,1-trichloroethane. In: *Biological Monitoring of Exposure to Chemicals. Organic Compounds.* Eds.: M. Ho, H. Dillon. Wiley Interscience, New York, p. 207, 1987.

R. Nolan, N. Freshour and D. Rick. Kinetics and metabolism of inhaled methyl chloroform (1,1,1-trichloroethane) in male volunteers. *Fund. Appl. Toxicol.* 4:654, 1984.

G. Pezzagno, M. Imbriani and E. Capodaglio. Urinary concentration, environmental concentration and respiratory uptake of some solvents: effect of the work load. *Am. Ind. Hyg. Assoc. J.* 49:546, 1988.

V. Rowe, T. Wujikowski, M. Wolf et al. Toxicity of a solvent mixture of 1,1,1-trichloroethane and tetrachloroethylene as determined by experiments on laboratory animals and human subjects. *Am. Ind. Hyg. Assoc. J.* 24:541, 1963.

Y. Seki, Y. Urashina, H. Aikawa et al. Trichloro-compounds in the urine of humans exposed to methylchloroform at sub-threshold levels. *Int. Arch. Arbeitsmed.* 34:39, 1975.

R. Stewart and V. Rowe. Quinze ans d'étude sur le 1,1,1-trichloroéthane. *Arch. Mal. Prof.* 28:194, 1967.

R. Stewart, H. Gay, A. Schaffer et al. Experimental exposure to methyl chloroform vapor. *Arch. Environ. Health* 19:467, 1969.

L. Wallace, E. Pelizzari, T. Hartwell et al. Personal exposures, indoor-outdoor relationships, and breath levels of toxic air pollutants measured for 355 persons in New Jersey. *Atmosph. Environ.* 19:1651, 1985.

L. Wallace, E. Pellizari, T. Hartwell et al. Concentrations of 20 volatile organic compounds in the air and drinking water of 350 residents of New Jersey compared with concentrations in their exhaled breath. *J. Occup. Med.* 28:603, 1986.

3.7. Trichloroethylene

Toxicokinetics

Like many other solvents, trichloroethylene is a ubiquitous pollutant. Trichloroethylene vapors are easily absorbed by the lungs; the retention ranges from 45 to 75% (Fernandez et al 1975; Monster et al 1979). Trichloroethylene liquid can also be absorbed through the skin (Stewart et al 1970; Sato and Nakajima 1978). Cutaneous absorption of trichloroethylene vapor is negligible. Unlike trichloroethane and tetrachloroethylene, only small amounts (less than 10%) of trichloro-

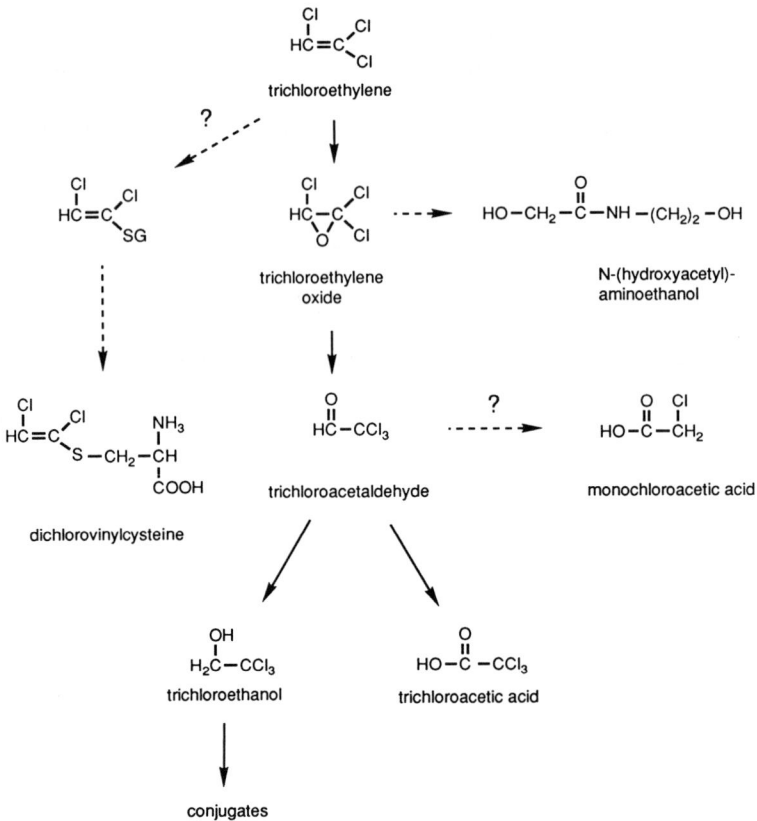

FIGURE 18 Main metabolic pathways of trichloroethylene.

ethylene are eliminated unchanged in expired air (Fernandez et al 1975). The major part of absorbed trichloroethylene is metabolized to trichloroethanol, trichloroethanol glucuronide, and trichloroacetic acid, which are excreted in urine. Very small amounts of trichloroethanol are also excreted by the lung (Monster et al 1979). The metabolic pathway of trichloroethylene in human is summarized in Figure 18.

The existence of other routes of elimination or of other metabolic pathways have been hypothesized, however, because the total recovery (trichloroethylene in air plus trichloroethanol and trichloroacetic acid in urine) usually does not exceed 60% (Fernandez et al 1975). Reported but unconfirmed minor metabolites in trichloroethylene-exposed humans are monochloroacetic acid (Soucek and Vlachova 1960) and N-(hydroxyacetyl)aminoethanol (Dekant et al 1984). Another metabolite might be the mercapturic acid N-acetyl-dichlorovinyl-cysteine. Its cleavage by a renal β-lyase would lead to the formation of (a) nephrotoxic metabolite(s). However, this pathway has only been observed in rats orally treated with high doses of trichloroethylene (Dekant et al 1986).

Investigation on workers has confirmed the high metabolic rate of trichloroethylene in comparison with the low one of tetrachloroethylene (Skender et al

1991). It should be noted that trichloroacetic acid and trichloroethanol are nonspecific metabolites of trichloroethylene. Other chlorine-containing solvents present in the workplace or in household products are metabolized in these products (trichloroethane, tetrachloroethane, perchloroethylene, etc.).

The biological half-lives in urine and blood of trichloroethanol and trichloroacetic acid are different; that of the latter (70-100 h) is greater than that of the former (10-15 h) because of the binding of trichloroacetic acid to plasma proteins (Müller et al 1972). This protein binding also explains that the concentration of trichloroacetic acid in plasma is higher by about a factor of two than in total blood (Monster 1986). Nomiyama and Nomiyama (1971) found a sex difference in trichloroethylene metabolism. Within 24 h after the end of the exposure, urinary excretion of trichloroacetic acid was greater in women than in men; the opposite was true for trichloroethanol. This observation was confirmed by Kimmerle and Eben (1973) and Triest and Lauwerys (1976).

Workload during exposure increases the uptake of trichloroethylene and therefore the blood concentrations (Vesterberg and Astrand 1976). Ingestion of ethanol inhibits the metabolism of trichloroethylene and therefore increases its concentration in blood (Müller et al 1975). Using a physiological simulation model, Sato et al (1991) estimated that ingestion of moderate amounts of ethanol before the start of work or at lunch time, but not at the end of the work, causes pronounced increase in blood trichloroethylene concentrations and decrease in the urinary excretion rates of trichloroethylene metabolites, this effect lasting until the next day. The consumption of ethanol the previous evening causes only small changes in trichloroethylene metabolism at an exposure level of 50 ppm, but appreciable changes at 500 ppm. The production of trichloroethylene metabolites seem also suppressed by co-exposure to tetrachloroethylene (Seiji et al 1989) and toluene (Ikeda 1974).

Biological monitoring

Breath analyses

Trichloroethylene. The concentration of trichloroethylene in alveolar air decreases rapidly during the first half hour after exposure from about 25% of the inhaled concentration at the end of exposure to about 5% (Monster 1986).

The results of several studies on volunteers indicate that the concentration of trichloroethylene in expired alveolar air collected during exposure reflects the current atmospheric concentration, whereas that in alveolar air 16 h after the end of exposure correlates with the average airborne exposure during the preceding day (Fernandez et al 1975; Kimmerle and Eben 1973; Monster et al 1979; Stewart et al 1970, 1974). An 8-h exposure at rest to 100 ppm trichloroethylene produces a concentration in expired alveolar air of approximately 25 ppm during exposure and a concentration below 1 ppm (0.4-0.9 ppm) 16 h after the end of exposure. The corresponding values for an exposure to 50 ppm are 12 and less than 0.5 (0.2-0.4) ppm, respectively. The observed values are in agreement with the predicted values using a mathematical model (Fernandez et al 1977). Workload and alcohol consumption during exposure increase the concentration in expired air.

The ACGIH (1991-1992) recommends 0.5 ppm trichloroethylene in end-exhaled air prior to the last shift of the week as BEI corresponding to a TWA of 50 ppm.

Trichloroethanol. Monster and co-workers (1979, 1986) have found an excellent correlation between trichloroethanol in blood and exhaled air, but during exposure the concentration in exhaled air is about 10,000 times lower than in blood.

Blood analysis

Trichloroethylene. Like the trichloroethylene concentration in expired air, the concentration of the solvent in blood reflects either the most recent exposure when blood is taken during exposure or the TWA exposure when blood is collected 16 h after exposure. At rest, an exposure to 100 ppm trichloroethylene leads to a concentration in blood of approximately 0.12 mg/100 ml during exposure and 0.007 mg/100 ml 16 h later (Müller et al 1974; Pfäffli and Backman 1972). According to Monster (1984, 1986), the mean blood trichloroethylene concentrations to be expected at a TWA exposure (8 h/d, 5 d/week) of 50 ppm are about 0.09 mg/100 ml at the end of exposure, 0.02 mg/100 ml a half hour after the exposure, and 0.002 mg/100 ml 16 h after.

Trichloroacetic acid, trichloroethanol. Müller et al (1972) have proposed determining trichloroacetic acid in plasma and trichloroethanol in whole blood rather than in urine. As in urine, trichloroacetic acid in blood increases progressively during the week. Trichloroethanol level also increases, but to a far lesser extent.

After exposure of sedentary volunteers to 50 ppm trichloroethylene for 5 d (6 h a day), trichloroacetic acid in plasma reaches a level of 5 mg/100 ml and that of trichloroethanol in whole blood, 0.23 mg/100 ml. Ertle et al (1972) reported a trichloroethanol concentration in blood of 0.5 mg/100 ml following exposure to 100 ppm trichloroethylene (5 h daily for 5 successive days).

Monster (1984, 1986) has estimated that a mean TWA exposure to 50 ppm entails approximate blood trichloroethanol and trichloroacetic acid concentrations of 0.46 mg/100 ml and 6-7 mg/100 ml at the end of exposure; 0.1-0.2 mg/100 ml and 6-7 mg/100 ml the next morning.

In a study conducted in four dry-cleaning shops where the trichloroethylene air concentration varied between 25 and 40ppm, median blood levels of 0.38 (0.15-3.58) μmol/l (or 4.99 μg/100 ml) trichloroethylene, 3.02 (0-10.71) μmol /l (or 45.12 μg/100 ml) trichloroethanol, 165.26 (6.12-302.4) μmol/l (or 2.7 mg/100 ml) trichloroacetic acid were observed on Monday morning, and 3.39 (0.46-12.71) μmol/l (or 45.55 μg/10 ml) trichloroethylene, 7.70 (0-26.10) μmol/l (or 0.115 mg/100 ml) trichloroethanol, 194.03 (13.47-393.56) μmol/l (or 3.17 mg/100 ml) trichloroacetic acid on Wednesday after work (Skender et al 1991).

From its assessment of published results, an ACGIH committee (1991-1992) has estimated that occupational exposure to 50 ppm trichloroethylene would lead to the following concentrations of free trichloroethanol in blood: prior to the first shift of the workweek: 0.02 mg/100 ml, end of the first shift of the workweek: 0.27 mg/100 ml, prior to the last shift of the workweek: 0.15 mg/100 ml, end of the last shift of the workweek: 0.38 mg/100 ml. The BEI recommended is 4 mg trichloroethanol/l (27 µmol/l) end-of-shift, end-of-workweek blood sample (ACGIH 1991-1992). The DFG (1991) recommends 5 mg/l, as biological tolerance value, for trichloroethanol (end-of-shift, after several weeks).

Urine analyses

Trichloroethanol, trichloroacetic acid. The determination of both metabolites in urine for monitoring trichloroethylene exposure has been extensively investigated (Ertle et al 1972; Ikeda et al 1972; Imamura and Ikeda 1973; Kimmerle and Eben 1973; Monster et al 1979; Müller et al 1972, 1974; Tanaka and Ikeda 1968; Landrigan et al 1987). The following practical conclusions can be drawn. The major portion of trichloroethanol produced is excreted within the first 24 h after the start of exposure, with a slight accumulation during the week, whereas the amount of trichloroacetic acid excreted in urine at the end of the week is much higher than at the first day.

In ten people exposed to trichloroethylene (25-40 ppm) in dry-cleaning shops, Skender et al (1991) report trichloroethanol and trichloroacetic median levels of 9.70 (0.38-35.65) mmol/mol creatinine (or 12.8 mg/g creatinine) and 32.47 (1.33-61.21) mmol/mol creatinine (or 46.9 mg/g creatinine), respectively, in Monday morning samples, and 54.89 (5.30-177.67) mmol/mol creatinine (or 72.5 mg/g creatinine) and 37.15 (1.92-77.35) mmol/mol creatinine (or 53.66 mg/g creatinine) in end-of-shift samples on Wednesday.

In workers exposed to a TWA of 205 mg/m^3 trichloroethylene (117-375), the mean urinary excretions of total metabolites and trichloroethanol were 480 mg/l and 155 mg/l at the end of the shift (Landrigan et al 1987).

The relationship between the degree of exposure and the metabolites excretion seems to remain linear up to 50 ppm for trichloroacetic acid and 1000 ppm for trichloroethanol. Following the estimation of Monster (1984, 1986), the approximate urinary concentrations of trichloroethanol and trichloroacetic corresponding to a TWA exposure to 50 ppm trichloroethylene are 180 and 100 mg/g creatinine at the end of the workshift, 120 and 80 mg/g creatinine 16 h after the end of exposure.

Droz et al (1989a,b) have developed a physiological model simulating a repeated constant exposure at 50 ppm with a constant light physical workload (50 watts). The average values of trichloroethylene and its metabolites that would be obtained in a group of workers at the end of the shift on Thursday are the following: trichloroethylene in expired air: 310 µmol/m^3 (7.5 ppm), in blood: 6 µmol/l (78.84 µg/100 ml), in urine: 1.4 µmol/l (184 µg/l) ; trichloroacetic acid in blood: 840 µmol/l (13.7 mg/100 ml), in urine: 61 mmol/mol creatinine (88 mg/g creatinine); trichloroethanol in blood: 32 µmol/l (0.478 mg/100 ml), in urine: 175

mmol/mol creatinine (231 mg/g creatinine). On Friday before the shift: trichloro-ethylene in expired air: 78 μmol/m^3 (1.84 ppm), in blood: 0.72 μmol/l (9.46 μg/100 ml), in urine: 0.12 μmol/l (15.76 μg/l); trichloroacetic acid in blood: 815 μmol/l (13.317 mg/100 ml), in urine: 81 mmol/mol creatinine (118 mg/g creatinine); trichloroethanol in blood: 16 μmol/l (0.239 mg/100 ml), in urine: 110 mmol/mol creatinine (146.56 mg/g creatinine).

For a TWA of 50 ppm (269 mg/m^3), the BEI recommended by the ACGIH (1991-1992) are 100 mg/l for trichloroacetic acid and 300 mg/l or 320 mg/g creatinine for trichloroacetic acid and trichloroethanol (end-of-shift at end of workweek). The DFG (1991) recommends the same value for trichloroacetic acid.

In summary, for a TWA exposure of 50 ppm trichloroethylene, 8 h/d, 5 d a week, we suggest the following limit values (sample collected at the end of the workweek): about 0.25 mg/100 ml blood and 150 mg/g urinary creatinine for trichloroethanol (which reflect the exposure level on the last day(s)); about 5 mg/100 ml blood and 75 mg/g urinary creatinine for trichloroacetic acid (which reflect the exposure level over the workweek). The use of trichloroethylene as biological indicator is more problematic since the time of sampling is critical. However, it could be used as a confirmatory test in case of doubt about the origin of the metabolites. A TWA exposure to 50 ppm would probably lead to mean blood and alveolar concentrations below 1 mg/l blood (end of shift) and 0.5 ppm in expired air 16 h after exposure) or 12 ppm at the end of the shift.

REFERENCES

ACGIH. American Conference of Governmental Industrial Hygienists. Threshold Limit Values for Chemical Substances and Physical Agents and Biological Exposure Indices. Cincinnati, 1991-1992.

W. Dekant, M. Metzler and D. Henschler. Novel metabolites of trichloroethylene through dechlorination reactions in rats, mice and humans. *Biochem. Pharmacol.* 33:2021, 1984.

W. Dekant, S. Vamvakas, K. Berthold et al. Bacterial β-lyase mediated cleavage and mutagenicity of cysteine conjugates derived from the nephrocarcinogenic alkenes trichloroethylene, tetrachloroethylene, hexachlorobutadiene. *Chem. Biol. Interact.* 60:31, 1986.

DFG. Deutsche Forschungsgemeinschaft. *Maximum Concentrations at the Workplace and Biological Tolerance Values for Working Materials.* Report No. XXVII. Commission for the Investigation of Health Hazards of Chemical Compounds in the Work Area. VCH, Weiheim, 1991.

P. Droz, M. Wu, W. Cumberland and M. Berode. Variability in biological monitoring of solvent exposure. I. Development of a population physiological model. *Br. J. Ind. Med.* 46:447, 1989a.

P. Droz, M. Wu and W. Cumberland. Variability in biological monitoring of organic solvent exposure. II. Application of a population physiological model. *Br. J. Ind. Med.* 46:547, 1989b.

T. Ertle, D. Henschler, G. Müller and M. Spassowski. Metabolism of trichloroethylene in man. I. The significance of trichloroethanol in long-term exposure conditions. *Arch. Toxicol.* 29:171, 1972.

J. Fernandez, B. Humbert, P. Droz and J. Caperos. Exposition au trichloréthylène. Bilan de l'absorption, de l'excrétion et du métabolisme chez des sujets humains. *Arch. Mal. Prof.* 36:397, 1975.

J. Fernandez, P. Droz, B. Humbert and J. Caperos. Trichloroethylene exposure. Simulation of uptake, excretion and metabolism using a mathematical model. *Br. J. Ind. Med.* 34:43, 1977.

M. Ikeda, H. Ohtsuji, T. Imamura and Y. Komoike. Urinary excretion of total trichlorocompounds, trichloroethanol and tirchloroacetic acid as a measure of exposure to trichloroethylene and tetrachloroethylene. *Br. J. Ind. Med.* 29:328, 1972.

M. Ikeda. Reciprocal metabolic inhibition of toluene and trichloroethylene in vivo and in vitro. *Int. Arch. Arbeitsmed.* 33:125, 1974.

T. Imamura and M. Ikeda. Lower fiducial limit of urinary metabolite level as an index of excessive exposure to industrial chemicals. *Br. J. Ind. Med.* 30:289, 1973.

G. Kimmerle and A. Eben. Metabolism, excretion and toxicology of trichloroethylene after inhalation. 2. Experimental human exposure. *Arch. Toxicol.* 30:127, 1973.

Ph. Landrigan, J. Kominsky, G. Stein et al. Common source community and industrial exposure to trichloroethylene. *Arch. Environ. Health* 42:327, 1987.

A. Monster and J. Houtkooper. Estimation of individual uptake of trichloroethylene, 1,1,1-trichloroethane and tetrachloroethylene from biological parameters. *Int. Arch. Occup. Environ. Health* 42:319, 1979.

A. Monster, G. Boersma and W. Duba. Kinetics of trichloroethylene in repeated exposure of volunteers. *Int. Arch. Occup. Environ. Health* 42:283, 1979.

A. Monster. Trichloroethylene. In: *Biological Monitoring and Surveillance of Workers Exposed to Chemicals.* Eds.: A. Aitio, V. Riihimari, H. Vaïnio. Hemisphere, Washington, D.C., 1984.

A. Monster. Biological monitoring of chlorinated hydrocarbon solvents. *J. Occup. Med.* 28:583, 1986.

G. Müller, M. Spassovski and D. Henschler. Trichloroethylene exposure and trichloroethylene metabolites in urine and blood. *Arch. Toxicol.* 29:335, 1972.

G. Müller, M. Spassovski and D. Henschler. Metabolism of trichloroethylene in man. II. Pharmacokinetics of metabolites. *Arch. Toxicol.* 32:283, 1974.

G. Müller, M. Spassovski and D. Henschler. Metabolism of trichloroethylene in man. III. Interaction of trichloroethylene and ethanol. *Arch. Toxicol.* 33:173, 1975.

K. Nomiyama and H. Nomiyama. Metabolism of trichloroethylene on human: sex difference in urinary excretion of trichloroacetic acid and trichloroethanol. *Int. Arch. Arbeitsmed.* 28:37, 1971.

P. Pfäffli and A. Backman. Trichloroethylene concentrations in blood and expired air as indicators of occupational exposure. A preliminary report. *Work Environ. Health* 9:140, 1972.

A. Sato and T. Nakajima. Differences following skin or inhalation exposure in the absorption and excretion kinetics of trichloroethylene and toluene. *Br. J. Ind. Med.* 35:48, 1978.

A. Sato, K. Endom, T. Kanebo and G. Johanson. Effects of consumption of ethanol on the biological monitoring of exposure to organic solvent vapours: a simulation study with trichloroethylene. *Br. J. Ind. Med.* 48:548, 1991.

K. Seiji, O. Inoue, Jin Chiu et al. Dose excretion relationship in tetrachloroethylene exposed workers and the effect of tetrachloroethylene co-exposure on trichloroethylene metabolism. *Am. J. Ind. Med.* 16:675, 1989.

L. Skender, V. Karacic and D. Prpic-Majic. A comparative study of human levels of trichloroethylene and tetrachloroethylene after occupational exposure. *Arch. Environ. Health* 46:174, 1991.

B. Soucek and D. Vlachova. Excretion of trichloroethylene metabolites in human urine. *Br. J. Ind. Med.* 17:60, 1960.

R. Stewart, H. Dodd, H. Gay and D. Erley. Experimental human exposure to trichloroethylene. *Arch. Environ. Health* 20:64, 1970.

R. Stewart, C. Hake, A. Lebrun et al. Biological standards for the industrial worker by breath analysis: trichloroethylene. NIOSH publication No. 74-133, U.S. Government Printing Office, Washington, D.C. 1974.

S. Tanaka and M. Ikeda. A method for determination of trichloroethanol and trichloroacetic acid in urine. *Br. J. Ind. Med.* 25:214, 1968.

A. Triest and R. Lauwerys. La surveillance médicale des travailleurs: contrôle du degré d'exposition avant et après l'instauration de mesures de prévention. *Cah. Méd. Trav.* 13:273, 1976.

O. Vesterberg and I. Astrand. Exposure to trichloroethylene monitored by analysis of metabolites in blood and urine. *J. Occup. Med.* 18:224, 1976.

3.8. Tetrachloroethylene

Toxicokinetics

As for many other volatile halogenated hydrocarbons, the exposure to tetrachloroethylene is ubiquitous (Hajimiragha et al 1986; Wallace et al 1986; Monster and Smolders 1984; Kido et al 1989). The pulmonary uptake increases with the duration of exposure and the pulmonary ventilation, hence the workload and this should be taken into account when interpreting the biological indicators. The absorption of the liquid form through the skin may also occur (Stewart and Dodd 1964). Aitio et al (1984) have determined the blood concentrations of tetrachloroethylene in both forearms of two subjects after one hand was soaked for 5 min in the solvent. The concentration on the soaked side was 130 times higher, i.e., up to 9 μmol/l (1.5 μg/ml) than in the other side. However, the risk of excessive dermal absorption is negligible since on the basis of physicochemical data (Fiserova-Bergerova and Pierce 1989) a skin penetration rate of only 0.11 mg/cm^2/h can be calculated for tetrachloroethylene in liquid form (ATSDR 1988). The capacity of human to metabolize tetrachloroethylene is limited and the compound is mainly excreted unchanged in exhaled air. Less than 3% is excreted in urine as trichloroacetic acid (Fernandez et al 1976). Ohtsuki et al (1983) have confirmed that the urinary concentration of the total trichloro-compounds is no longer proportional to the amount inhaled when exposure is above 100 ppm tetrachloroethylene. Moreover they have calculated (on the basis of results obtained in workers) that at the end of an 8-h shift with exposure to tetrachloroethylene at 50 ppm, 38% of the absorbed dose would be exhaled unchanged and less than 2% would be excreted as metabolites into the urine, the rest remaining in the body to be eliminated later. Elimination from the body is slow due to progressive release from adipose tissue. Tetrachloroethylene has a half-life in blood of approximately 4 to 6 d in short-term experiments and approximately 6 to 8 d in long-term exposure. The concentration of trichloroacetic acid in blood increases up to 20 h after a single exposure; thereafter the concentration decreases with a half-life of about 80 h (Monster 1984, 1986). Hence, in repeated daily exposure, there is a progressive accumulation of the solvent in the body. An increased trend of excretion of thioethers throughout the working week has been detected in workers employed in a dry-cleaning shop. However, the levels were within the range found in controls (2-12 mmol/mol creatinine) and the significance of this finding is not clear (Lafuente and Mallol 1986).

Alcohol consumption during exposure does not seem to have an effect on the concentration of tetrachloroethylene in blood (Monster 1986).

Biological monitoring

Breath analysis

Tetrachloroethylene. In persons exposed at rest to 100 ppm tetrachloroethylene for 7 to 8 h, the solvent concentration in alveolar air amounts to approximately 60 ppm at the end of exposure and to 4 ppm 16 h after the end of exposure (Fernandez et al 1976; Stewart et al 1970). When the same exposure is repeated for 5 d, the concentration 16 h after the end of exposure is increased twofold.

The concentration of tetrachlorethylene in alveolar air during normal breathing seems to provide a reasonable estimate of the time-weighted concentration in arterial blood (Opdam and Smolders 1986, 1987).

In dry-cleaning shop workers exposed to a TWA of 20.8 ppm (8.9-37.5), the mean concentration of tetrachloroethylene in alveolar air 0.5 and 16 h after the end of exposure amounted to 5.1 (0.2-10) ppm and 1.9 (0.1-5.5) ppm (Lauwerys et al 1983).

Monster and co-workers (1983, 1984, 1986) have estimated the mean alveolar air concentration of tetrachloroethylene to be 23 ppm (160 mg/m^3) 5-15 min after an average exposure to 50 ppm tetrachloroethylene (8 h/d, 5 d/week).

In a field study, Droz (1984) found that there was a highly significant correlation between preshift (3rd day of the workweek) perchloroethylene concentration in end-exhaled air and the TWA exposure measured on the previous day. A level of 9.5 ppm was extrapolated to correspond to a TWA exposure to 50 ppm.

Imbriani et al (1988) exposed ten volunteers to tetrachloroethylene vapor at 3.6-316 mg/m^3 for 2-4 h at rest. The alveolar concentration corresponding to a TWA exposure at 50 ppm was about 11.5 ppm at the end of exposure (11.17 mg/m^3) (extrapolated from the regression line).

The DFG (1991) recommends 9.5 ppm, and the ACGIH (1991-1992) 10 ppm in end-exhaled air collected prior to the shift after at least two consecutive workdays, as biological exposure indice corresponding to a TWA of 50 ppm (340 mg/m^3).

Blood analysis

Tetrachloroethylene. The data on the blood levels of tetrachloroethylene in volunteers exposed at rest to the solvent (Monster et al 1979; Stewart et al 1970) are limited. Appreciably raised blood tetrachloroethylene concentrations have been detected in neighbors of dry-cleaning shops (Popp et al 1992).

A study conducted on workers exposed to tetrachloroethylene in dry-cleaning shops suggests that if the blood concentration of tetrachloroethylene does not exceed 0.1 mg/100 ml, 16 h after the end of exposure, the TWA exposure is likely to have been below 50 ppm (Lauwerys et al 1983). For a mean exposure to 20.8 ppm (8.9-37.5) these authors observed mean tetrachloroethylene blood concentrations of 0.04 (0.01-0.08) mg/100 ml before work, and 0.12 (0.04-0.31) 30 min after work.

The mean blood concentrations corresponding to a TWA exposure of 50 ppm (8 h/d, 5 d/week) were estimated to be 0.23 mg/100 ml tetrachloroethylene and 0.58 mg/100 ml trichloroacetic acid 5-15 min after the exposure (Monster et al 1983; Monster 1984, 1986).

In 18 workers exposed to tetrachloroethylene concentrations ranging between 33 and 53 ppm, the median blood concentrations found on Monday morning before work and Thursday afternoon after work, were 3.74 (1.75-31.78) µmol/l (62 µg/100 ml) and 8.93 (4.76-79.97) µmol/l (148 µg/100 ml) tetrachloroethylene; 7.10 (1.65-17.08) µmol/l (116 µg/100 ml) and 7.78 (1.71-20.93) µmol/l (127 µg/ 100 ml) trichloroacetic acid (Skender et al 1991).

The DFG (1991) and the ACGIH (1991-1992) recommend for tetrachloroethylene 1 mg/l at the beginning of the next shift as biological limit value corresponding to a TWA of 50 ppm (340 mg/m³).

Urine analyses

Trichloroacetic acid, trichloroethanol. Trichloroacetic acid has been detected in urine of workers or volunteers exposed to tetrachloroethylene, but trichloroethanol is not usually present (Fernandez et al 1976; Ogata et al 1971; Monster et al 1979).

According to Ohtsuki et al (1983), when the lower 95% confidence limit is selected, the screening mean levels (end-of-shift) for a TWA exposure of 50 and 100 ppm will be 30 and 61 mg/l for total trichloro-compounds values, respectively, or 19 and 35 mg/l for values adjusted for a specific gravity of urine of 1016.

According to Monster and co-workers (1983, 1984, 1986) exposure to a TWA of 50 ppm (5 h/d, 5 d/week) would lead to mean urinary concentrations of 9.7 mg/ g creatinine for trichloroacetic acid and 0.5 mg/g creatinine for trichloroethanol.

In none of the urine samples of workers exposed to 20 ppm tetrachloroethylene on the average was trichloroacetic detected (Lauwerys et al 1983).

Median (range) urinary concentrations of 0.08 (0-2.56) mmol trichloroethanol/ mol creatinine (about 0.1 mg/g creatinine) and 2.68 (0.71-15.76) mmol trichloroacetic acid/mol creatinine (about 3.87 mg/g creatinine) were found in Monday morning samples of workers exposed to tetrachloroethylene concentrations ranging from 33 to 53 ppm. The corresponding Thursday afternoon concentrations were 0.13 mmol trichloroethanol (0-0.73)/mol creatinine (about 0.17 mg/g creatinine) and 2.34 mmol trichloroacetic acid (0.81-10.81)/mol creatinine (about 3.38 mg/g creatinine) (Skender et al 1991). The BEI is 7 mg trichloroacetic acid/l (end-of-workweek) adopted by the ACGIH (1991-1992) corresponding to a TWA exposure of 50 ppm (340 mg/m³).

Tetrachloroethylene. In both experimentally and occupationally exposed subjects, the urinary excretion of tetrachloroethylene showed a linear relationship with the average environmental concentration. In workers performing moderate work, the urinary tetrachloroethylene concentration after 4 h exposure to a TWA air concentration of 50 ppm was 120 µg/l and its lower confidence limit (biological threshold) 100 µg/l (Imbriani et al 1988; Ghittori et al 1987). As already mentioned, the concentrations are ventilation-dependent and hence workload dependent. Imbriani et al (1988) have estimated that in subjects exposed to 4 h-50 ppm tetrachloroethylene, the concentrations of tetrachloroethylene in urine (end-of-shift sample) are 101, 206, 277, 418 µg/l for a pulmonary ventilation of 7.5, 15, 20, and 30 l/min, respectively.

According to the physiological model, developed by Droz et al (1989a,b), simulating repeated constant exposure to 50 ppm tetrachloroethylene with a constant light physical workload, the estimated biological indicators are the following: just after the end-of-work on Thursday: tetrachloroethylene in alveolar air: 1000 µmol/m^3 (165.85 mg/m^3, 25 ppm), in venous blood: 18 µmol/l (298.5 µg/100 ml), in urine: 1.3 µmol/l (215.6 µg/l); trichloroacetic acid in venous blood: 29 µmol/l (473.86 µg/100 ml), in urine: 2.9 mmol/mol creatinine (4.2 mg/g creatinine). On Friday morning; tetrachloroethylene in alveolar air: 435 µmol/m^3 (72 mg/m^3, 10.8 ppm), in venous blood: 6.1 µmol/l (101.17 µg/100 ml), in urine: 0.43 µmol/l (71.3 µg/l); trichloroacetic acid in venous blood: 28 µmol/l (457.5 µg/100 ml), in urine: 2.8 mmol/l (4 mg/g creatinine).

In view of the long biological half-life of tetrachloroethylene, the internal dose may be better estimated by measuring its concentration in blood or expired air 16 h after the end of exposure (i.e., before next shift). A TWA exposure to 50 ppm will probably lead to mean concentrations of 0.1 mg tetrachloroethylene/100 ml blood and about 8 ppm tetrachloroethylene in alveolar air. The corresponding mean concentration in urine amounts to about 70 µg/g creatinine. Due to its slow rate of excretion, trichloroacetic acid can also be used as an indicator of the integrated exposure over the workweek. A TWA exposure to 50 ppm tetrachloroethylene will probably entail a mean urinary concentration of about 5 mg/l at the end of the week.

REFERENCES

ACGIH. American Conference of Governmental Industrial Hygienists. Threshold Limit Values for Chemical Substances and Physical Agents and Biological Exposure Indices. Cincinnati, 1991-1992.

DFG. Deutsche Forschungsgemeinschaft. *Maximum Concentrations at the Workplace and Biological Tolerance Values for Working Materials.* Report No. XXVII. Commission for the Investigation of Health Hazards of Chemical Compounds in the Work Area. VCH, Weinheim, 1991.

A. Aitio, K. Pekari and J. Järvisalo. Skin absorption as a source of error in biological monitoring. *Scand. J. Work Environ. Health* 10:317, 1984.

P. Droz. Biological monitoring and health surveillance of workers exposed to various solvents. Report to Fonds National Suisse de la Recherche Scientifique, 1984. Quoted in "Documentation of the Biological Exposure Indices", ACGIH. 1991.

P. Droz, M. Wu, W. Cumberland and M. Berode. Variability in biological monitoring of solvent exposure. I. Development of a population physiological model. *Br. J. Ind. Med.* 46:447, 1989a.

P. Droz, M. Wu and W. Cumberland. Variability in biological monitoring of organic solvent exposure. II. Application of a population physiological model. *Br. J. Ind. Med.* 46:547, 1989b.

J. Fernandez, E. Guberan and J. Caperos. Experimental human exposure to tetrachloroethylene vapor and elimination in breath after inhalation. *Am. Ind. Hyg. Assoc. J.* 37:143, 1976.

V. Fiserova-Bergerova and J. Pierce. Biological monitoring. V. Dermal absorption. *Appl. Ind. Hyg.* 4:F14, 1989.

S. Ghittori, M. Imbriani, G. Pezzagno and E. Capodaglio. The urinary concentration of solvents as a biological indicator of exposure: proposal for the biological equivalent exposure limit for nine solvents. *Am. Ind. Hyg. Assoc. J.* 48:786, 1987.

H. Hajimiragha, U. Ewers, R. Jansen-Rosseck and A. Brockhaus. Human exposure to volatile halogenated hydrocarbons from the general environment. *Int. Arch. Occup. Environ. Health* 58:141, 1986.

M. Imbriani, S. Ghittori, G. Pezzagno and E. Capodaglio. Urinary excretion of tetrachloroethylene in experimental and occupational exposure. *Arch. Environ. Health* 43:292, 1988.

K. Kido, T. Shiratori, T. Watanabe et al. Correlation of tetrachloroethylene in blood and in drinking water: a case of well water pollution. *Bull. Environ. Contam. Toxicol.* 43:444, 1989.

A. Lafuente and J. Mallol. Thioethers in urine during occupational exposure to tetrachloroethylene. *Br. J. Ind. Med.* 43:68, 1986.

R. Lauwerys, J. Herbrand, J.-P. Buchet et al. Health surveillance of workers exposed to tetrachloroethylene in dry-cleaning shops. *Int. Arch. Occup. Environ. Health* 52:69, 1983.

A. Monster, G. Boersma and H. Steenweg. Kinetics of tetrachloroethylene in volunteers; influence of exposure concentration and work load. *Int. Arch. Occup. Environ. Health* 42:303, 1979.

A. Monster, W. Regouin-Peeters, A. Van Schijndel and J. Van Der Tuin. Biological monitoring of occupational exposure to tetrachloroethylene. *Scand. J. Work Environ. Health* 8:273, 1983.

A. Monster and J. Smolders. Tetrachloroethylene in exhaled air of persons living near pollution sources. *Int. Arch. Occup. Environ. Health* 53:331, 1984.

A. Monster. Biological monitoring of industrial solvents: tetrachloroethylene. In: *Biological Monitoring and Surveillance of Workers Exposed to Chemicals.* Eds.: A. Aitio, V. Riihimaki, H. Vaïnio. Hemisphere, Washington, D.C., 1984.

A. Monster. Biological monitoring of chlorinated hydrocarbon solvents. *J. Occup. Med.* 28:583, 1986.

M. Ogata, Y. Takatsuka and K. Tomokuni. Excretion of organic chlorine compounds in the urine of persons exposed to vapours of trichloroethylene and tetrachloroethylene. *Br. J. Ind. Med.* 28:386, 1971.

T. Ohtsuki, S. Kunihiko, A. Koizumi et al. Limited capacity of humans to metabolize tetrachloroethylene. *Int. Arch. Occup. Environ. Health* 51:384, 1983.

J. Opdam and J. Smolders. Alveolar sampling and fast kinetics of tetrachloroethene in man. I. Alveolar sampling. *Br. J. Ind. Med.* 43:814, 1986.

J. Opdam and J. Smolders. Alveolar sampling and fast kinetics of tetrachloroethene in man. *Br. Ind. J. Med.* 44:26, 1987.

W. Popp, G. Müller, B. Baltes-Schmitz et al. Concentrations of tetrachloroethylene in blood and neighbours of dry-cleaning shops. *Int. Arch. Occup. Environ. Health* 63:393, 1992.

L. Skender, V. Karacic, D. Prpic-Majic. A comparative study of human levels of trichloroethylene and tetrachloroethylene after occupational exposure. *Arch. Environ. Health* 46:174, 1991.

R. Stewart and H. Dodd. Absorption of carbon tetrachloride, trichloroethylene, tetrachloroethylene, methylene chloride and 1,1,1-trichloroethane through the human skin. *Am. Ind. Hyg. Assoc. J.* 25:439, 1964.

R. Stewart, E. Baretta, H. Dodd and T. Torkelson. Experimental human exposure to tetrachloroethylene. *Arch. Environ. Health* 20:224, 1970.

L. Wallace, E. Pellizari, T. Hartwell et al. Concentrations of 20 volatile organic compounds in the air and drinking water of 350 residents of New Jersey compared with concentrations in their exhaled breath. *J. Occup. Med.* 28:603, 1986.

U.S. Department of Health and Human Services. Agency for Toxic Substances and Disease Registry. Toxicological profile for tetrachloroethylene. ATSDR/TP-88/22, 1988.

3.9. Vinyl Chloride

Toxicokinetics

Vinyl chloride is absorbed by the lungs. Experiments with volunteers have shown that approximately 40% of an inhaled dose of the chemical is retained in the lungs (Krajewski et al 1980). Vinyl chloride is quickly eliminated either unchanged with expired air or as water-soluble metabolites in urine. Animal experiments suggest that a saturation of the metabolic pathway may rapidly occur. Hence, the proportion of vinyl chloride eliminated varies with the exposure level (Watanabe et al 1976).

The metabolic pathway of vinyl chloride has been extensively investigated in animals. One metabolite, thiodiglycolic acid, is known to be excreted in greater amounts by workers exposed to vinyl chloride (Müller et al 1978).

Biological monitoring

Breath analysis

Vinyl chloride. Baretta et al (1969) have demonstrated on volunteers and on workers exposed to vinyl chloride that there exists a relationship between environmental exposure and the concentration in expired air 16 h after the end of exposure. However, these observations were made under exposure conditions (greater than 25 ppm) that are much higher than the current acceptable exposure level.

Table 13
Relationship Between Vinyl Chloride Exposure and
Thiodiglycolic Acid in Urine

Vinyl Chloride in Air Mean (ppm)	n	Thiodiglycolic Acid in Urine Mean (mg/l)	SD
Control	78	0.83	0.56
<10	9	1.25	0.58
<30	23	2.84	1.68
>100	32	3.73	4.02

From Chen et al 1983.

Urine analysis

Thiodiglycolic acid. In control subjects, the normal concentration of this substance in urine is less than 2 mg/l (Müller et al 1979). Chen et al (1983) studied the relationship between thiodiglycolic acid concentration in urine and the intensity of occupational exposure to vinyl chloride. The results are summarized in Table 13.

It seems that individual variation renders this method unreliable at low exposure concentrations (5 ppm) (Tarkowski 1984). Since vinyl chloride is a human carcinogen, the exposure should be kept as low as possible, so that the urinary concentration of thiodiglycolic acid does not increase significantly above the background level.

REFERENCES

E. Baretta, R. Stewart and J. Mutchler. Monitoring exposures to vinyl chloride vapor: breath analysis and continuous air sampling. *Am. Ind. Hyg. Assoc. J.* 30:537, 1969.

Z. Chen, X. Gu, M. Cui and X. Zhu. Sensitive flame-photometric-detector analysis of thiodiglycolic acid in urine as a biological monitor of vinyl chloride. *Int. Arch. Occup. Environ. Health* 52:281, 1983.

J. Krajewski, M. Dobecki and J. Gromiec. Retention of vinyl chloride in the human lung. *Br. J. Ind. Med.* 37:373, 1980.

G. Müller, K. Norpoth, E. Kusters et al. Determination of thiodiglycolic acid in urine specimens of vinyl chloride exposed workers. *Int. Arch. Occup. Environ. Health* 41:199, 1978.

G. Müller, K. Norpoth and R. Wickramasinghe. An analytical method, using GC-MS for the quantitative determination of urinary thiodiglycolic acid. *Int. Arch. Occup. Environ. Health* 44:185, 1979.

S. Tarkowski. Preventive measure against occupational hazards in the PVC production industry. *Prog. Clin. Biol. Res.* 141:177, 1984.

P. Watanabe, G. McGowan and P. Gehring. Fate of (^{14}C)-vinyl chloride after single oral administration in rats. *Toxicol. Appl. Pharmacol.* 37:49, 1976.

3.10. Fluorinated Anesthetics

Operating room personnel may be exposed to low levels of volatile anesthetics. The metabolism in man of some modern fluorinated anesthetics has been investigated. The parent compound can usually be detected in expired air and in blood (Hallen et al 1970; Witcher et al 1971). Various metabolites can be found in urine (Holaday and Fiserova-Bergerova 1979; Van Dyke 1973; Van Stee 1976). Except for halothane, no biological threshold limit value has yet been proposed for the urinary metabolites of volatile fluorinated anesthetics.

3.10.1. Halothane (Trifluoro-2-bromo-2-chloroethane)

Toxicokinetics

The major route of exposure is via inhalation. Cohen et al (1975) have identified three metabolites in urine of individuals who were given labeled halothane: trifluoroacetic acid, N-trifluoroacetyl-2-aminoethanol, and N-acetyl-S-(2-bromo-2-chloro-1,1-difluoroethyl)-L-cysteine. Trifluoroacetic acid, the major metabolite, can also be detected in blood. Its biological half-life in blood is between 50 to 70 h (Dallmeier and Henschler 1981). At the end of a 1-week exposure, a steady-state concentration of trifluoroacetic acid is found in blood and urine and there is a linear relationship between the blood or urine concentration of the metabolite and the airborne concentration of halothane.

Biological monitoring

Breath analysis

Halothane. Stevens et al (1987) found a statistically significant, but low correlation (r: 0.39) between breath halothane concentration at the end of exposure and the average level of environmental exposure. As for all volatile solvents, expired air concentration (sample collected during exposure) is greatly affected by variations in exposure levels and mainly reflects the airborne concentration at the time of sampling. It seems, however, that in groups of occupationally exposed subjects, there is a satisfactory correlation between the level of environmental exposure and the alveolar concentration of halothane (at end of exposure) (NIOSH 1977; Pezzagno et al 1989). It has been proposed that the biological value of about 0.5 ppm (sample collected at the end of exposure) corresponds to an environmental value of 5 ppm halothane.

Urine analyses

Halothane. Stevens et al (1987) have investigated the relationship between the halothane levels in urine of subjects occupied in operating theaters and the airborne halothane concentrations (range: 11.3 to 1671 ppm; median 36 ppm). The end-of-shift urinary concentration halothane (r: 0.78; $p < 0.01$) was correlated with the TWA exposure level.

Imbriani et al (1991) also found a significant association between halothane environmental and urinary concentrations (r: 0.92; $p < 0.001$) in 58 operating theater personnel. The results tend to show that urinary halothane concentration can be used as an appropriate biological exposure index. The extrapolated values corresponding to mean (4 h) occupational exposure levels of 50 (404 mg/m³), 2 (16 mg/m³), and 0.5 ppm (4 mg/m³) are 101.3 µg/l (95% lower confidence limit: 92 µg/l), 7.4 µg/l (95% confidence limit: 6.5 µg/l), and 4.5 µg/l (95% confidence limit: 3.9 µg/l), respectively.

Trifluoroacetic acid. According to Dallmeier and Henschler (1981) an exposure to 5 ppm halothane for 5 d (8 h/d) entails a trifluoroacetic acid concentration in urine of about 10 mg/g creatinine. Another study has shown a mean increase of trifluoroacetic acid urinary level of about 40% during the working day (Gilli et al 1985).

Other. Stevens et al (1987) did not detect any increase in inorganic fluoride concentration in urine during the workday or the workweek in subjects exposed on average to 36 ppm halothane.

Blood analyses

Halothane. Halothane blood concentrations have been investigated in some studies (Hallen et al 1970; Nikki et al 1972) but there is not enough data to draw any conclusion.

Trifluoroacetic acid. Dallmeier and Henschler (1981) suggest the measurement of trifluoroacetic concentration in blood at the end of a workweek because the scatter is lower than in urine. They have estimated that an exposure to 10 or 5 ppm halothane for 5 d (8 h/d) will lead to a blood concentration of trifluoroacetic acid of 500 or 250 µg/100 ml, respectively. They consider the latter blood concentration as a risk-free value.

The DFG (1991) has proposed a biological limit value of 2.5 mg/l blood (at the end of the shift after several shifts), corresponding to a TWA exposure level of 5 ppm.

Other. Increased plasma bromide has also been found after exposure to halothane (Tinker et al 1976; Pezzagno et al 1989).

The following tentative biological values corresponding to a mean exposure level to 5 ppm halothane are suggested: halothane in alveolar air (during exposure): 0.5 ppm; halothane in urine (end-of-exposure): 10 µg/g creatinine; trifluoroacetic acid in blood (end-of-exposure): 0.25 mg/100 ml; trifluoroacetic acid in urine: 10 mg/g creatinine.

3.10.2. Enflurane

Difluoromethoxydifluoric acid (CHF_2OCF_2COOH), oxalic acid, and inorganic fluoride have been identified as metabolites of the anesthetic enflurane in human urine (Burke et al 1981; Hitt et al 1977). The observations of Carlsson et al (1985) indicate that inorganic fluoride cannot be considered as reliable biological indicators of occupational exposure to enflurane. This can be partly explained by the wide individual variability in enflurane metabolism (Strube et al 1987), the nonspecificity of this metabolite, and the influence of drinking water and other factors (e.g., toothpaste).

3.10.3. Fluoroxene

Trifluoroacetic acid and trifluoroethanol have been identified as urinary metabolites of fluoroxene (Holaday and Fiserova-Bergerova 1979).

3.10.4. Methoxyflurane

Methoxyflurane gives rise *in vivo* to oxalic acid and fluoride, which are excreted in urine (Dahlgren 1979; Holaday et al 1970; Vaughan et al 1978; Yoshimura et al 1976). Other metabolic products identified in urine are: methoxydifluoroacetic acid and dichloroacetic acid, formaldehyde, and chlorine (see the review by Pezzagno et al 1989).

3.10.5. Isoflurane

Imbriani et al (1988) observed in 45 occupationally exposed subjects a close relationship (r: 0.9) between the TWA environmental and urinary isoflurane concentrations. Mean exposure levels of 2 ppm and 10 ppm would lead at the end of a 4-h exposure to mean isoflurane concentrations of 4.3 µg/l (95% lower confidence limit: 3.4 µg/l) and 14.5 µg/l, respectively.

REFERENCES

T. Burke, R. Branchflower, D. Lees and L. Pohl. Mechanism of defluorination of enflurane. Identification of an organic metabolite in rat and man. *Drug Metab. Dispos.* 9:19, 1981.

P. Carlsson, J. Ekstrand and B. Hallen. Plasma fluoride and bromide concentrations during occupational exposure to enflurane or halothane. *Acta Anaesthesiol. Scand.* 29:669, 1985.

E. Cohen, J. Trudell, H. Edmunds and E. Watson. Urinary metabolites of halothane in man. *Anesthesiology* 43:392, 1975.

B. Dahlgren. Fluoride concentrations in urine of delivery ward personnel following exposure to low concentrations of methoxyflurane. *J. Occup. Med.* 21:624, 1979.

E. Dallmeier and D. Henschler. Halothan-Belastung am Arbeitzplaz im Operationssaal. *Dtsch. Med. Wochenschr.* 106:324, 1981.

DFG. Deutsche Forschungsgemeinschaft. *Maximum Concentrations at the Workplace and Biological Tolerance Values for Working Materials.* Report No. XXVII. Commission for the Investigation of Health Hazards of Chemical Compounds in the Work Area. VCH, Weinheim, 1991.

G. Gilli, G. Corrao, E. Scursatone et al. Significance of biological indexes in exposure to anaesthetic gases. Meeting of Italian Society of Occupational Medicine. 611 Pavia, 1985.

B. Hallen, H. Ehrner-Samuel and M. Thomason. Measurements of halothane in the atmosphere of an operating theatre and in expired air and blood of the personnel during routine anaesthetic work. *Acta Anaesthesiol. Scand.* 14:17, 1970.

B. Hitt, R. Mazze, W. Beppu et al. Enflurane metabolism in rats and man. *J. Pharmacol. Exp. Ther.* 203:193, 1977.

D. Holaday, S. Rudofsky and P. Threuhaft. The metabolic degradation of methoxyflurane in man. *Anesthesiology* 33:579, 1970.

D. Holaday and V. Fiserova-Bergerova. Fate of fluorinated metabolites of inhalation anesthetics in man. *Drug Metab. Rev.* 9:61, 1979.

M. Imbriani, S. Ghittori, G. Pezzagno and E. Capodaglio. Evaluation of exposure to isoflurane (forane) environmental and biological measurements in operating room personnel. *J. Toxicol. Environ. Health* 25:393, 1988.

M. Imbriani, S. Ghittori, P. Zadra and R. Imberti. Biological monitoring of the occupational exposure to halothane (fluothane) in operating room personnel. *Am. J. Ind. Med.* 20:103, 1991.

P. Nikki, P. Pfaffli, K. Ahiman and R. Raili. Chronic exposure to anesthetic gases in the operation theatre and recovery room. *Ann. Clin. Res.* 4:273, 1972.

NIOSH Publ. No. Dhew. Criteria for a recommended standard. Occupational exposure to waste anaesthetic gases and vapors. 77:140, 1977.

G. Pezzagno, M. Imbriani, S. Ghittori and E. Capodaglio. Biological indicators for the assessment of human exposure to industrial chemicals: Inhalation anaesthetics. Eds.: L. Alessio, A. Berlin, M. Boni and R. Roi. Commission of the European Communities, Luxembourg, Eur 12174 En, 1989.

M. Stevens, J. Walrand, J.-P. Buchet and R. Lauwerys. Evaluation de l'exposition à l'halothane et au protoxyde d'azote en salle d'opération par des mesures d'ambiance et des mesures biologiques. *Cah. Méd. Trav.* 34:41, 1987.

P. Strube, G. Hulands and M. Halsey. Serum fluoride levels in morbidly obese patients: enflurane compared with isoflurane anaesthesia. *Anaesthesia* 42:685, 1987.

J. Tinker, A. Gandolfi and R. Van Dyke. Elevation of plasma bromide levels in patients following halothane anesthesia. *Anesthesiology* 44:194, 1976.

R. Van Dyke. Biotransformation of volatile anaesthetics with special emphasis on the role of metabolism in the toxicity of anaesthetics. *Can. Anaesth. Soc. J.* 20:21, 1973.

E. Van Stee. Toxicology of inhalation anesthetics and metabolites. *Annu. Rev. Pharmacol. Toxicol.* 16:67, 1976.

R. Vaughan, I. Sipes and B. Brown. Role of biotransformation in the toxicity of inhalation anesthetics. *Life Sci.* 23:2447, 1978.

C. Whitcher, E. Cohen and J. Tridell. Chronic exposure to anesthetic gases in the operating room. *Anesthesiology* 35:348, 1971.

N. Yoshimura, D. Holaday and V. Fiserova-Bergerova. Metabolism of methoxyflurane in man. *Anesthesiology* 44:372, 1976.

3.11. Chlorofluorocarbons

Toxicokinetics

Studies designed to assess the retention and elimination of trichlorofluoromethane (CFC-11) (Angerer et al 1985), chlorodifluoromethane (CFC-22) (Woollen et al 1990a) and 1,1,2-trichloro-1,2,2-trifluoroethane (CFC-113) (Woollen et al 1990b) have been carried out in human volunteers.

Pulmonary retention has been estimated to be 14% for CFC-113 and on average 18% for CFC-11.

During exposure, blood concentrations rapidly reached a plateau (after 1 h exposure for CFC-22, 30 min for CFC-113). Peak CFC-22 and CFC-113 concentrations in blood and breath were related to the exposure level. The average amounts expired in the postexposure period represented only 2%, for CFC-22, and 2.6 to 4.3%, for CFC-113, of the dose inhaled. After the termination of exposure, the concentrations in alveolar air and blood declined rapidly. The blood elimination process could be followed through a two-phase model for CFC-11 (half-times: 7 min, 1.8 h) and CFC-22 (half-times: 0.23, 2.8 h). The breath-elimination curves were described by a two-compartment model for CFC-11 (half-times: 11 min, 1 h) and a three-compartment model for CFC-22 (half-times: 0.005 h, 0.17 h, 2.4 h) and for CFC-113 (half-times: 0.22, 2.3, 29 h). This suggests that CFC-113 may accumulate in the body following repeated exposure.

A peak concentration of CFC-22 in urine was measured 0.5 h after the end of exposure (average elimination half-time: 2.7 h). But less than 0.01% (of the amount inhaled) of CFC-11 and less than 0.02% of CFC-113 were excreted in urine unchanged. A very slight increase (<0.1% of the amount inhaled) was detected as fluoride in urine in case of exposure to CFC-113, but not in case of exposure to CFC-111 or CFC-22.

Biological monitoring

These studies indicate that occupational exposure to CFC-111 could be assessed by measuring CFC-111 concentration in alveolar air, provided that the samples are taken at a well-defined time (e.g., 30 min) after the termination of exposure. The values quoted by Angerer et al (1985) for an estimation of recent exposure are 75 ml/m^3 in alveolar air and 1.1 mg/l blood, corresponding to an average exposure to 1000 ppm CFC-111.

Breath analysis could also be used to monitor exposure to CFC-22. Sample should be taken at least 30 min after the end of exposure. Alternatively, measurements of CFC-22 in urine samples collected at the end of an exposure period could be measured.

Breath CFC-113 concentration on the morning after exposure seems an appropriate method to estimate mean CFC-113 exposure during the previous day. Woollen et al (1990b) stressed the fact that the predictive value is improved if the results are normalized to the body fat content of individual workers. Data, however, are still lacking to propose any biological threshold.

REFERENCES

J. Angerer, B. Schröder and R. Heinrich. Exposure to fluorochloromethane. *Int. Arch. Occup. Environ. Health* 56:67, 1985.
B. Woollen, T. Auton, P. Blain et al. A study on the rate of uptake and elimination of chlorodifluoromethane in humans by inhalation. In: *Occupational Health in the Chemical Industry.* XVII Medichem Congress. Cracow, Poland 26-29 Sept. 1989. WHO, Regional Office for Europe, Copenhagen 1990a.
B. Woollen, E. Guest, W. Howe et al. Human inhalation pharmacokinetics of 1,1,2-trifluoroethane FC 113. *Int. Arch. Occup. Environ. Health* 62:73, 1990b.

3.12. Polychlorinated Biphenyls

Toxicokinetics

Polychlorinated biphenyls (PCB) are ubiquitous contaminants of low biodegradability. The more highly chlorinated compounds of the family known as polychlorinated biphenyls are cumulative toxic chemicals that can be absorbed by all routes. In industry, skin absorption seems to play an important role (Maroni et al 1981a; Lees et al 1987; Perkins and Knight 1989; Wester et al 1983; Wolff 1985; Grandjean 1990).

Being highly lipophilic, their distribution is mainly into fat. Moreover in lactating women, milk may also be an important excretory route (Rogan et al 1983, 1986; Yakushiji et al 1984; Mes et al 1984; Slorach and Vaz 1985).

In workers exposed to various PCB (20 to 54% chlorine), PCB concentration in adipose tissue was found to be proportional to that in plasma with a partition for total PCB of approximately 190:1, the adipose-plasma partition varying with the congeneer and both concentrations were related to the duration and intensity of exposure (Wolff et al 1982).

The different components undergo variable biotransformation patterns, depending upon the number and position of chlorine atoms on the biphenyl rings. Steele et al (1986) have estimated the biological half-life in human serum for the highly chlorinated congeneers to be 33 to 34 months and 6 to 7 months for the less chlorinated ones.

Phillips et al (1989) have followed in a group of 58 workers previously exposed to PCB, the changes in serum concentrations of PCB between 1977 when the factory stopped the use of PCB and 1985. Using another half-time estimation procedure than Steele et al, they calculated a median half-life of 2.6 years for less-chlorinated PCB and 4.8 years for more highly chlorinated components. Moreover, their data suggest that the half-life varies inversely with the initial serum concentration.

Biological monitoring

Blood and adipose tissue analyses

The intensity of human occupational exposure to PCB has been assessed by measurement of blood and adipose levels. According to Letz (1983), the "background" levels in human serum are typically less than 2 µg/100 ml and in adipose tissues range from 1 to 2 ppm. The serum and adipose tissue concentrations are related to the duration and the magnitude of exposure. However, the range of concentrations varies considerably (from nondetectable levels to more than 300 µg/100 ml serum and from nondetectable levels to more than 40 µg/g adipose tissue) and overlaps the range of values seen in nonoccupationally exposed subjects (Wolff 1985; Phillips et al 1989; Emmett et al 1988a,b; Luotamo et al 1985a,b, 1988; Mes et al 1984; Takamatsu et al 1984). PCB with lower chlorine (less than 40%) are less cumulative than higher chlorinated biphenyls and their levels in blood seem to be more related to recent exposure than to body burden.

In a group of workers exposed for many years to PCB mixtures with a 42% chlorine content, it was found that the PCB concentration in the blood was closely correlated with the duration of occupational exposure (Maroni et al 1981a). This suggests that the PCB concentration in whole blood is a reflection of the amount stored in the body. A similar observation has been made by Wolff et al (1982b). Maroni et al (1981b) found a significant positive association between the prevalence of liver abnormalities (mainly increased serum gamma-glutamyltrans-peptidase, transaminases, and ornithine-carbamoyltransferase) and blood PCB concentrations. However, the dose-response relationship is not sufficiently precise to define a biological threshold for blood PCB. A tentative biological limit of 20 µg PCB/100 g blood has been suggested by Ouw et al (1976). Maroni et al (1981b) suggest that at this level some workers may still exhibit abnormal hepatic findings but this certainly needs further confirmation. No correlation has yet been demonstrated between blood PCB level and the prevalence of chloracne.

Wolff et al (1982a) and Luotamo et al (1988, 1991) have shown that elevated serum concentrations of specific isomers of different chlorinated levels are typical for the different sources and types of exposure. Fait et al (1989) have also demonstrated that the distribution of the PCB peak in a population of transformer-repair workers differs from that in capacitor workers (exposed to less highly chlorinated PCB). Therefore, as the toxicity differs among the various PCB congeners caution should be observed in using total PCB levels for risk assessment.

No biological threshold value for blood nor adipose tissue can currently be proposed.

REFERENCES

E. Emmett, M. Maroni, J. Schmith et al. Studies of transformer repair workers exposed to PCBs. I. Study design, PCBs concentrations, questionnaire and clinical examination results. *Am. J. Ind. Med.* 13:415, 1988a.

E. Emmett, M. Maroni, J. Jeffreys et al. Studies of transformer repair workers exposed to PCBs. II. Results of clinical laboratory investigations. *Am. J. Ind. Med.* 14:47, 1988b.

A. Fait, E. Grosman, S. Self et al. Polychlorinated biphenyl congeners in adipose tissue lipid and serum of past and present transformer repair workers and a comparison group. *Fundam. Appl. Toxicol.* 12:42, 1989.

Ph. Grandjean. *Skin penetration: hazardous chemicals at work.* Report prepared for the Commission of the European Communities, Luxembourg. Eds.: Taylor and Francis, London, Eur 12599 En, 1990.

P. Lees, M. Corn and P. Breysse. Evidence for dermal absorption as the major route of body entry during exposure of transformer maintenance and repairmen to PCBs. *Am. Ind. Hyg. Assoc. J.* 48:257, 1987.

G. Letz. The toxicology of PCBs: an overview for clinicians. *West. J. Med.* 13:534, 1983.

M. Luotamo, J. Jarvisalo, A. Aitio et al. Biological monitoring of workers exposed to polychlorinated biphenyl compounds in capacitor accidents. In: *Monitoring Human Exposure to Carcinogenic and Mutagenic Agents.* Eds.: A. Berlin, M. Draper, K. Hemminki, H. Vainio. IARC Scientific Publications Lyon, 1985a, No. 59, p. 307.

M. Luotamo, J. Järvisalo and A. Aitio. Analysis of polychlorinated biphenyls PCB in human serum. *Environ. Health Perspect.* 60:27, 1985b.

M. Luotamo. Isomer. Specific biological monitoring of polychlorinated biphenyls in human serum. *Scand. J. Work Environ. Health* 14:60, 1988.

M. Luotamo, J. Järvisalo and A. Aitio. Assessment of exposure to PCB: analysis of selected isomers in blood and adipose tissue. *Environ. Res.* 54:121, 1991.

M. Maroni, A. Colombi, S. Cantoni et al. Occupational exposure to polychlorinated biphenyls in electrical workers. I. Environmental and blood polychlorinated biphenyls concentrations. *Br. J. Ind. Med.* 38:49, 1981a.

M. Maroni, A. Colombi, G. Arbosti et al. Occupational exposure to polychlorinated biphenyls in electrical workers. II. Health effects. *Br. J. Ind. Med.* 38:55, 1981b.

J. Mes, J. Boyle, B. Adams et al. Polychlorinated biphenyls and organochlorine pesticides in milk and blood of Canadian women during lactation. *Arch. Environ. Contam. Toxicol.* 13:217, 1984.

H. Ouw, G. Simpson and D. Siyali. Use and health effects of Aroclor 1242, a polychlorinated biphenyl in an electrical industry. *Arch. Environ. Health* 31:189, 1976.

J. Perkins and B. Knight. Risk assessment of dermal exposure to PCB permeating a polyvinyl chloride glove. *Am. Ind. Hyg. Assoc. J.* 50:171, 1989.

D. Phillips, A. Smith, V. Burse et al. Half-life of polychlorinated biphenyls in occupationally exposed workers. *Arch. Environ. Health* 44:351, 1989.

W. Rogan, B. Gladen, J. McKinney and Ph. Albro. Chromatographic evidence of polychlorinated biphenyl exposure from a spill. *JAMA* 249:1057, 1983.

W. Rogan, B. Gladen, J. McKinney et al. Polychlorinated biphenyls (PCBs) and dichlorodiphenyl dichloroethene (DDE) in human milk effects of material factors and previous lactation. *Am. J. Public Health* 76:172, 1986.

S. Slorach and R. Vaz. PCB levels in breast milk; data from the UNEP/WHO pilot project on biological monitoring and some other recent studies. *Environ. Health Perspect.* 60:121, 1985.

G. Steele, P. Stehr-Green and E. Welty. Estimation of the biological half-life of polychlorinated biphenyls in human serum. *N. Engl. J. Med.* 314:926, 1986.

M. Takamatsu, M. Oki, K. Maeda et al. PCBs in blood of workers exposed to PCB and their
health status. *Am. J. Ind. Med.* 5:59, 1984.

R. Wester, D. Bucks, H. Maibach and J. Anderson. Polychlorinated biphenyls: dermal
absorption, systemic elimination and dermal wash efficiency. *J. Toxicol. Environ.
Health* 12:511, 1983.

M. Wolff, J. Thornton, A. Fischbein et al. Disposition of polychlorinated biphenyl congeners in occupationally exposed persons. *Toxicol. Appl. Pharmacol.* 62:294, 1982a.

M. Wolff, A. Fischbein, J. Thornton et al. Body burden of polychlorinated biphenyls
among persons employed in capacitor manufacturing. *Int. Arch. Occup. Environ.
Health* 49:199, 1982b.

M. Wolff. Occupational exposure to polychlorinated biphenyls (PCBs). *Environ. Health
Perspect.* 60:133, 1985.

T. Yakushiji, I. Watanabe, K. Kuwabara et al. Postnatal transfer of PCBs from exposed
mothers to their babies: influence of breast-feeding. *Arch. Environ. Health* 39:368,
1984.

3.13. Polychlorodibenzo-*p*-Dioxins (PCDD)

Toxicokinetics

Exposure to these compounds in the general population occurs mainly through
ingestion. In case of occupational exposure inhalation and skin contact are the
primary routes of concern, but there are no data concerning the rates of absorption.
Due to their liposolubility and high stability, polychlorinated dibenzo-*p*-dioxins
are predominately stored in fat and can also be detected in breast milk (Fiedler et
al 1990).

Limited human data indicate half-lives in the range of 2-7 years (Gorski et al
1984; Poiger and Schlatter 1986; CDC 1987; Pirkle et al 1989).

Biological monitoring

Blood and adipose tissue analyses

Blood and adipose tissue levels have been suggested as indicators of body burden
(Pirkle et al 1989; Strechter and Ryan 1988; Beck et al 1989; Zober et al 1990;
Sweeney et al 1990). Very high levels (up to 56 µg/100 ml serum) have been
recorded after accidental exposure of the general population (Mocarelli et al
1991).

The adipose tissues of nine workers from a former trichlorophenol manufac-
turer showed a mean 2,3,7,8-tetrachlorodibenzo-*p*-dioxin (TCDD) level of 246
pg/g compared with a mean value of 8.7 pg/g for the control subjects. The lipid
adjusted mean level in serum was 363 pg/g (range: 61-1090; n: 8) compared
with a value of 47.1 pg/g for the nonexposed workers (Patterson et al 1989).
Workers with a history of moderate and light chloracne 17 years after the end
of occupational exposure to PCDDs showed, respectively, median levels of 498

pg 2,3,7,8-TCDD pg/g blood lipid (n: 3; range: 193-498; calculated median level at the end of exposure: 2682 pg/g blood lipid) and 305 pg/g blood lipid (n: 6; range: 98-659; calculated median level at the end of exposure: 1640 pg/g blood lipid). The median levels in control groups ranged from 13 to 20 pg/g (Neuberger et al 1991).

The relationship between exposure, internal dose, and effects has not been sufficiently characterized to propose meaningful biological limit values.

REFERENCES

H. Beck, K. Eckart, W. Mather and R. Wittkowski. Levels of PCDDs and PCDs in adipose tissue of occupationally exposed workers. *Chemosphere* 18:507, 1989.

CDC. Centers for Disease Control. Comparison of serum levels of 2,3,7,8-TCDD with indirect estimates of agent orange exposure in Vietnam veterans. Public Health Service, U.S. Department of Health and Human Services, Atlanta, 1987.

H. Fiedler, O. Hutzinger and C. Timms. Dioxins: sources of environmental load and human exposure. *Toxicol. Environ. Chem.* 29:157, 1990.

T. Gorski, L. Konopka and M. Brodzki. Persistence of some polychlorinated dibenzo-p-dioxins and polychlorinated debenzofurans of pentachlorophenol in human adipose tissue. *Pocz. Pzh.* XXXV:297, 1984.

P. Mocarelli, L. Needham, A. Marocchi et al. Serum concentrations of 2,3,7,8- tetrachloro-dibenzo-p-dioxin and test results from selected residents of Seveso, Italy. *J. Toxicol. Environ. Health* 32:357, 1991.

M. Neuberger, W. Landvoigt and F. Derntl. Blood levels of 2,3,7,8-tetrachlorodibenzo-p-dioxin in chemical workers after chlorane and in comparison groups. *Int. Arch. Occup. Environ. Health* 63:325, 1991.

D. Patterson, M. Fingerhut, D. Roberts et al. Levels of polychlorinated dibenzo-p-dioxins and dibenzofurans in workers exposed to 2,3,7,8-tetrachlorodibenzo-p-dioxin. *Am. J. Ind. Med.* 16:135, 1989.

J. Pirkle, W. Wolfe, D. Patterson et al. Estimates of the half-life of 2,3,7,8-tetrachlorodibenzo-p-dioxin in Vietnam veterans of operation ranch hand. *J. Toxicol. Environ. Health* 27:165, 1989.

H. Poiger and R. Schlatter. Pharmacokinetics of 2,3,7,8-TCDD in man. *Chemosphere* 15:9, 1986.

A. Strechter and J. Ryan. PCDD and PCDF levels in human adipose tissues from workers 32 years after occupational exposure to 2,3,7,8-TCDD in 1953 in Germany. *Chemosphere* 17:915, 1988.

M. Sweeney, M. Fingerhut, D. Patterson et al. Serum levels of 2,3,7,8-tetrachlorodibenzo-p-dioxin in New Jersey and Missouri chemical workers exposed to dioxin contamined processes. *Chemosphere* 30:993, 1990.

A. Zober, P. Messerer and P. Huber. Thirty-four-year mortality follow-up of BASF employees exposed to 2,3,7,8-TCDD after the 1953 accident. *Int. Arch. Occup. Environ. Health* 62:139, 1990.

4. AMINO AND NITRO DERIVATIVES
4.1. Aliphatic Amino Compounds

Toxicokinetics

Inhalation seems to constitute the main route of absorption of these compounds. A few metabolic studies have been carried out on volunteers and subjects occupationally exposed to various aliphatic amines (Åkesson et al 1988, 1989a,b; Lundh et al 1991; Al Waiz et al 1987a,b; Stählbom et al 1991b).

The amines are excreted in urine as such and after oxidation, which appears to be the major route of biotransformation. The average rate of oxidation of dimethylamine was about 90%, which was considerably larger than that of triethylamine (about 24%) but somewhat lower than that of trimethylamine (>95%). Large interindividual variations were observed. The half-lives of dimethylethylamine (DMEA) and dimethylethylamine-*N*-oxide (DMEAO) in plasma were 1.3 and 3 h respectively (Stählbom et al 1991b). The urinary excretion of DMEA and DMEAO followed a two-phase pattern. The half-lives were 1.5 h for DMEA and 2.5 h for DMEAO for the first phase and 7 and 8 h for the second phase, which started about 9 h after the end of exposure. These half-lives are somewhat shorter than for triethylamine (TEA) and triethylamine-*N*-oxide (TEAO). Elimination of DMEA and TEA by exhalation is minimal (Stählbom et al 1991b; Åkesson et al 1989b).

The results of Åkesson and Skerfving (1990) provide strong evidence of the inhibiting effect of ethanol intake on the rate of oxidation of triethylamine.

Biological monitoring

Blood and urine analyses

Triethylamine, triethylamine-N-oxide. According to Åkesson et al (1988), an exposure to TWA of 40 mg triethylamine/m^3 air would lead to an average TEA urinary concentration of about 160 mmol/mol creatinine (about 145 mg/g creatinine) at rest and 320 mmol/mol creatinine (about 290 mg/g creatinine) after a moderate workload (samples taken within 2 h of the end-of-exposure).

To prevent the occurrence of visual disturbances, however, a TWA of 10 mg/m^3 would be more appropriate (Åkesson et al 1985, 1986, 1988; Lundh et al 1991; Stählbom et al 1991a; Warren and Selchan 1988). According to experimental studies, this level of exposure would correspond to an average urinary TEA concentration of 40 mmol/mol creatinine (about 36 mg/g creatinine) (at rest) and 80 mmol/mol creatinine (about 72 mg/g creatinine) (moderately heavy work) (Åkesson et al 1988). On the basis of a field study, they concluded that an air level of 10 mg/m^3 corresponds to a urinary excretion of 65 mmol TEA/mol creatinine (about 58.5 mg/g creatinine) and a plasma level of 1.9 μmol/l (about 19.25 μg/100 ml) (Åkesson et al 1989b).

The same authors have suggested that due to the interindividual variation of the oxidation reaction the sum of TEA and TEAO (in plasma or urine) constitutes a better biological indicator of occupational exposure (Åkesson et al 1989a,b).

Dimethylethylamine, dimethylethylamine-N-oxide. An experimental study conducted on volunteers exposed during 8 h to different air DMEA concentrations, leads to the conclusion that an 8-h exposure to 10 mg DMEA/m³ corresponds to a postexposure plasma concentration and 2 h postexposure urinary excretion of 4.9 µmol/l (about 36 µg/100 ml) and 75 mmol/mol creatinine (about 50 mg/g creatinine), respectively (Stählbom et al 1991b).

The observations of Åkesson et al (1989a,b) also showed that postshift concentration of the sum of DMEA and DMEAO in plasma and urine might be used to assess exposure to dimethylethylamine. Air concentrations of 40 and 10 mg/m³ would correspond to average concentrations of the summed amount of DMEA and DMEAO of 570 mmol/mol creatinine and 23 µmol/l plasma and 135 mmol/mol creatinine and 5.7 µmol/l plasma, respectively (during light to moderate work).

Hexamethylenediamine. In the urine of volunteers, who were given hexamethylenediamine orally, Brorson et al (1990) have identified the compound itself, 6-aminohexanoic acid and *N*-acetyl-1,6-hexamethylene diamine.

4.2. Aliphatic Nitrates

Toxicokinetics

The aliphatic nitrates (ethylene glycol dinitrate, propyleneglycol dinitrate and nitroglycerin) are absorbed through the lungs and the skin. In practice, skin absorption is the most important route of entry during occupational exposure (Gjesdal et al 1985; Grandjean 1990). These aliphatic nitrates give rise to methemoglobin *in vivo*, but only under high exposure conditions. Therefore, methemoglobin determination is not sufficiently sensitive for monitoring exposure of workers.

Biological monitoring

Blood analyses

These compounds can be detected in blood (Fukuchi 1981; Gottel 1976; Williams et al 1966). A biological threshold limit value of 0.2 and 0.4 µg/100 ml of blood has been proposed by Gottel (1976) for ethylene glycol dinitrate and nitroglycerin, respectively. Since the biological half-life of both compounds is short (about 30 min), blood sampling should be done during exposure. According to Hogstedt and Stahl (1980), however, venous blood levels seem to reflect almost exclusively the concentration of locally absorbed ethylene glycol dinitrate from the part of the arm distal to the sampling spot. This has been confirmed by Gjesdal et al (1985), who observed much higher plasma glyceryl trinitrate in blood samples obtained from the cubital vein than in those taken from the femoral vein. Therefore, biological monitoring of this compound through blood samples cannot be used for quantitative estimations of the total uptake. However, it may be of value for monitoring efforts made towards reducing dermal uptake.

Urine analyses

Ethylene glycol dinitrate can also be measured in urine and, according to Fukuchi (1981), its concentration in urine appears to be a more stable index of the amount entering the body than its concentration in the blood. According to Ahonen et al (1989), it is possible to detect exposure to isopropyl nitrate by expired air, blood, and urine analysis. Although its elimination is rapid, urinary monitoring seems the most promising method for simple sampling and sensitive analysis.

4.3. Aromatic Amino and Nitro Compounds

Toxicokinetics

The cutaneous and the pulmonary routes represent the main routes of exposure to aromatic amino and nitro compounds in industry. Skin exposure to many of these compounds (aniline, nitroaniline, 4,4′-methylenedianiline, toluidine, 4,4′-methylene-bis(2-chloroaniline) or MOCA, phenylenediamine, benzidine, (di)nitrobenzene, (di-, tri-)nitrotoluene, etc.) may effectively result in considerable absorption (Grandjean 1990; Hopkins and Manoharan 1985; Bronaugh and Maibach 1985; Hansen and Andersen 1988; Fiserova-Bergerova and Pierce 1989; Woollen et al 1985, 1986; Chin et al 1983; Scansetti et al 1987).

These compounds undergo a variety of biotransformation reactions. Figure 19 illustrates the biotransformation of aniline and nitrobenzene. The main detoxification pathway of the aromatic compounds occurs through N-acetylation and oxidation at ring-carbons by cytochrome P450-dependent reactions with the formation of phenolic compounds that are excreted in the bile and urine as O-glucuronides and O-sulfates (Lauwerys 1988a). This explains that slow acetylators are more at risk of toxic effects than fast acetylators. Aromatic nitro-compounds undergo transformations such as reduction of the nitro group, oxidation of the aromatic ring and the lateral chains, and conjugation. These metabolites are also mainly excreted via the urine (Lauwerys 1988b).

For more information on the metabolic handling of these substances see Lauwerys 1988a,b; Rickert et al 1984; Rickert 1987; Gorrod and Manson 1986; Land et al 1989; Turner et al 1985; Lewalter and Korallus 1985; IARC 1987).

Biological monitoring

Various biological tests have been proposed for evaluating the intensity of exposure to these chemicals.

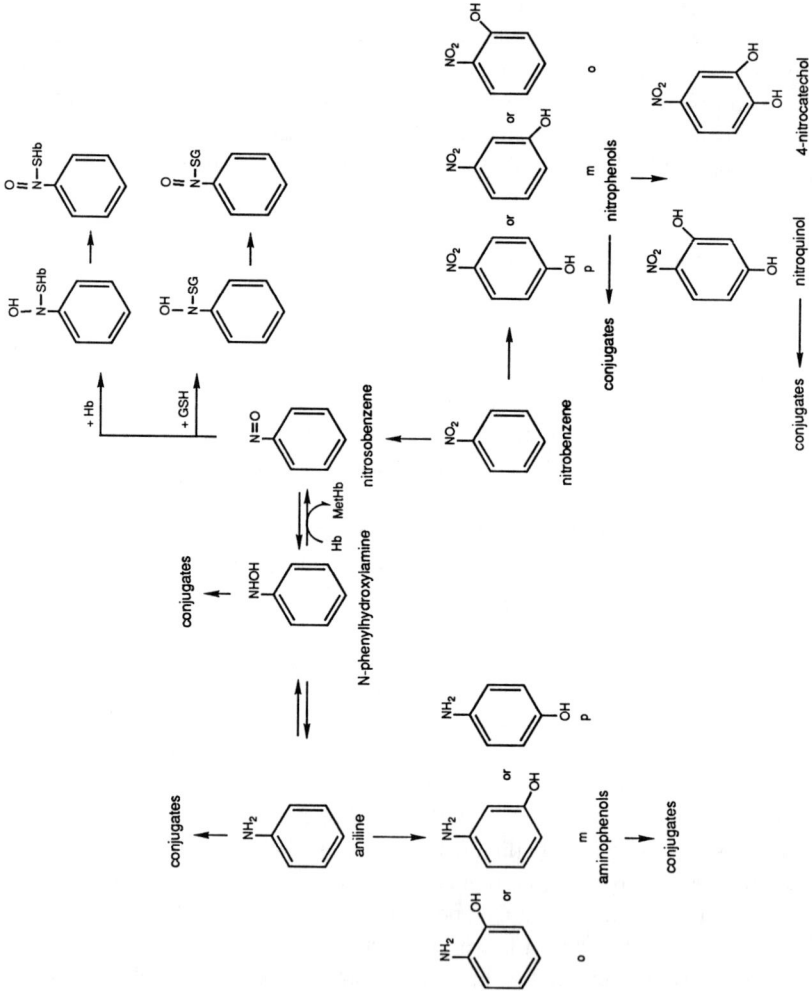

FIGURE 19 Main metabolic pathways of aniline and nitrobenzene.

Blood analyses

Methemoglobin level. Several single-ring aromatic nitro and amino compounds (e.g., aniline, nitrobenzene, dinitrobenzene, and *o*-, *m*-, and *p*-chloroaniline) are metabolized *in vivo* to methemoglobin-forming agents (e.g., phenylhydroxylamine following exposure to aniline). The determination of the methemoglobin level may serve as a routine biological method for detecting excessive absorption of these compounds. Approximately 1 to 2% methemoglobin exists in erythrocytes of normal individuals. Because of the rapid reduction of methemoglobin in erythrocytes, blood must be analyzed soon after withdrawal. Storage of blood for more than a few hours will lead to erroneous results. A biological threshold limit value of 5% methemoglobin has been proposed (Linch 1974a). This seems acceptable when methemoglobin formation is the critical effect. For specific substances (e.g., carcinogenic aromatic amines), it may be justified to prevent any increase above background value.

It should be pointed out that other industrial chemicals are also methemoglobin-forming agents, such as aliphatic nitrates (e.g., ethylene glycol dinitrate), aliphatic nitrites, inorganic nitrites, and chlorates. Exposure to some of these agents may also be monitored by methemoglobin analysis.

Sulfhemoglobin level. In addition to producing methemoglobin, aromatic amino and nitro compounds can also induce the formation of sulfhemoglobin, which is much more stable than methemoglobin (Jope 1946). The practical application of this test has not yet been investigated.

Heinz bodies. Heinz bodies (denatured hemoglobin) can be found in erythrocytes in workers exposed to various aromatic amino and nitro compounds (aniline, nitrobenzene), but this parameter is not a reliable indicator of the amount of chemical absorbed.

Hemoglobin adducts. Several aromatic amines undergo metabolic activation leading to electrophilic metabolites, which bind to nucleophilic sites on biological macromolecules such as DNA or proteins.

Only free, nonacetylated active metabolites can lead to the formation of conjugates with hemoglobin (Hb). The degree of adduct formation correlates inversely with the magnitude of the acetylation potential, depending on the availability of free, nonacetylated compounds and hence the "acetylating capacity" of the subject (Lewalter and Korallus 1985).

The analysis of Hb adducts has been recommended to assess exposure of individuals to these potentially genotoxic chemicals (Bolt et al 1985; Birner and Neumann 1988; Birner et al 1990; Neumann 1988). The method offers the advantage of assessing the internal exposure over the lifetime of the protein (120 d for hemoglobin).

So far in human, Hb adducts determination has been reported for aniline (Lewalter and Korallus 1985), MOCA (Braselton et al 1988), 4-aminobiphenyl (ABP) (Bryant et al 1987; Skipper et al 1986) and benzidine (Birner et al 1990).

Skipper et al (1986) have observed "background levels" of Hb bound 4-aminobiphenyl of 0 to 0.25 mg/10 ml blood.

Bryant et al (1987) have shown that cigarette smokers have a significantly higher mean value of ABP adducts (154 pg ABP/g Hb) than the nonsmokers (28 pg ABP/g Hb) with no overlap between the two groups. 4-ABP adduct was found in all nonsmokers, indicating a low level ubiquitous contamination by this aromatic amine.

Urine analyses

Diazo-positive metabolites. An increased concentration of diazo-positive metabolites has been found in urine of workers exposed to various aromatic amino and nitro compounds (Ikeda et al 1977; Linch 1974b; Tabuchi et al 1983; Ahlborg et al 1988a). Despite its lack of specificity, this test may be useful to detect groups of workers at risk.

In 133 persons nonoccupationally exposed to aromatic amines, Armeli et al (1980) found a mean total urinary amine concentration of 5.14 mg/l (range: 3.2-10.5). No differences were found due to sex or smoking. The mean ($\pm$SD) normal concentration reported by Ahlborg et al (1988a) was about 0.5 ± 0.3 mmol/mol creatinine (as 2,4-diaminotoluene equivalent). There was a great interindividual variation, but no difference between smokers and nonsmokers. Yoshida et al (1988) monitored 10 workers exposed to *p*-nitrochlorobenzene (PNCB) in a dye factory. The urinary diazo-positive metabolites level reflected the PNCB exposure level at least in autumn and winter. In summer, most of the workers showed high urinary values irrespective of the ambient PNCB level. Increased skin absorption during the summer due to perspiration and less efficient protection with gloves and clothes (caused by the hot and humid weather) was probably responsible for this observation. The mean (geometric) diazo-positive metabolite concentrations, before and after shift, corresponding to TWA values of 0.306 mg/m³ (autumn), 0.158 (winter), and 0.380 (summer), were 0.358 and 0.776 mg *p*-aminophenol equivalent/mg creatinine, 0.413 and 0.624 mg *p*-aminophenol equivalent/mg creatinine, 0.434 and 0.944 mg *p*-aminophenol equivalent/mg creatinine, respectively.

Buchet et al (1988) have also assessed the usefulness of measuring diazo-positive compounds in urine as a biological index of exposure to aromatic amino compounds in four groups of workers. On a group basis, the geometric mean was significantly increased in the urine samples collected in the 45 dye factory workers: +40% (5.6 mg equivalent aniline/g creatinine; P95: 20.4; range: 2.9-79.9), in the 42 coke oven workers: +70% (6.8 mg equivalent aniline/g creatinine; P95: 33.7; range: 3.2-42.4) and in the 19 rubber factory workers: +48% (5.9 mg equivalent aniline/g creatinine; P95: 61.7; range: 3.5-61.7). The increase in the 32 hairdressers was not statistically significant (5.1 mg equivalent aniline/g creatinine, P95: 24.2; range 2.8-38.7). In the control group (n = 149) the geometric mean value was 4 mg equivalent aniline/g creatinine (P95: 6.63, range: 0.7-20.7). Age, sex, smoking habits, and drug consumption had no significant effect on the urinary excretion of these compounds. However, consumption of certain drugs (e.g., some analgesics) may lead to a marked increased urinary excretion of diazo-positive metabolites (Lauwerys 1988a).

Parent compounds or their metabolites. Since aromatic amines are partly excreted unchanged in urine, the analysis of urine collected at the end of a working day can be used to confirm absorption of these chemicals (e.g., 4,4'-methylene-bis(2-chloroaniline) or MOCA, 4-aminobiphenyl, 2-naphthylamine, benzidine, 3,3'-dimethylbenzidine, 3,3'-dichlorobenzidine, 3,3'-dimethoxybenzidine, 4,4'methyl-ene-dianiline or MDA) (Linch 1974b; Nony and Bowman 1978; Ichikawa et al 1990; Thomas and Wilson 1984; Cocker et al 1986, 1988; Ward et al 1986; Ducos et al 1985; London et al 1989; Clapp et al 1991; Gristwood et al 1984; Smith et al 1989; Neumeister 1991). Although there are no data available on the relation-ship between intake and concentration in urine, such analyses seem very useful to detect exposure to carcinogenic aromatic amines.

For aniline and nitrobenzene, the determination of the main urinary metabo-lites, p-aminophenol and p-nitrophenol, is also a sensitive method of monitoring. After 6 h of exposure to an airborne vapor concentration of 1 ppm of nitrobenzene, the urinary concentration of p-nitrophenol reaches approximately 5 mg/l (Piotrowski 1967). A dangerous exposure to aniline is suspected when the p-aminophenol concentration in urine exceeds 50 mg/l (Michaux et al 1971). A reasonable biological threshold limit value should probably be on the order of 30 mg/l.

The elimination of dinitrotoluene (DNT) and its metabolites in the urine of exposed workers has been studied by several investigators (Levine et al 1985; Turner et al 1985; Woollen et al 1985). Several metabolites were identified: 2,4- and 2,6-dinitrotoluene, 2,4- and 2,6-dinitrobenzoic acid, 2,4- and 2,6-dinitrobenzyl glucuronide, 2-amino-4-nitrobenzoic acid, 4-N-(acetyl)amino-2-nitrobenzoic acid. There is a sex-difference in the proportion of the different metabolites produced: the ratio of dinitrobenzyl glucuronide to dinitrobenzoic acids is higher in females than in males. There is also a large interindividual difference in the rate of excretion of the different metabolites. According to Levine et al (1985), most urinary metabolites related to exposure during an 8-hour shift are excreted by the start of work the following day. The major metabolite is dinitrobenzoic acid and on a group basis its measurement in urine seems a useful tool to monitor the exposure. Levine et al (1985) advise a 24-hour urine collection but for Woollen et al (1985), an end-of-shift sample is sufficiently adequate for assessing the effectiveness of measures taken to limit the absorption of DNT.

Trinitrotoluene (TNT) absorption has also been assessed by detecting TNT and its metabolites in urine (Woollen et al 1986; Ahlborg et al 1988b). The major metabolites are 2,4-dinitro-6-aminotoluene and 2,6-dinitro-4-aminotoluene. As for DNT, there is also a large interindividual variation in the rate of clearance of the metabolites. Woollen et al (1986) found that in most cases, excretion was highest at the end of the workshift, but a few workers had still high urinary concentrations in the morning the day after exposure. In some cases the metabo-lites could still be detected 17 d after the end of the exposure.

In the urine of workers exposed to methylenedianiline (MDA), Cocker et al (1986) identified N-acetylmethylenedianiline to be the major metabolite. While a significant proportion of MDA is excreted as the acetylated compounds, N-acetyl MOCA is a relatively minor metabolite of MOCA (Cocker et al 1986, 1988; Ducos et al 1985).

4-Chloro-o-toluidine has been detected in the urine of workers occupationally exposed to chlordimeform (Geyer and Fattal 1987) and N-mononitrosopiperazine in the urine of workers exposed to piperazine (Bellander et al 1987, 1988).

El Bayoumi et al (1986a,b) have identified and quantified aniline and o-toluidine in the urine of nonoccupationally exposed subjects. The concentrations (mean ± SD) in smokers (3.1 ± 2.6 µg/24 h of aniline; 6.3 ± 3.7 µg/24 h of o-toluidine) were not significantly different from those in nonsmokers (2.8 ± 2.5 µg/24 h of aniline; 4.1 ± 3.2 µg/24 h of o-toluidine). However, as the intra- and interindividual variations in the amounts excreted were relatively large, it is suggested that source(s) other than cigarette smoke contribute to their concentrations in urine.

Tests reflecting exposure to mutagenic and potentially carcinogenic substances

Since several aromatic amino and nitro compounds are genotoxic, the aspecific methods which have been proposed for detecting exposure to organic mutagenic compounds (e.g., Ames test on urine, thioethers in urine, etc.) may also be considered (see Section 17).

4.4. Azo Compounds

These compounds can be absorbed by all the routes, but in industry exposure mainly occurs through skin contact. Benzidine and its acetylated metabolites (monacetyl- and diacetylbenzidine) were found in the urine of workers exposed to benzidine-based azo dyes such as Direct Black 38 (Dewan et al 1988; Lowry et al 1980; Meal et al 1981; NIOSH 1980). Not only the liver but also the intestinal flora may catalyze the reduction of the azo bound (Cerniglia et al 1982).

In practice, the same monitoring strategies developed for the aromatic amino and nitro-compounds can be applied to assess the exposure to benzidine-derived azo dyes.

For these substances mainly absorbed through the skin, biological monitoring certainly represents the best approach for assessing the risk of excessive exposure. Unfortunately, with a few exceptions, the knowledge of the relationship between the biological marker(s) and the exposure intensity or the risk of adverse effects is still too fragmentory to propose meaningful biological limit values. Nevertheless biological monitoring is still very useful to assess the efficiency of preventive measures (e.g., skin protection) mainly for genotoxic chemicals (e.g., benzidine, MDA, MOCA, etc.) for which the internal dose should be kept as low as possible.

For aniline, the following relationship has been reported between exposure and urinary excretion of p-aminophenol (WHO 1986): an 8-hour exposure to 5 and 19 mg aniline/m³ air results in an excretion of 35 and 150 mg p-aminophenol within the first 24 h following the beginning of exposure. At the above exposure levels, the rates of p-aminophenol excretion in urine during the 4th and 6th hours of exposure are 1.5 mg/h and 13 mg/h, respectively.

The ACGIH (1991-1992) has adopted as BEI for aniline a total p-aminophenol concentration of 50 mg/g creatinine (end-of-shift) and a methemoglobin concentration of 1.5% (end-of-shift). The same level of methemoglobin has been adopted for nitrobenzene. Its urinary metabolite, p-nitrophenol, should not exceed 5 mg/g creatinine (end-of-shift, end-of-workweek).

The DFG (1991) recommends 1 mg aniline/l urine (end-of-shift) as biological tolerance value, and 100 µg/l blood has been proposed for the amount of aniline released from aniline hemoglobin conjugate. This last value has also been adopted for nitrobenzene.

REFERENCES

ACGIH. American Conference of Governmental Industrial Hygienists. Threshold Limit Values for Chemical Substances and Physical Agents and Biological Exposure Indices. Cincinnati, 1991-1992.

G. Ahlborg, A. Ulander, B. Bergstrom and A. Oliv. Diazo-positive metabolites in urine from workers exposed to aromatic-nitro-amino compounds. *Int. Arch. Occup. Environ. Health* 60:51, 1988a.

G. Ahlborg, P. Einisto and M. Sorsa. Mutagenic activity and metabolites in the urine of workers to trinitrotoluene TNT. *Br. J. Ind. Med.* 45:353, 1988b.

I. Ahonen, P. Oksa and S. Rantanen. Biological monitoring of isopropyl nitrate. *Toxicol. Lett.* 47:205, 1989.

B. Åkesson, I. Florén and S. Skerfving. Visual disturbances after experimental exposure to triethylamine. *Br. J. Ind. Med.* 42:848, 1985.

B. Åkesson, M. Bengtsson and I. Florén. Visual disturbances by industrial triethylamine exposure. *Int. Arch. Occup. Environ. Health* 57:297, 1986.

B. Åkesson, S. Skerfving and L. Mattiasson. Experimental study on the metabolism of triethylamine in man. *Br. J. Ind. Med.* 45:262, 1988.

B. Åkesson, S. Skerfving, B. Ståhlbom and T. Lundh. Metabolism of triethylamine in polyurethane-foam workers. *Am. J. Ind. Med.* 16:255, 1989a.

B. Åkesson, E. Vinge and S. Skerfving. Pharmacokinetics of triethylamine and triethylamine-N-oxide in man. *Toxicol. Appl. Pharmacol.* 100:529, 1989b.

B. Åkesson and S. Skerfving. Effects of ethanol ingestion and urinary acidity on the metabolism of triethylamine in man. *Int. Arch. Occup. Environ. Health* 62:89, 1990.

M. Al Waiz, S. Mitchell, J. Iddle and R. Smith. The metabolism of 14c-labelled trimethylamine and its n-oxide in man. *Xenobiotica* 17:551, 1987a.

M. Al Waiz, S. Mitchell, J. Iddle and R. Smith. The relative importance of n-oxidation and n-demethylation in the metabolism of triethylamine in man. *Toxicology* 43:117, 1987b.

G. Armeli, G. Baldratti, F. Cazzoli and E. Rabbi. Determination of total urinary aromatic amines in non-occupationally exposed subjects. *G. Ital. Med. Lav.* 2:117, 1980 (in Italian).

T. Bellander, B. Oesterdahl and L. Hagmar. N-mononitrosopiperazine in urine after occupational exposure to piperazine. In: *Relevance of n-Nitroso Compounds to Human Cancer: Exposures and Mechanisms.* IARC Scientific Publication No. 84(2):553, Lyon, 1987.

T. Bellander, B. Oesterdahl, L. Hagmar and S. Skerfving. Excretion of *n*-mononitrosopiperazine in urine in workers manufacturing piperazine. *Int. Arch. Occup. Environ. Health* 60:25, 1988.

G. Birner and H. Neumann. Biological monitoring of aromatic amines. Hemoglobin binding of some monocyclic aromatic amines. *Arch. Toxicol.* 62:110, 1988.

G. Birner, W. Albrecht and H. Neumann. Biomonitoring of aromatic amines. Hemoglobin binding of benzidine and some benzidine congeners. *Arch. Toxicol.* 64:97, 1990.

H. Bolt, G. Neumann and J. Lewalter. Zur Problematik von Bat-Werten für Aromatische Amine. *Arbeitsmed. Sozialmed. Präventivmed.* 20:197, 1985.

W. Braselton, T. Chen and B. Kuslikis. Dose monitoring of exposure to 4,4-methylene bis (2-chloroaniline) MBOCA by determination of hemoglobin adducts. *Toxicologist* 8:183, 1988.

R. Bronaugh and H. Maibach. Percutaneous absorption of nitroaromatic compounds: in vivo and in vitro studies in the human and the monkey. *J. Invest. Dermat.* 84:180, 1985.

T. Brorson, G. Skarping, J. Sandström and M. Stenberg. Biological monitoring and related amines. I. Determination of 1,6-hexamethylene diamine in hydrolysed human urine after oral administration. *Int. Arch. Occup. Environ. Health* 62:79, 1990.

M. Bryant, P. Skipper, S. Tannenbaum and M. Maclure. Hemoglobin adducts of 4-aminobiphenyl in smokers and non-smokers. *Cancer Res.* 47:602, 1987.

J.-P. Buchet, D. Verschelde, A. Devroede et al. Excretion urinaire des composés diazopositifs chez des travailleurs exposés aux aminés aromatiques. *Cah. Méd. Trav.* 25:4, 1988.

C. Cerniglia, J. Freeman, W. Franklin and L. Pack. Metabolism of benzidine-congener based dyes by human, monkey and rat intestinal bacteria. *Biochem. Biophys. Res. Commun.* 107:1224, 1982.

B. Chin, M. Tobes and S. Han. Absorption of 4,4′-methylenebis (2-chloroaniline) by human skin. *Environ. Res.* 32:167, 1983.

D. Clapp, G. Piacitelli, D. Zaebst and E. Ward. Assessing exposure to 4,4′-methylenebis (2-chloroaniline)(MBOCA) in the workplace. *Appl. Occup. Environ. Hyg.* 6:125, 1991.

J. Cocker, W. Gristwood and H. Wilson. Assessment of occupational exposure to 4,4′-diaminodiphenylmethane (methylene dianiline) by gas chromatography-mass spectrometry analysis or urine. *Br. J. Ind. Med.* 43:620, 1986.

J. Cocker, A. Boobis and D. Davies. Determination of the n-acetyl metabolites of 4,4′-methylenebis (2-chloroaniline) in urine. *Biomed. Environ. Mass. Spectrometry* 17:161, 1988.

A. Dewan, J. Jani, J. Patel et al. Benzidine and its acetylated metabolites in the urine of workers exposed to direct black 38. *Arch. Environ. Health* 43:269, 1988.

DFG. Deutsche Forschungsgemeinschaft. *Maximum Concentrations at the Workplace and Biological Tolerance Values for Working Materials.* Report No. XXVII. Commission for the Investigation of Health Hazards of Chemical Compounds in the Work Area. VCH, Weinheim, 1991.

P. Ducos, C. Maire and R. Gaudin. Assessment of occupational exposure to 4,4′-methylenebis (2-chloroaniline) "MOCA" by a new sensitive method for biological monitoring. *Int. Arch. Occup. Environ. Health* 55:159, 1985.

K. El Bayoumi, J. Donahue, S. Hecht and D. Hoffmann. Identification and quantitative determination of aniline and toluidines in human urine. *Cancer Res.* 46:6064, 1986a.

K. El Bayoumi, J. Donahue, S. Hecht and D. Hoffmann. Analysis of aniline and *o*-toluidine in human urine. Branbury Report 23. Mechanisms in tobacco carcinogenesis, Cold Spring Harbor Laboratory, Cold Spring Harbor, NY, 0-87969-223-5-8/86:77, 1986b.

V. Fiserova-Bergerova and J. Pierce. Biological monitoring. V. Dermal absorption. *Appl. Ind. Hyg.* 4:F14, 1989.

Y. Fukuchi. Nitroglycol concentrations in blood and urine of workers engaged in dynamite production. *Int. Arch. Occup. Environ. Health* 48:339, 1981.

R. Geyer and F. Fattal. HPLC determination of the metabolite 4-chloro-o-toluidine in the urine of workers occupationally exposed to chlodimeform. *J. Anal. Toxicol.* 11:24, 1987.

K. Gjesdal, S. Bille, J. Bredesen et al. Exposure to glyceryl trinitrate during gun powder production: plasma glyceryl trinitrate concentrations, elimination kinetics and discomfort among production workers. *Br. J. Ind. Med.* 42:27, 1985.

J. Gorrod and D. Manson. The metabolism of aromatic amines. *Xenobiotica*, 16:933, 1986.

P. Gotell. Environmental and clinical aspects of nitroglycol and nitroglycerin exposure. *Occup. Health Safety* 45:50, 1976.

Ph. Grandjean. Skin penetration: hazardous chemicals at work. A report prepared for the Commission of the European Communities, Luxembourg. Taylor and Francis, London, Eur 12599 En, 1990.

W. Gristwood, S. Robertson, and H. Wilson. The determination of 4,4'-methylenebis (2-chlorosys aniline) in urine by electron capture gas chromatography. *J. Anal. Toxicol.* 8:101, 1984.

C. Hansen and B. Andersen. The affinities of organic solvents in biological systems. *Am. Ind. Hyg. Assoc. J.* 49:301, 1988.

C. Hogstedt and R. Stahl. Skin absorption and protective gloves in dynamite work. *Am. Ind. Hyg. Assoc. J.* 41:367, 1980.

J. Hopkins and A. Manoharan. Severe aplastic anemia following the use of hair dye: report of two cases and review of literature. *Postgrad. Med. J.* 61:1003, 1985.

IARC. International Agency for Research on Cancer. The Relevance of n-Nitroso Compounds to Human Cancer. Exposure and Mechanisms. Eds.: H. Bartsch, I. O'Neill, R. Schulte-Hermann. IARC Scientific Publications, No. 84, Lyon 1987.

Y. Ichikawa, M. Yoshida, A. Okayama et al. Biological monitoring for workers exposed to 4,4'-methylenebis (2-chloroaniline). *Am. Ind. Hyg. Assoc. J.* 51:5, 1990.

M. Ikeda, T. Watanabe, I. Hara et al. A field survey on the health status of workers in dye-producing factories. *Int. Arch. Occup. Environ. Health* 39:219, 1977.

E. Jope. The disappearance of sulphaemoglobin from the blood of TNT workers in relation to the dynamics of red cell destruction. *Br. J. Ind. Med.* 3:136, 1946.

S. Land, K. Zukowski, M. Lee et al. Metabolism of aromatic amines: relationships of N-acetylation, O-acetylation, N,O-acetyl transfer and deacetylation in human liver and urinary bladder. *Carcinogenesis* 10:727, 1989.

R. Lauwerys. Biological indicators for the assessment of human exposure to industrial chemicals: Aromatic amines. Eds.: L. Alessio, A. Berlin, M. Boni and R. Roi. Commission of the European Communities, Luxembourg, Eur 11478 En, 1988a.

R. Lauwerys. *Biological Indicators for the Assessment of Human Exposure to Industrial Chemicals: Aromatic Nitro Compounds.* Eds.: L. Alessio, A. Berlin, M. Boni and R. Roi. Luxembourg, Eur 11478 En, 1988b.

R. Levine, M. Turner, Y. Crume et al. Assessing exposure to dinitrotoluene using a biological monitor. *J. Occup. Med.* 27:627, 1985.

J. Lewalter and U. Korallus. Blood protein conjugates and acetylation of aromatic amines: new findings on biological monitoring. *Int. Arch. Occup. Environ. Health* 56:179, 1985.

A. Linch. Biological monitoring for industrial exposure to cyanogenic aromatic nitro- and amino-compounds. *Am. Ind. Hyg. Assoc. J.* 35:426, 1974a.

A. Linch. *Biological Monitoring for Industrial Chemical Exposure Control*. CRC Press, Boca Raton, FL, 1974b.

M. London, J. Boiano and S. Lee. Exposure assessment of 3,3-dichlorobenzidine at two chemical plants. *Appl. Ind. Hyg.* 4:101, 1989.

L. Lowry, W. Tolos, M. Boeniger et al. Chemical monitoring of urine from workers potentially exposed to benzidine-derived azo-dyes. *Toxicol. Lett.* 7:29, 1980.

T. Lundh, B. Stählbom and B. Akesson. Dimethylethylamine in mould core manufacturing exposure, metabolism and biological monitoring. *Br. J. Ind. Med.* 48:230, 1991.

P. Meal, J. Cocker, H. Wilson and J. Gilmour. Search for benzidine and its metabolites in urine of workers weighing benzidine-derived dyes. *Br. J. Ind. Med.* 38:191, 1981.

P. Michaux, H. Boiteau and F. Tolot. Valeur et limite du dépistage clinique et biologique en pathologie professionnelle. *Arch. Mal. Prof.* 32:1, 1971.

H. Neumann. Biomonitoring of aromatic amines and alkylating agents by measuring Hb adducts. *Int. Arch. Occup. Environ. Health* 60:151, 1988.

Ch. Neumeister. Analysis of urine to monitor exposure to benzidine, o-dianisidine, o-toluidine, and 4,4'-methylene dianiline. *Appl. Occup. Environ. Hyg.* 6:953, 1991.

NIOSH. Technical Report. Carcinogenicity and metabolism of azo-dyes, especially those derived from benzidine. National Institute for Occupational Safety and Health, NIOSH Publication No. 80-119, U.S. Government Printing Office, Washington, D.C., 1980.

C. Nony and M. Bowman. Carcinogens and analogs: trace analysis of thirteen compounds in admixture in wastewater and human urine. *Int. J. Environ. Anal. Chem.* 5:203, 1978.

J. Piotrowski. Further investigations on the evaluation of exposure to nitrobenzene. *Br. J. Ind. Med.* 24:60, 1967.

D. Rickert, B. Butterworth and J. Popp. Dinitrotoluene: acute toxicity, oncogenicity, genotoxicity and metabolism. *CRC Crit. Rev. Toxicol.* 13:217, 1984.

D. Rickert. Metabolism of nitroaromatic compounds. *Drug Metab. Rev.* 18:23, 1987.

G. Sabbioni, H. Neumann. Biomonitoring of arylamines: hemoglobin adducts of urea and carbamate pesticides. *Carcinogenesis* 11:111, 1991.

G. Scansetti, E. Byglione, M. Massiccio et al. Excretion kinetics of the rubber anti-oxidant n-isopropyl-n'-phenyl-p-phenylenediamine. *Int. Arch. Occup. Environ. Health* 59:537, 1987.

P. Skipper, M. Bryant, S. Tannenbaum and J. Groopman. Analytical methods for assessing exposure to 4-aminobiphenyl based on protein adduct formation. *J. Occup. Med.* 28:643, 1986.

A. Smith, C. Jefferies and M. Hall. Urine monitoring of an aromatic amine in the work-place. In: *Occupational Health in the Chemical Industry*. XVII Medichem Congress, Cracow, Poland 26-29/9 1989. WHO, Regional Office for Europe, Copenhagen, p. 181, 1989.

B. Stählbom, T. Lundh, I. Floren and B. Åkesson. Visual disturbances in man as a result of experimental and occupational exposure to dimethylethyl amine. *Br. J. Ind. Med.* 48:26, 1991a.

B. Stählbom, T. Lundh and B. Åkesson. Experimental study on the metabolism of dimethylethylamine in man. *Int. Arch. Occup. Environ. Health* 63:305, 1991b.

T. Tabuchi, I. Hara and M. Minami. Evaluations of diazo-positive metabolite in urine as index of aromatic nitro and amino compounds exposure. *Proc. Osaka Pref. Public Health Ind. Health* 21:27, 1983.

J. Thomas and H. Wilson. Biological monitoring of workers exposed to 4,4'-methylenebis (2-chloroaniline) MBOCA. *Br. J. Ind. Med.* 41:547, 1984.

M. Turner, R. Levine, D. Nystrom et al. Identification and quantification of urinary metabolites of dinitrotoluenes in occupationally exposed humans. *Toxicol. Appl. Pharmacol.* 80:166, 1985.

E. Ward, D. Clapp, W. Tolos and D. Groth. Efficacity of urinary monitoring for 4-4'-methylenebis (2-chloroaniline). *J. Occup. Med.* 28:637, 1986.

D. Warren and D. Selchan. An industrial hygiene appraisal of triethylamine and dimethylamine exposure limits in the foundry industry. *Ind. Hyg. Assoc. J.* 49:630, 1988.

A. Williams, W. Murray and B. Gibb. Determination of traces of ethyleneglycol dinitrate (and nitroglycerine) in blood and urine. *Nature* 210:816, 1966.

B. Woollen, M. Hall, R. Craig ang G. Steel. Trinitrotoluene: assessment of occupational absorption during manufacture of explosives. *Br. J. Ind. Med.* 43:465, 1986.

B. Woollen, M. Hall, R. Craig and G. Steel. Dinitrotoluene: an assessment of occupational absorption during the manufacture of blasting explosives. *Int. Arch. Occup. Environ. Health* 55:319, 1985.

M. Yoshida, M. Sunaga and I. Hara. Urinary diazo-positive metabolites levels of workers handling p-nitrochlorobenzene in a dye producing factory. *Ind. Health* 26:87, 1988.

WHO. World Health Organization. Early detection of occupational diseases. WHO, Geneva, 1986.

5. ALCOHOLS
5.1. Methanol

Toxicokinetics

Exposure to methanol in industry occurs mainly by inhalation of vapor. The retention is independent from the lung ventilation and averages 58% (Sedivec et al 1981) of the amount inhaled. Because of the volatility of methanol, severe dermal intoxications have only been documented in cases where the exposed skin has been occluded (Grandjean 1990). When exposure is stable, the excretion reaches its maximum at the end of the exposure and then decreases exponentially with a biological half-life of about 1.5 to 2.0 h. Hence, about 12 h following termination of exposure the urinary concentration returns to normal level (Sedivec et al 1981).

Methanol is a natural constituent of blood, urine, and expired air. It is probably formed by the activities of the intestinal microflora or by other enzymatic processes (Sedivec et al 1981).

Biological monitoring

Measurement of methanol itself or its metabolites, formaldehyde and formic acid, in blood and in urine have been considered for the assessment of occupational exposure to methanol. It is clearly established that the determination of unchanged methanol in urine represents the best biological monitoring method (Leaf and Zatman 1952; Ferry et al 1980; Sedivec et al 1981).

Urine analyses

Methanol. According to Sedivec et al (1981), the mean normal level of methanol in urine of control subjects (n: 31) is 0.73 mg/l, with extreme values ranging from 0.32 to 2.61 mg/l.

When exposure is more or less constant during the workshift, the concentration of methanol in end-of-shift urine samples correlates with the intensity of exposure. According to Sedivec et al (1981), after an 8-h exposure to 200 ppm (260 mg/m³) methanol, the concentration of methanol in urine collected at the end of the exposure period amounts approximately to 7 mg/l (Sedivec et al 1981). A much higher estimate (40 mg/l) has been reported by Heinrich and Angerer (1982) and Kawai et al (1991). The study of Kawai et al (1991) also demonstrates a quantitative relationship between the intensity of exposure to methanol and methanol concentration in urine samples obtained at the end of the shift from methanol vapor-exposed workers. But, the calculated regression line indicates that the methanol concentration in urine samples collected at the end of the shift following 8 h exposure to 200 ppm methanol would be 42 mg/l (95% confidence range: 28-60 mg/l). The arithmetic and geometric mean concentrations found in the urine of nonexposed subjects were 1.90 mg/l and 1.73 mg/l.

In workers exposed to about 120 ppm methanol, Ogata and Iwamoto (1990) found, in the end-of-shift urine sample, a mean concentration of 35.8 mg methanol/l (SD: 41.5) (1.12 mmol/l ± 1.30). The mean value observed in nonexposed subjects was 1.34 mg/g creatinine (SD: 1.66).

When methanol concentration in air changes markedly during the shift, it is preferable to measure methanol concentration in whole-shift urine samples for estimating the methanol dose absorbed by the organism (Sedivec et al 1981). The biological limit value recommended by the ACGIH (1991-1992) and the DFG (1991) are 15 mg/l and 30 mg/l, respectively (end-of-shift sample), corresponding to a TWA of 200 ppm (260 mg/m³).

Formic acid. Formic acid in urine has been considered as a possible parameter for estimating methanol exposure (Baumann and Angerer 1979; Angerer and Lehnert 1977). Heinzow and Ellrott (1992) determined in healthy occupationally unexposed adults (20-80 years) a mean urinary formic acid concentration of 21 mg/l (SD:30) (95 percentile: 60 mg/l). Smoking and dietary habits had no influence on formic acid excretion but age was positively correlated with increased concentrations. An oral methanol intake of 10 mg/kg body weight had no significant impact on urine excretion of formic acid. Baumann and Angerer (1979) have studied workers from a printing shop exposed to about 100 ppm methanol. The corresponding concentrations in urine were 13.1 and 20.2 mg/l, respectively. According to the field study of Liesivuori and Savolainen (1987), an exposure to an average methanol vapor concentration of 200 ppm produces a mean urinary excretion of 80 mg formic acid/g creatinine, 16 h after the end of shift. The urinary formic acid concentration (mean ± SD) in the morning samples taken from control subjects was 15.1 ± 6.1 mg/g creatinine. No correlations were found between methanol exposure and urinary formic acid or methanol concentrations in samples taken immediately after the workshift.

Ogata and Iwamoto (1990) have recorded a geometric mean concentration of urinary formic acid in healthy subjects of 7.82 mg/g creatinine. The end-of-shift mean concentration found in workers exposed to about 120 ppm methanol was 123.3 mg formic acid/l (SD: 168.1) (2.68 mmol/l ± 3.64).

The ACGIH (1991-1992) recommends 80 mg/g creatinine before the shift, at the end of the workweek as biological exposure indice.

Blood analysis

Formic acid. Formic acid in blood has also been considered as a possible parameter for estimating methanol exposure. The concentration of formic acid in blood of workers exposed to about 100 ppm methanol increased on the average from 0.32 mg/100 ml before to 0.79 mg/100 ml after the shift (Baumann and Angerer 1979).

In view of the high background level of formic acid in blood and in urine, this analysis does not offer any advantage over methanol analysis. Urinary concentrations of methanol rarely exceed 3 mg/l in subjects nonoccupationally exposed to methanol. It seems that an 8-hour exposure to a mean air concentration of 200 ppm would lead at the end of the shift to a mean urinary concentration of around 35-40 mg/l. However, a urinary concentration of 25 mg/g creatinine cannot totally exclude an exposure to this level.

5.2. Ethanol

An experimental study on one volunteer suggests that exposure to ethanol vapor at 1900 mg/m^3 (1000 ppm) does not produce detectable amounts of ethanol in venous blood, even after 3 h of exposure. The ethanol content in all the blood samples remained consistently below the detection limit of 2 mg/l (Campbell and Wilson 1986).

5.3. *n*-Butyl Alcohol

Astrand et al (1976) have exposed volunteers to *n*-butyl alcohol (300 and 600 mg/m^3 or 100 and 200 ppm) for 2 hours at rest or during exercise. The concentration of butyl alcohol in venous blood never exceeded 0.1 mg/100 ml. When the exposure was decreased to 150 mg/m^3 (50 ppm), no butyl alcohol could be detected in the blood with their assay technique (limit of detection, 0.008 mg/100 ml).

214 INDUSTRIAL CHEMICAL EXPOSURE

5.4. Isopropanol

Toxicokinetics

2-Propanol is rapidly absorbed and distributed throughout the body after inhalation or ingestion. The main metabolite is acetone which has been detected in alveolar air, blood and urine (Daniel et al 1981; Kelner and Bailey 1983; Brugnone et al 1983; Martinez et al 1986; Kawai et al 1990).

Isopropanol half-lives of 2.5, 3 and 6.4 h have been determined in the blood of two alcoholics and one nonalcoholic, respectively, after ingestion of rubbing alcohol. In the first two subjects the acetone blood level declined slowly over the next 30 h and in the nonalcoholic subject the acetone blood half-life was calculated to be 22 h (Daniel et al 1981; Natowicz et al 1985).

Biological monitoring

Breath analysis

Isopropanol. Brugnone et al (1983, 1985) have analyzed the alveolar air, blood, and urine of 12 printing workers exposed to 2-propanol at air concentrations ranging between 7 and 645 mg/m³. The alveolar 2-propanol concentration was highly correlated with the level of exposure at any time of exposure (ratio: 0.418). Isopropanol was not detected in the blood or urine. Before the workshift the mean acetone concentration was 3.7 mg/m³ (SD: 3.8) in alveolar air, 0.14 mg/100 ml (SD: 0.05) in blood, and 3.9 mg/l (SD: 7.6) in urine. During the shift, acetone ranged between 3 and 93 mg/m³ in alveolar air, 0.076 and 1.56 mg/100 ml in blood, 0.9 and 18.2 mg/l in urine. The concentration of acetone in urine collected from the end of the work to the next morning ranged between 0.85 and 53.7 mg/l. Alveolar and blood acetone concentrations were highly correlated with alveolar isopropanol concentrations at any time during exposure.

According to their results, a TWA exposure to 400 mg/m³ (164 ppm) and 980 mg/m³ (400 ppm) corresponds to isopropanol alveolar concentrations of about 200 and 515 mg/m³, respectively.

Urine analysis

Acetone. Kawai et al (1990) have also studied the urinary excretion of isopropanol and acetone in workers exposed to isopropanol. Isopropanol itself was not found in control subjects and was detectable only in the urine of those exposed to isopropanol above 5 ppm. Acetone was detected in the urine of most of the control subjects (up to 1.6 mg/l) and increased in proportion to the alcohol exposure intensity. The authors claim that urinary acetone is a valuable index for biological monitoring of occupational exposure to isopropanol as low as 70 ppm. This exposure level would lead to a urinary acetone excretion of 20.3 mg/l (at the end of the shift) and a TWA exposure of 400 ppm to 109 mg acetone/l urine.

It should be noted, however, that following exposure to acetone itself, a biological limit value of 30 mg/g creatinine has been suggested (see below). The DFG (1991) recommends 50 mg acetone/l blood and 50 mg acetone/l urine (end-of-shift samples) as biological tolerance values for isopropyl alcohol corresponding to a TWA of 400 ppm.

5.5. Furfuryl Alcohol

Its principal urinary metabolite is furoylglycine measured as furoic acid after alkaline hydrolysis. Its determination might serve as exposure index (Pfäffli et al 1985). Furoic acid is also a metabolite of furfural (see below).

REFERENCES

ACGIH. American Conference of Governmental Industrial Hygienists. Threshold Limit Values for Chemical Substances and Physical Agents and Biological Exposure Indices. Cincinnati, 1991-1992.

J. Angerer and G. Lehnert. Occupational exposure to methanol. *Acta Pharmacol. Toxicol.* 41:551, 1977.

I. Astrand, P. Ovrum, T. Lindqvist and M. Hultengren. Exposure to butyl alcohol. Uptake and distribution in man. *Scand. J. Work Environ. Health* 3:165, 1976.

K. Baumann and J. Angerer. Occupational chronic exposure to organic solvents. IV. Formic acid concentration in blood and urine as an indicator of methanol exposure. *Int. Arch. Occup. Environ. Health* 42:241, 1979.

F. Brugnone, L. Perbellini, P. Apostoli et al. Isopropanol exposure: environmental and biological monitoring in a printing works. *Br. J. Ind. Med.* 40:160, 1983.

F. Brugnone and L. Perbellini. Biological monitoring of occupational exposure to solvents by analysis of alveolar air and blood. Working Group on Chronic Effects of Organic Solvents on Central Nervous System and Diagnostic Criteria. WHO, Copenhagen, June 1985.

L. Campbell and H. Wilson. Blood alcohol concentrations following the inhalation of ethanol vapour under controlled conditions. *J. Foren. Sci. Soc.* 26:129, 1986.

D. Daniel, B. McNalley and J. Garriott. Isopropyl alcohol metabolism after acute intoxication in humans. *J. Anal. Toxicol.* 5:110, 1981.

DFG. Deutsche Forschungsgemeinschaft. *Maximum Concentrations at the Workplace and Biological Tolerance Values for Working Materials.* Report No. XXVII. Commission for the Investigation of Health Hazards of Chemical Compounds in the Work Area. VCH, Weinheim, 1991.

D. Ferry, W. Temple and E. McQueen. Methanol monitoring. Comparison of urinary methanol concentration with formic acid excretion rate as a measure of occupational exposure. *Int. Arch. Occup. Environ. Health* 47:155, 1980.

Ph. Grandjean. Skin penetration: hazardous chemicals at work. A report prepared for the Commission of the European Communities, Luxembourg. Taylor and Francis, London, Eur 12599 En, 1990.

R. Heinrich and J. Angerer. Occupational chronic exposure to organic solvents. X. Biologi-
    cal monitoring parameters for methanol exposure. *Int. Arch. Occup. Environ. Health*
    50:131, 1982.
B. Heinzow and T. Ellrott. Formic acid in urine. A useful indicator in environmental
    medicine. *Zentralbl. Hyg. Umweltmed.* 192:455, 1992.
T. Kawai, T. Yasugi, S. Horiguchi et al. Biological monitoring of occupational exposure
    to isopropyl alcohol vapor by urinalysis for acetone. *Int. Arch. Occup. Environ. Health*
    62:409, 1990.
T. Kawai, T. Yasugi, K. Mizunuma et al. Methanol in urine as a biological indicator of
    occupational exposure to methanol vapour. *Int. Arch. Occup. Environ. Health* 63:311,
    1991.
M. Kelner and D. Bailey. Isopropanol ingestion: interpretation of blood concentrations and
    clinical findings. *J. Toxicol. Clin. Toxicol.* 20:497, 1983.
G. Leaf and L. Zatman. A study of the conditions under which methanol may exert a toxic
    hazard in industry. *Br. J. Ind. Med.* 9:19, 1952.
J. Liesivuori and H. Savolainen. Urinary formic acid as an indicator of occupational
    exposure to formic acid and methanol. *Am. Ind. Hyg. Assoc. J.* 48:32, 1987.
T. Martinez, R. Jaeger, F. de Castro et al. A comparison of the absorption and metabolism
    of isopropyl alcohol by oral, dermal and inhalation routes. *Vet. Hum. Toxicol.* 28:233,
    1986.
M. Natowicz, J. Donahve, L. Gorman et al. Pharmacokinetic analysis of a case of isopropanol
    intoxication. *Clin. Med.* 31:326, 1985.
M. Ogata and T. Iwamoto. Enzymatic assay of formic acid and gas chromatography of
    methanol for urinary biological monitoring to methanol. *Int. Arch. Occup. Environ.
    Health* 62:227, 1990.
P. Pfäffli, A. Tossovainen and H. Savolainen. Comparison of inhaled furfuryl alcohol
    vapour with urinary furoic acid excretion in exposed foundry workers by chromato-
    graphic techniques. *Analyst* 110:337, 1985.
V. Sedivec, M. Mzaz and J. Flek. Biological monitoring of persons exposed to methanol
    vapours. *Int. Arch. Occup. Environ. Health* 48:257, 1981.

6. GLYCOLS AND DERIVATIVES
6.1. Ethylene Glycol

Toxicokinetics

Ethylene glycol has a low volatility. The risk of excessive exposure by inhalation
is therefore moderate unless it is heated or aerosolized. It gives rise to various
metabolites *in vivo*: glycol aldehyde, glycolic acid, glyoxylic acid, oxalic acid,
formic acid, and carbon monoxide (Figure 20) (Clay and Murphy 1977; Wiener
and Richardson 1988).

Biological monitoring

The determination of oxalic acid in urine and ethylene glycol itself in serum can
be used for monitoring exposure, but no biological threshold limit values have yet
been proposed. Nonexposed subjects usually excrete less than 50 mg oxalic acid/
g of urinary creatinine.

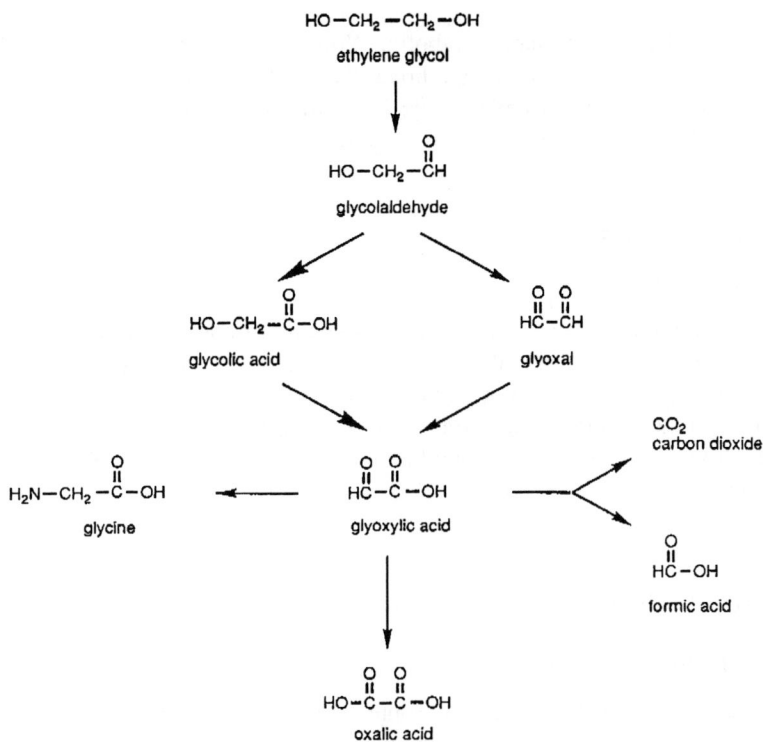

FIGURE 20 Main metabolic pathways of ethylene glycol.

According to Hewlett and McMartin (1986), glycolic acid determination in serum and urine might be of interest for diagnosis and evaluation of patients poisoned by ethylene glycol.

6.2. Diethylene Glycol

Diethylene glycol gives rise *in vivo* to the same metabolites as ethylene glycol. In view of its low volatility, the risk of inhalation is limited. In rats, diethylene glycol is progressively oxidized by the alcohol and aldehyde dehydrogenase into 2-hydroxyethoxyacetaldehyde and 2-hydroxyethoxyacetic acid. The ether bond seems relatively stable (Wiener and Richardson 1989).

6.3. Dioxane

Toxicokinetics

Exposure to dioxane results from inhalation of vapor and probably also through skin contact. In man exposed to dioxane, β-hydroxyethoxyacetic acid was found to represent the main urinary metabolite (Young et al 1976). Young et al (1977) exposed four volunteers to 50 ppm dioxane vapor for 6 h at rest. The half-life for elimination of dioxane was 59 min; 99.3% of the elimination was by biotransformation of dioxane to β-hydroxyethoxyacetic acid, and 0.7% was by excretion of dioxane in the urine.

Biological monitoring

Urine analysis

β-*Hydroxyethoxyacetic acid.* In volunteers exposed to 50 ppm dioxane during 6 h, the amount of β-hydroxyethoxyacetic acid excreted in urine within 24 h after the start of exposure was 620 mg (Young et al 1977). In workers exposed for 7.5 h to an average airborne concentration of dioxane of 1.6 ppm, the concentration of β-hydroxyethoxyacetic acid found in urine collected at the end of the shift was 36.5 mg/l (Young et al 1976).

6.4. Glycol Ethers

Toxicokinetics

In the occupational setting, the main routes of absorption of these compounds are the respiratory tract and the skin. The pulmonary retention rate seems to increase when the molecular weight decreases and the pulmonary uptake increases with physical exercise (Groeseneken et al 1986b,c, 1987a,b; Johanson 1986; Johanson et al 1986).

Skin absorption results not only from direct skin contact with the liquid form but also from exposure to vapors; therefore it is not surprising that it represents an important part, sometimes a major part of the total uptake. There is a trend of reduced dermal absorption rates with increasing molecular weight or reduced volatility within the monoethylene glycol ethers series and the diethylene glycol compounds. The diethylene glycol liquids, however, are less rapidly absorbed through human epidermis than the corresponding monoethylene glycol members (Dugard et al 1984).

The acetate derivatives are probably first converted to the corresponding ethers and then follow the same metabolic pathway as the latter (Stott and McKenna 1985; Groeseneken et al 1987b). The main biotransformation of the ethylene glycol ethers is the oxidation of the hydroxy group with production of the corresponding alkoxyacetic acids which are excreted in urine. Other metabolites have been identified in laboratory animals, but not yet in humans. It seems that there is an inverse relationship between the elimination half-life of the alkoxyacetic acids and the length of the alkyl group in the glycol ethers. Propylene glycol ethers seem to undergo another metabolic pathway yielding via *o*-dealkylation propylene glycol, CO_2, sulfate derivatives, and glucuronide conjugates.

6.4.1. Ethylene Glycol Monomethyl Ether (Methylcellosolve, 2-Methoxyethanol)

Toxicokinetics

Ethylene glycol monomethyl ether (EGMME) is a volatile liquid that may enter the organism by inhalation of vapor, ingestion, or through skin contact (Ohi and Wegman 1978; Cohen 1984; Grandjean 1990). It may be detected in expired air during or shortly after exposure. Methoxyacetic acid (MAA) is the main urinary metabolite, but the relationship between its concentration and the intensity of exposure has not been sufficiently investigated (Figure 21).

Groeseneken et al (1989a) studied the uptake of EGMME and the urinary excretion of MAA in volunteers during exposure (4 h) to EGMME vapor at rest. A high pulmonary retention rate was noted (76%). MAA was detected in urine up to 120 h after the start of exposure: the elimination half-life was on average 77.1 h. The total amount of MAA excreted was calculated to average 85.5% of the retained EGMME.

According to the National Institute for Occupational Safety and Health (NIOSH 1990), in the absence of skin exposure, a urinary MAA of 0.8 mg/g creatinine approximates the concentration that would result from exposure to 0.1 ppm EGME during an 8-h workshift (sample taken at the beginning of the week).

6.4.2. Ethylene Glycol Monoethyl Ether

Toxicokinetics

Ethylene glycol monoethyl ether (EGMEE) may enter the organism by inhalation, ingestion, or through skin contact (liquid, vapor). This last route of absorption is of great concern in occupational settings.

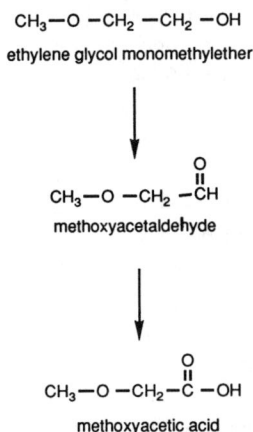

$$CH_3-O-CH_2-CH_2-OH$$

ethylene glycol monomethylether

$$CH_3-O-CH_2-\overset{\overset{O}{\|}}{C}H$$

methoxyacetaldehyde

$$CH_3-O-CH_2-\overset{\overset{O}{\|}}{C}-OH$$

methoxyacetic acid

FIGURE 21 Main metabolic pathway of ethylene glycol monomethylether.

Groeseneken et al (1986b,c, 1987a,b) carried out short-term inhalation studies on volunteers with EGMEE and EGMEE acetate. The time course of ethoxyacetic acid (EAA) excretion during and after exposure to the acetate derivative was similar to that found after ethylene glycol monoethylether exposure. The maximum excretion rate of EAA was reached 3 to 4 h after the end of the 4-hour exposure. Afterwards the excretion declined with a biological half-life of 21-24 h. On the average, 23% of the absorbed compound was recovered as EAA within 42 h. EAA excretion increased with the intensity of exposure and during physical exercise.

Respiratory elimination of the unmetabolized compound is only a minor route of excretion; the total amount of the unchanged EGMEE or EGMEE acetate eliminated through the lungs accounted for less than 0.5% of the total body uptake.

The long biological half-life of EAA and hence the risk of accumulation of the metabolite through repetitive exposure was confirmed by a field study. Veulemans et al (1987) showed that urinary EAA excretion clearly increased during the workweek, the elimination was far from complete over the weekend, and traces were still detectable 12 d after exposure. The half-life might even be longer than 24 h and the maximal urinary concentration might not be reached after a single week of exposure.

Biological monitoring

Urine analysis

Ethoxyacetic acid. Veulemans et al (1987) obtained a good linear correlation between the average exposure over 5 days (14.4 mg/m^3) and the EAA excretion at the end of the week (105.7 mg/g creatinine). An EAA estimate of 150 ± 35 mg/g creatinine (mean $\pm$ SD) was found to correspond with repeated 5 d full-shift exposure to 19 mg/m^3 EGMEE or 27 mg/m^3 EGMEE acetate. The same exposure level would lead to a concentration of 32 mg/g creatinine, 16 h after the end of exposure (INRS, France, Internal Report 1992).

In workers exposed to average concentrations of ethoxyethanol, ethoxyethyl-acetate, butoxyethanol, 1-methoxy-2-propanol, 2-methoxypropyl-1-acetate, and xylene of 2.8, 2.7, 1.1, 7, 2.8, and 1.7 ppm, Angerer et al (1990) found mean ethoxyacetic levels in pre- and postshift urine samples of 128.5 mg/l (4.6-423.2) and 167.8 mg/l (49.9-497). However, no significant correlation between the airborne concentration of the glycolethers and their concentration in blood or their metabolites concentration in urine was observed. Comparing their results with those reported in the literature, Angerer et al (1990) have suggested that a tolerable limit value for the concentration of EAA in urine might be in the order of 100 to 200 mg/l (urine collected at the end of the workweek).

According to the National Institute for Occupational Safety and Health (NIOSH 1990a), in the absence of skin exposure, a urinary EAA excretion of 5 mg/g creatinine approximates the concentration that would result from an exposure to 0.5 ppm EGEE during an 8-h workshift (sample taken before the shift, last day of the week).

6.4.3. Ethylene Glycol Monobutyl Ether (2-Butoxyethanol)

Toxicokinetics

Ethylene glycol monobutyl ether (EGMBE) is readily absorbed through the lung and the skin. The studies of Johanson and co-workers (1988, 1991) provide evidence of the important contribution of dermal absorption to the total uptake of 2-butoxyethanol vapor. Their results indicate that percutaneous absorption of 2-butoxyethanol (liquid or vapor) may cause toxic effects in man. According to Bartnik et al (1987), because of the ability of butoxyethanol to readily penetrate the skin the potential for toxic uptake is more likely due to excessive skin exposure than to vapor inhalation. However, as stressed by the authors, evaporative loss from the skin may considerably reduce the amount of substance available for skin penetration. Butoxyacetic acid (BAA) is the main urinary metabolite. In a case of acute intoxication by ingestion, Rambourg-Schepens et al (1987) found not only butoxyacetic acid in urine, but also a small amount of oxalic acid.

Simulated inhalation exposure to EGMBE agreed with results from human experimental exposure under identical conditions (Johanson 1986; Johanson et al 1986). Blood concentration of EGMBE increased with increasing workload and with coexposure to ethanol. The concentration in blood rose rapidly, reaching an apparent plateau level within 1-2 h. It could no longer be detected in blood 2-4 h after the end of the exposure. The amount of EGMBE excreted in urine was less than 0.03% of the total uptake while that of butoxyacetic acid ranged from 17 to 55%, the excretion rate increasing for 5 h and then declining with a half-life of 4 h.

Biological monitoring

In workers exposed to average concentrations of butoxyethanol, ethoxyethanol, ethoxyethyl acetate, 1-methoxy-2-propanol, 2-methoxypropyl acetate, and xylene of 1.1, 2.8, 2.7, 7, 2.8, and 1.7 ppm, Angerer et al (1990) found a mean butoxyethanol postshift blood concentration of 121.3 µg/l (<5-570). The mean pre- and postshift butoxyacetic acid values in urine were 3.3 mg/l (0.1-14.3) and 10.5 (0.6-30.3) mg/l, respectively.

According to the National Institute for Occupational Safety and Health (NIOSH 1990b), in the absence of skin exposure, a urinary BAA concentration of 60 mg/g creatinine approximates the concentration that would result from exposure to 5 ppm EGMBE or EGMBEA during an 8-h workshift (sample taken at the end of the shift).

6.4.4. Propylene Glycol Monomethyl Ether (1-Methoxy-2-Propanol)

Toxicokinetics

The monomethyl ether of propylene glycol is a volatile liquid. It can probably give rise to methanol *in vivo*. In contrast to the ethylene glycol ethers, the 1-methoxy-

2-propanol (α isomere of propylene glycol monomethyl-ether) cannot be oxidized to a corresponding acid and is mainly metabolized into propylene glycol.

It should be noted, however, that animal experiment has shown that 2-methoxy-1-propanol (β isomers of 2-propylene glycol monomethyl ether) give rise to methoxypropionic acid (Miller et al 1986).

Biological monitoring

In volunteers exposed to 240 ppm of propylene glycol monomethylether for 0.25 to 3 h, Stewart et al (1970) found on the average 4 ppm of the compound in expired air 15 s after the end of exposure.

As it is not as easily subject to metabolic degradation, the secondary alcohol, 1-methoxy-2-propanol, might be a candidate for monitoring in blood or urine samples (Johanson 1988).

Since glycol ethers can be absorbed by all the routes, biological monitoring should represent the best approach for assessing the risk of overexposure. The best approach for the biological monitoring of exposure to ethylene glycol ethers is the measure of the corresponding alkoxyacetic acids in urine (Groeseneken et al 1986a,b,c, 1987a,b, 1989a,b; Johanson 1988; Clapp et al 1984, 1987; Smallwood et al 1984, 1988; Veulemans et al 1987; Angerer et al 1990; Piacitelli et al 1990), i.e.: MAA (exposure to ethylene glycol monomethyl ether (acetate)), ethoxycetic acid (exposure to ethylene glycol monoethyl ether (acetate)), butoxyacetic acid (exposure to ethylene glycol monobutyl ether (acetate)).

The data are, at present, still too limited to propose BEI. For EAA a tentative maximum permissible concentration of 150 mg/g creatinine can be proposed, corresponding to a TWA of 5 ppm.

REFERENCES

J. Angerer, E. Lichterbeck, J. Bergerow et al. Occupational chronic exposure to organic solvents. XIII. Glycolether exposure during the production of varnishes. *Int. Arch. Occup. Environ. Health* 62:123, 1990.

F. Bartnik, A. Reddy, G. Klecak et al. Percutaneous absorption, metabolism and hemolytic activity of n-butoxyethanol. *Fundam. Appl. Toxicol.* 8:59, 1987.

D. Clapp, D. Zaebst and R. Herrick. Measuring exposures to glycol ethers. *Environ. Health Perspect.* 57:91, 1984.

D. Clapp, A. Smallwood, C. Moseley and K. Debord. Workplace assessment of exposure to 2-ethoxyethanol. *Appl. Ind. Hyg.* 2:1983, 1987.

K. Clay and R. Murphy. On the metabolic acidosis of ethylene glycol intoxication. *Toxicol. Appl. Pharmacol.* 39:39, 1977.

R. Cohen. Reversible subacute ethylene glycol monomethyl ether toxicity associated with microfilm production, a case report. *Am. J. Ind. Med.* 6:441, 1984.

P. Dugard, M. Walker, S. Mawdsley and R. Scott. Absorption of some glycol ethers through human skin in vitro. *Environ. Health Perspect.* 57:193, 1984.

Ph. Grandjean. Skin penetration: hazardous chemicals at work. A report prepared for the Commission of the European Communities, Luxembourg. Taylor and Francis, London, Eur 12599 En, 1990.

D. Groeseneken, E. Van Vlem, H. Veulemans and R. Masschelein. Gas chromatographic determination of methoxyacetic and ethoxyacetic acid in urine. Br. J. Ind. Med. 43:62, 1986a.

D. Groeseneken, H. Veulemans and R. Masschelein. Respiratory uptake and elimination of ethylene glycol monoethyl ether after experimental human exposure. Br. J. Ind. Med. 43:544, 1986b.

D. Groeseneken, H. Veulemans and R. Masschelein. Urinary excretion of ethoxyacetic acid after experimental human exposure to ethylene glycol monomethyl ether. Br. J. Ind. Med. 13:615, 1986c.

D. Groeseneken, H. Veulemans, R. Masschelein and E. Van Vlem. Pulmonary absorption and elimination of ethylene glycol monoethyl ether acetate in man. Br. J. Ind. Med. 44:309, 1987a.

D. Groeseneken, H. Veulemans, R. Masschelein and E. Van Vlem. Ethoxyacetic acid: a metabolite of ethylene glycol monoethyl ether acetate in man. Br. J. Ind. Med. 44:488, 1987b.

D. Groeseneken, H. Veulemans, R. Masschelein and E. Van Vlem. Experimental human exposure to ethylene glycol monomethyl ether. Int. Arch. Occup. Environ. Health 61:243, 1989a.

D. Groeseneken, H. Veulemans, R. Masschelein and E. Van Vlem. An improved method for the determination in urine of alkoxyacetic acids. Int. Arch. Occup. Environ. Health 61:249, 1989b.

T. Hewlett and K. McMartin. Ethylene glycol poisoning. The value of glycolic acid determinations for diagnosis and treatment. Clin. Toxicol. 24:389, 1986.

G. Johanson. Physiologically based pharmacokinetic modeling of inhaled 2-butoxyethanol in man. Toxicol. Lett. 34:23, 1986.

G. Johanson, H. Kronborg, Ph. Näslund and M. Nordqvist. Toxicokinetics of inhaled 2-butoxyethanol (ethylene glycol monobutyl ether) in man. Scand. J. Work Environ. Health 12:594, 1986.

G. Johanson, A. Boman and B. Dynesius. Percutaenous absorption of 2-butoxyethanol in man. Scand. J. Work Environ. Health 4:101, 1988.

G. Johanson. Aspects of biological monitoring of exposure to glycol ethers. Toxicol. Lett. 43:5, 1988.

G. Johanson and A. Boman. Percutaneous absorption of 2-butoxyethanol vapour in human subjects. Br. J. Ind. Med. 48:788, 1991.

R. Miller, P. Langvardt, L. Calhoun and M. Yahrmarkt. Metabolism and disposition of propylene glycol monomethyl ether (PGME) beta isomer in male rats. Toxicol. Appl. Pharmacol. 83:170, 1986.

NIOSH. The National Institute for Occupational Safety and Health. Criteria for a recommended standard. Occupational Exposure to Ethylene Glycol Monomethyl Ether, Ethylene Glycol Monoethyl Ether and Their Acetates. U.S. Department of Health and Human Services. Public Health Service. Center for Disease Control, Washington, D.C., 1990a.

NIOSH. The National Institute for Occupational Safety and Health. Criteria for a recommended standard. Occupational Exposure to Ethylene Glycol Monobutyl Ether and Ethylene Glycol Monobutyl Ether Acetate. U.S. Department of Health and Human Services. Public Health Service. Center for Disease Control, Washington, D.C., 1990b.

G. Ohi and D. Wegman. Transcutaneous ethyleneglycol monomethyl ether poisoning in the work setting. *J. Occup. Med.* 20:675, 1978.

G. Piacitelli, D. Votaw and E. Radha Krishnan. An exposure assessment of industries using ethylene glycol ethers. *Appl. Occup. Environ. Hyg.* 5:107, 1990.

M. Rambourg-Schepens, M. Buffet, R. Bertault et al. Aspects métaboliques de l'intoxication aigüe par l'ingestion de butylglycol. *Arch. Mal. Prof.* 48:121, 1987.

A. Smallwood, K. Bebord and L. Lowry. Analyses of ethylene glycol monoalkyl ethers and their proposed metabolites in blood and urine. *Environ. Health Perspect.* 57:249, 1984.

A. Smallwood, K. Debord, J. Burg et al. Determination of urinary 2-ethoxyacetic acid as an indicator of occupational exposure to 2-ethoxyethanol. *Appl. Ind. Hyg.* 3:47, 1988.

R. Stewart, E. Baretta, H. Dodd and T. Torkelson. Experimental human exposure to vapor of propylene glycol monomethyl ether. *Arch. Environ. Health* 20:218, 1970.

W. Stott and M. McKenna. Hydrolysis of several glycol ether acetates and acrylate esters by nasal mucosal carboxyl esterase in vitro. *Fundam. Appl. Toxicol.* 5:399, 1985.

A. Veulemans, D. Groeseneken, R. Masschelein and E. Van Vlem. Field study of the urinary excretion of ethoxyacetic acid during repeated daily exposure to the ethyl ether of ethylene glycol and the ethyl ether of ethylene glycol acetate. *Scand. J. Work Environ. Health* 13:239, 1987.

H. Wiener and K. Richardson. The metabolism and toxicity of ethylene glycol. *Res. Commun. Substances Abuse* 9:77, 1988.

H. Wiener and K. Richardson. Metabolisme of diethylene glycol in male rats. *Biochem. Pharmacol.* 38:8539, 1989.

J. Young, W. Braun, P. Gehring et al. 1,4-dioxane and β-hydroxyethoxyacetic acid excretion in urine of humans exposed to dioxane vapors. *Toxicol. Appl. Pharmacol.* 38:643, 1976.

J. Young, W. Braun, L. Rampy et al. Pharmacokinetics of 1,4-dioxane in humans. *J. Toxicol. Environ. Health* 3:507, 1977.

7. KETONES

The ketones of industrial importance are usually volatile liquids that may enter the organism by inhalation of vapors or through direct skin contact with the liquid form. They are rapidly eliminated from the body either unchanged in urine and expired air or after biotransformation. The parent compound or its metabolites can also be detected in blood. Only a few ketones for which human data are available will be considered in this chapter.

7.1. Methyl Ethyl Ketone (Butanone)

Toxicokinetics

Methyl ethyl ketone (MEK) is a volatile liquid that is absorbed by all routes in man. Its metabolism in man has not yet been fully characterized. In workers exposed to methyl ethyl ketone, Perbellini et al (1984) have estimated the alveolar retention at around 70%. But, Liira et al (1988a,b) determined in volunteers a relative pulmonary uptake of about 41 to 53% throughout 4-hour exposure periods at 200 ppm MEK. The concentration of MEK in blood increased steadily until the end of the exposure: a slower increase taking place after a more rapid phase. Approximately 3% of the total uptake was excreted unchanged in expired air. Elimination of MEK from blood occurred in two phases with half times of 30 min and 81-95 min, respectively (Liira et al 1988a,b). According to Liira et al (1990a), the kinetics of MEK are dose-dependent.

Miyasaka et al (1982) have shown that only 0.1% of absorbed methyl ethyl ketone is excreted unchanged in urine. These authors have found that in occupationally exposed workers the urinary MEK level increased rapidly to reach a plateau in the first quarter of the workshift. In the field study conducted by Ong et al (1991a), the urinary MEK excretion rose steadily until the end of exposure, whereas the MEK concentration in exhaled breath varied markedly throughout the day. About 2-3% of the total uptake seemed to be excreted as 2,3-butanediol; however, great individual variations were observed (Liira et al 1988a,b).

Perbellini et al (1984) have identified the metabolite 3-hydroxy-2-butanone (acetyl methyl carbinol) in the end-of-shift urine of workers exposed to methyl ethyl ketone. It was unquantifiable in the preshift urine samples in nonexposed individuals. The urinary excretion of MEK and this metabolite together accounted for not more than 0.1% of the uptake. The concentration of the solvent in urine collected at the end of the exposure period is related to the intensity of exposure. During exposure, the ratio between alveolar and environmental concentration amounts on the average to 0.36 (Brugnone et al 1980).

Exercise increased the overall blood MEK level (Liira et al 1988a), but combined exposure to MEK (200ppm) and m-xylene (100 ppm) did not cause any change in the blood concentration of MEK or in the excretion of 2,3-butanediol (Liira et al 1988b). Ingestion of ethanol inhibited the primary oxydative metabolism of MEK and hence caused an increase in MEK blood concentration and in the pulmonary excretion of MEK and the urinary excretion of MEK and 2,3-butanediol (Liira et al 1990b). A study on volunteers has shown that following combined exposure to toluene and MEK (at 98.3 ± 0.6 and 50 ± 1 ppm, respectively), the results of breath analysis were not significantly different from those obtained after exposure to each substance separately (Tolos et al 1987).

Biological monitoring

According to Miyasaka et al (1982), a TWA exposure of 100 ppm methyl ethyl ketone leads to a mean postshift urinary concentration of the solvent of about 2.6 mg/l.

For Perbellini and co-workers (Perbellini et al 1984; Brugnone and Perbellini 1985) a TWA exposure to 200 ppm (590 mg/m³) MEK would lead to MEK concentrations of about 177 µg/l in alveolar air and 2 mg/l in urine and to a 3-hydroxy-2-butanone concentration in urine of 2.9 mg/l. This is in agreement with the result of Ghittori et al (1987), who estimated that a 4-h exposure to 200 ppm MEK corresponds to a urinary value of 2.2 mg/l (95% lower confidence limit: 2 mg/l).

In volunteers exposed by inhalation to an average (±SD) concentration of 189 ± 5.2 ppm MEK during 4 h, Tolos et al (1987) recorded MEK mean concentrations (±SD) of 3.7 mg/l (±1.6) in blood and 9.1 ppm (±1.3) in breath.

In field conditions, Ong et al (1991a) found a good correlation between MEK TWA and MEK concentrations in blood, breath, and urine (all samples taken at the end of the shift). After 8-h exposure to 200 ppm MEK, the corresponding end-of-shift breath and urinary concentrations were 9.7 ppm and 5.1 µmol/l (or 4.11 mg/g creatinine).

Discrepancies between the different studies can probably be explained by differences in the intensity of physical activities and the importance of skin contact.

The BEI recommended by the ACGIH (1991-1992) and the DFG (1991) for MEK in urine are 2 mg/l and 5 mg/l (urine collected at the end of the shift), corresponding to a TWA of 200 ppm (590 mg/m^3).

On the basis of the reported results we propose that in workers exposed to a TWA of 200 ppm MEK, its urinary concentration should not exceed 2.5 mg/g creatinine.

7.2. Methyl-*n*-Butyl Ketone (2-Hexanone)

Toxicokinetics

This volatile liquid is easily absorbed by all routes in man (Di Vincenzo et al 1978). Following an epidemic outbreak of peripheral neuropathy in workers exposed to methyl-*n*-butyl ketone (Allen et al 1975), the metabolism of this compound was investigated in humans and in various animal species. Its main *in vivo* metabolites are 2-hexanol, 5-hydroxy-2-hexanone, 2,5-hexanedione, and 2,5-hexanediol. In volunteers exposed to 50 and 100 ppm methyl-*n*-butyl ketone for 7 h, Di Vincenzo et al (1978) detected the presence of 2,5-hexanedione in serum, but neither methyl-*n*-butyl ketone nor its metabolites were detected in urine during or after exposure.

Biological monitoring

As recently stated by Bos et al (1991), data are still insufficient and more studies are necessary to determine whether or not the concentration of methyl-*n*-butyl ketone or one of its metabolites in biological samples may be useful for biological monitoring.

The DFG (1991) proposes the sum of 2,5-hexanedione and 4,5-dihydroxy-2-hexanone in the end-of-shift urine as biological exposure indice: this sum should not exceed 9 mg/l urine, corresponding to a TWA of 200 ppm.

Since 2,5-hexanedione is probably the main active metabolite of methyl-*n*-butylketone, the same biological limit value as proposed for *n*-hexane can be recommended i.e., 4 mg/g creatinine (end-of-week).

7.3. Acetone

Toxicokinetics

Acetone is a highly volatile solvent that enters the organism mainly through inhalation of vapor. It is also formed in the organism as a product of fat metabolism. Wigaeus Hjelm et al (1981) exposed volunteers to acetone vapor and estimated the retention at about 45% of the amount administered. Pezzagno et al (1986) estimated the relative uptake to average 53%. Higher values (71-80%) were found by Di Vincenzo et al (1973) and by Brugnone et al (1978). Wigaeus Hjelm et al (1981) found that approximately 79% of the amount absorbed is metabolized, 20% is excreted unchanged with expired air, and about 1% is excreted via urine. Kawai et al (1990) have estimated that 3% of the acetone absorbed is excreted into the urine. The half-time of acetone in alveolar air is about 4 h, in venous and arterial blood it is about 6 and 4 h, respectively. The highest concentration of acetone in urine occurs 3 to 3.5 h after exposure. Some experiments tend to indicate that high levels of blood acetone could result in transformation to isopropyl alcohol (Lewis et al 1984).

Biological monitoring

Breath analysis

Acetone. During exposure of volunteers to 1309 mg/m^3 (545 ppm) at rest, the average concentration of acetone in alveolar air amounts to approximately 500 mg/m^3. In workers exposed to 240 mg/m^3 (100 ppm), Brugnone et al (1980) found a mean alveolar air concentration of 53 mg/m^3 during exposure. In 1985, this author proposed as biological exposure indice 498 µg/l (end-of-shift) corresponding to a TWA of 150 ppm (1780 mg/m^3) (Brugnone and Perbellini 1985). At the end of a 4-h exposure to 250 ppm and 125 ppm acetone, Lowry (1987) observed a mean acetone concentration of 25 ppm (58 mg/m^3) and 13 ppm (30 mg/m^3) in expired air.

Blood analysis

Acetone. The endogenous acetone concentration in blood and urine does not normally exceed 2 mg/l (Pezzagno et al 1986; Teitz 1983). After a 2-h exposure of volunteers to 545 ppm acetone, the acetone concentration in blood increased throughout the exposure to reach a value of about 1 mg/100 ml (venous blood) (Wigaeus Hjelm et al 1981). In workers exposed to approximately 30 ppm acetone for 8 h, Baumann and Angerer (1979) found a mean blood acetone concentration of 0.33 mg/100 ml at the end of the shift. In workers exposed to a mean environmental concentration of 24.1 mg/m^3 (10 ppm) (range: 1.1 to 58.1 mg/m^3), Brugnone et al (1978) found a mean blood acetone concentration of 0.15 mg/100 ml (range: 0.05 to 0.3 mg/100 ml). At the end of a 4-h exposure to 250 ppm and 125 ppm acetone, Lowry (1987) observed a mean acetone concentration of 1.5 mg/100 ml and 1 mg/100 ml in blood. The blood BEI proposed by Brugnone and Perbellini (1985) is 5.5 mg/100 ml (end-of-shift) for a TWA of 750 ppm (1780 mg/m³).

Urine analyses

Acetone. According to Wigaeus Hjelm et al (1981), the urinary concentration of acetone following 2-h exposure to 545 ppm acetone is 10 mg/l. Pezzagno et al (1986) also found a close relationship between the environmental concentration of acetone in the breathing zone and its urinary concentration (samples collected after 4 h of exposure). The urinary concentration corresponding to a TWA of 750 ppm (1780 mg/m³) was about 58 mg/l (1000 µmol/l) and the corresponding biological equivalent exposure limit (95% lower confidence limit of the regression line) amounted to 54 mg/l (930 µmol/l) (Pezzagno et al 1986; Grampella et al 1988). They reported similar results in a further study (Ghittori et al 1987) and recommended 56 mg acetone/l as biological exposure limit.

Kawai et al (1990) also determined acetone concentration in end-of-shift urine samples (collected during the second half of the week) from workers exposed to acetone. But, according to their estimation, an 8-h exposure to 50 ppm would lead to an acetone mean urinary concentration of about 20 mg/l.

Formic acid. Baumann and Angerer (1979) consider that the determination of the urinary excretion rate of formic acid, or the total amount excreted, is also suitable for monitoring acetone exposure. The amount of formic acid excreted during 24 h by workers exposed to 30 ppm acetone for 8 h is about 13 mg.

7.4. Methyl Isobutyl Ketone (MIBK)

Toxicokinetics

Wigaeus Hjelm et al (1990) exposed volunteers during 2 h to different concentrations of MIBK vapor (10-100-200 mg/m³). The relative pulmonary uptake was about 60% and the uptake increased linearly with the intensity of exposure. The concentration of MIBK in blood rose rapidly after the onset of exposure without tendency to saturation kinetics. After the exposure, the elimination of MIBK from blood was biphasic (with elimination half-times of about 12 and 70 min). Only a very small amount (0.04%) was eliminated unchanged via the kidneys within 3 h postexposure. 4-Hydroxy-4-methyl-2-pentanone and 4-methyl-2-pentanol, two metabolites identified in animal experiments, were not detected in human exposed to MIBK.

The inhalation study carried out by Dick et al (1990) suggests that a steady-state blood level is attained during exposure to 410 mg MIBK/m³ (100 ppm) for 4 h. Blood and breath samples collected 90 min after exposure indicated essentially complete clearance of the absorbed MIBK.

Biological monitoring

For a TWA of 25 ppm, Ogata et al (1990) have estimated that the mean urinary concentration of methyl isobutyl ketone (end-of-shift sample) amounts to 0.9 mg/l.

The ACGIH (1991-1992) proposes as biological exposure indice 2 mg MIBK/ l urine (0.02 mmol/mol creatinine) in sample taken at the end of the shift corresponding to a TWA of 50 ppm (205 mg/m^3).

Since toxic effects can still be induced by an exposure to 25 ppm methyl isobutyl ketone (Wigaeus Hjelm et al 1990), we propose a biological limit value of 0.5 mg/ g creatinine for the concentration of the solvent in urine (end-of-shift sample).

7.5. Cyclohexanone

Cyclohexanone is readily absorbed by inhalation and by skin contact. Cyclohexanediols and cyclohexanol have been identified in the urine and blood of subjects exposed to cyclohexanone (Flek and Sedivec 1989; Mills and Walker 1990; Sakata et al 1989; Ong et al 1991b).

Flek and Sedivek (1989) exposed volunteers for 8 h to 415 mg/m^3 cyclohexanone and observed that only 3.5% of the cyclohexanone found in the urine was conjugated to glucuronic acid, the rest being present as trans-1,2-cyclohexanediol glucuronide and as trans-1,4-cyclohexanediol.

The usefulness of urine analysis of cyclohexanediols (free or conjugated) as a biological monitoring method should be further investigated. According to Ong et al (1991b) the measurement of cyclohexanol in urine might be used for the surveillance of cyclohexanone occupational exposure, but the available data are insufficient to propose a BEI.

In conclusion, the measurement of the ketones themselves in urine and/or in blood collected at the end of the workshift appears the most practical approach for assessing exposure.

For MEK, exposure to a TWA of 200 ppm will probably lead to a mean end-of-shift urinary excretion of about 2 mg/l, provided the physical activity and skin contact are not too important.

An average exposure to 750 ppm acetone probably corresponds to an average acetone urinary concentration of 58 mg/l and a mean acetone blood concentration of 5.5 mg/100 ml.

For methyl isobutyl ketone, a biological limit value of 0.5 mg/g creatinine (end-of-shift) is proposed.

REFERENCES

ACGIH. American Conference of Governmental Industrial Hygienists. Threshold Limit Values for Chemical Substances and Physical Agents and Biological Exposure Indices. Cincinnati, 1991-1992.

N. Allen, J. Mendell, D. Billmaier et al. Toxic polyneuropathy due to methyl n-butyl ketone. *Arch. Neurol.* 32:209, 1975.

K. Baumann and J. Angerer. Biological monitoring in occupational exposure to acetone. *Dtsch. Ges. Arbeitsmed.* 19:403, 1979 (in German).

P. Bos, G. de Mik and P. Bragt. Critical review of the toxicity of methyl n-butyl ketone: risk from occupational exposure. *Am. J. Ind. Med.* 20:175, 1991.

F. Brugnone, L. Perbellini, L. Grigolini and P. Apostoli. Solvent exposure in a shoe upper factory. I. n-Hexane and acetone concentration in alveolar and environmental air and in blood. *Int. Arch. Occup. Environ. Health* 42:51, 1978.

F. Brugnone, L. Perbellini, E. Gaffuri and P. Apostoli. Biomonitoring of industrial solvent exposure in workers' alveolar air. *Int. Arch. Occup. Environ. Health* 47:245, 1980.

F. Brugnone and L. Perbellini. Biological monitoring of occupational exposure to solvents by analysis of alveolar air and blood. Working Group on Chronic Effects of Organic Solvents on Central Nervous System and Diagnostic Criteria. WHO, Regional Office for Europe, Copenhagen, June 1985.

DFG. Deutsche Forschungsgemeinschaft. *Maximum Concentrations at the Workplace and Biological Tolerance Values for Working Materials.* Report No. XXVII. Commission for the Investigation of Health Hazards of Chemical Compounds in the Work Area. VCH, Weinheim, 1991.

R. Dick, D. Dankovic, J. Setzer et al. Body burden profiles of methyl ethyl ketone and methyl isobutyl ketone exposure in human subjects. *Toxicologist* 10:122, 1990.

G. Di Vincenzo, F. Yanno and B. Astill. Exposure of man and dog to low concentrations of acetone vapor. *Am. Ind. Hyg. Assoc. J.* 34:329, 1973.

G. Di Vincenzo, M. Hamilton, C. Kaplan et al. Studies on the respiratory uptake and excretion and the skin absorption of methyl-*n*-butyl ketone in humans and dogs. *Toxicol. Appl. Pharmacol.* 44:593, 1978.

J. Flek and V. Sedivec. Identifikace a stanoveni metabolitu cyclohexanonu v lidske moci. *Prac. Lek.* 41:259, 1989.

S. Ghittori, M. Imbriani, G. Pezzano and E. Capodaglio. The urinary concentration of solvents as a biological indicator of exposure: proposal for the biological equivalent exposure limit for nine solvents. *Am. Ind. Hyg. Assoc. J.* 48:786, 1987.

D. Grampella, L. Garavaglia, G. Pezzagno and M. Imbriani. Biological monitoring of workers exposed to acetone. In: *Occupational Health in the Chemical Industry.* Papers presented at the XXII ICOH Congress, Sydney, Australia 27 Sept-2 Oct 1987. WHO, Regional Office for Europe, Copenhagen, 1988.

T. Kawai, T. Yasugi, Y. Uchida and O. Iwami. Urinary excretion of unmetabolized acetone as an indicator of occupational exposure to acetone. *Int. Arch. Occup. Environ. Health* 62:165, 1990.

G. Lewis, A. Kaufman, B. McAnalley and J. Garriott. Metabolism of acetone to bipropyl alcohol in rats and humans. *J. Foren. Sci.* 29:541, 1984.

J. Liira, V. Riihimaki and P. Pfäffli. Kinetics of methyl ethyl ketone in man: absorption, distribution, elimination in inhalation exposure. *Int. Arch. Occup. Environ. Health* 60:195, 1988a.

J. Liira, V. Riihimäki, K. Engström and P. Pfäffli. Coexposure of man to *m*-xylene and methyl ethyl ketone: kinetics and metabolism. *Scand. J. Work Environ. Health* 14:322, 1988b.

J. Liira, G. Johanson, and V. Riihimäki. Dose-dependent kinetics of inhaled methyl ethylketone in man. *Toxicol. Lett.* 50:195, 1990a.

J. Liira, V. Riihimäki and K. Engström. Effects of ethanol on the kinetics of methyl ethyl ketone in man. *Br. J. Ind. Med.* 47:325, 1990b.

L. Lowry. The biological exposure index: its use in assessing chemical exposures in the workplace. *Toxicology* 47:55, 1987.

G. Mills and V. Walker. Urinary excretion of cyclohexanediol, a metabolite of the solvent cyclohexanone, by infants in a speical care unit. *Clin. Chem.* 36:870, 1990.

M. Miyasaka, M. Kumai, A. Koizumi et al. Biological monitoring of exposure to methylethylketone by means of urinalysis for methylethylketone itself. *Int. Arch. Occup. Environ. Health* 50:131, 1982.

M. Ogata, T. Taguchi, N. Hirota et al. A database for biological monitoring of aromatic solvents. In: *Biological Monitoring of Exposure to Industrial Chemicals.* Eds.: V. Fiserova-Bergerova, M. Ogata. Proceedings of the U.S.-Japan Cooperative Seminar on Biological Monitoring. ACGIH, Cincinnati, 1990.

C. Ong, G. Sia, H. Ong et al. Biological monitoring of occupational exposure to methyl ethyl ketone. *Int. Arch. Occup. Environ. Health* 63:319, 1991a.

C. Ong, G. Sia, S. Chia et al. Determination of cyclohexanol in the urine and its use in environmental monitoring of cyclohexanone exposure. *J. Anal. Toxicol.* 15:13, 1991b.

L. Perbellini, F. Brugnone, P. Mozzo et al. Methyl ethyl ketone exposure in industrial workers. Uptake and kinetics. *Int. Arch. Occup. Environ. Health* 54:73, 1984.

G. Pezzagno, M. Imbriani, S. Ghittori et al. Urinary elimination of acetone in experimental and occupational exposure. *Scand. J. Work Environ. Health* 12:603, 1986.

M. Sakata, J. Kikuchi, M. Haga et al. Disposition of acetone, methyl ethylketone and cyclohexanone in acute poisoning. *J. Toxicol. Clin. Toxicol.* 27:67, 1989.

N. Teitz. *Clinical Guide to Laboratory Tests.* W. B. Saunders, Philadelphia, 1983.

W. Tolos, J. Setzer, B. Mac Kenzie et al. Biological monitoring of experimental human inhalation exposures of methyl ethyl ketone and toluene. In: *Biological Monitoring of Exposure to Chemicals. Organic Compounds.* Eds.: M. Ho, M. Dillon. Wiley Interscience, New York, p. 133, 1987.

E. Wigaeus Hjelm, S. Holm and I. Astrand. Exposure to acetone. Uptake and elimination in man. *Scand. J. Work Environ. Health* 7:84, 1981.

E. Wigaeus Hjelm, M. Hagberg, A. Iregren and A. Löf. Exposure to methyl isobutyl ketone: toxicokinetics and occurrence of irritative and central nervous system symptoms in man. *Int. Arch. Occup. Environ. Health* 62:19, 1990.

8. ALDEHYDES
8.1. Formaldehyde

Toxicokinetics

Occupational exposure to formaldehyde occurs usually by inhalation of the gas or by skin contact with an aqueous solution.

Biological monitoring

Attempts have been made to monitor worker exposure to formaldehyde by measuring its concentration in blood and that of its oxidation product, formic acid, in urine.

Blood analysis

Formaldehyde. Piotrowski (1977) reported that workers exposed to approximately 5 ppm formaldehyde showed blood levels of 0.06 to 0.4 mg formaldehyde/ 100 ml, but much higher values have been reported by Einbrodt et al (1976). They found that men exposed to 0.9 ppm formaldehyde for 3 h had blood formaldehyde levels of about 0.85 mg/100 ml at the end of the exposure. An investigation on human volunteers, however, showed that an exposure to an average air concentration of 1.9 ppm formaldehyde during 40 min did not significantly increase its concentration in venous blood (before exposure: 2.61 ± 0.14 µg/g of blood; after exposure: 2.77 ± 0.28).

Urine analysis

Formic acid. The average urinary formate concentration reported by Boeniger (1987) in non-exposed individuals ranges from 11.7 to 18 mg/l. Heinzow and Ellrott (1992) measured a mean concentration in urine of occupational unexposed adults (20-80 years) of 21 mg/l (SD: 30; 95 percentile: 60 mg/l). According to these authors, excretion in the general population is determined by endogenous metabolism of amino acids, purine, and pyrimidine bases rather than the uptake and metabolism of precursors like formaldehyde. Formate precursor compounds encountered in industry include besides formaldehyde, methanol, halomethanes, and acetone. Some pharmaceuticals can also be transformed into formate (e.g., ephedrine, methylephedrine) (Boeniger 1987).

Einbrodt et al (1976) found that exposure to 0.93-1.19 ppm formaldehyde for 8 h increases the formic acid level in urine by a factor of 3 to 7.

In veterinary medicine students exposed to low levels of formaldehyde (20.5 ppm), no significant increase was seen in the urinary formic acid levels (Gottschling et al 1984).

In practice, it seems that moderate exposure to formaldehyde cannot be adequately assessed by measuring formaldehyde in blood or formic acid in urine, their levels depending on too many variables.

8.2. Furfural

Toxicokinetics

The metabolism of furfural, a heterocyclic aldehyde that is used as a solvent, has been investigated in man (Flek and Sedivec 1978; Sedivec and Flek 1978). Furfural vapor can enter the organism not only through the lung, but also through the intact skin. Percutaneous absorption of liquid furfural has also been demonstrated (Flek and Sedivec 1978). In the body, furfural is metabolized very rapidly; its biological half-life is about 2 to 2.5 h. Only a very small proportion (less than 1%) of the retained furfural is eliminated by the lung. The major part is oxidized to furoic acid, which is conjugated with glycine to give furoylglycine and eliminated in the urine. Less than 5% of the absorbed furfural condenses with acetic acid to give rise to 2-furanacrylic acid, which is excreted in urine after conjugation with glycine (i.e., as furanacryluric acid) (Figure 22).

Biological monitoring

Urine analysis

Furoic acid. Furoic acid is a natural constituent of human urine. Its average concentration (free plus conjugated) is on the order of 15 mg/g creatinine (range 4 to 65).

Sedivec and Flek (1978) exposed volunteers to concentrations of furfural vapors ranging from 7 to 30 mg/m³ and found that a stable exposure to 2.5 ppm (10 mg/m³) for 8 h leads to an average concentration of 109 mg furoic acid/g creatinine (range: 61 to 208) in urine collected during the last 2 h of the shift. The corresponding values for an exposure to 5 ppm (20 mg/m³) are mean, 204, range 120-353 mg furoic acid/g creatinine. In view of the very short half-life of furfural *in vivo*, these biological limits only apply when exposure is rather stable during the shift.

ACGIH (1991-1992) has proposed a biological exposure indice for formic acid in urine (end-of-shift) of 200 mg/g creatinine.

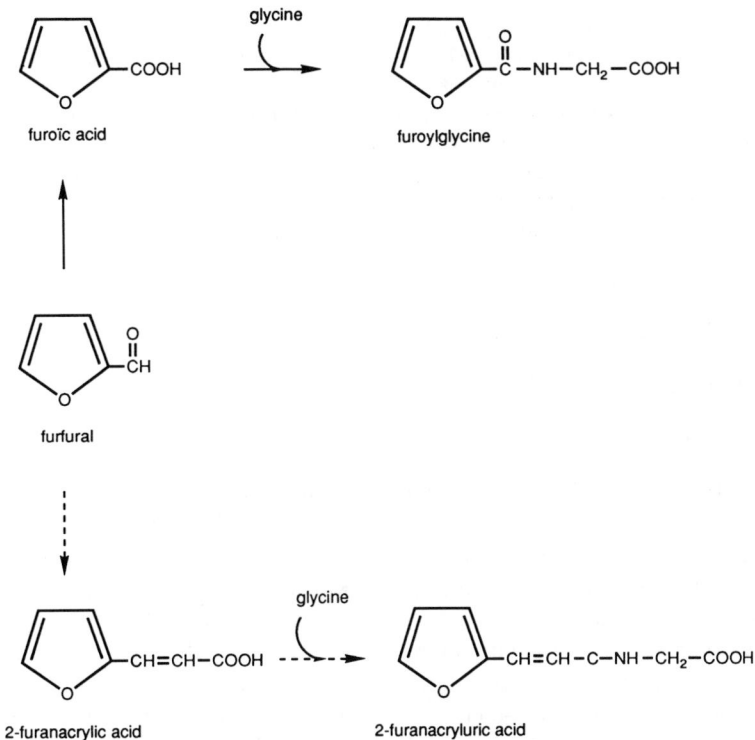

FIGURE 22 Main metabolic pathways of furfural.

In view of the results obtained by Sedivec and Flek (1978) on volunteers exposed to furfural, we are inclined to propose that the furoic acid concentration in urine collected at the end of the shift should not exceed 80 mg/g creatinine.

However, a field study conducted by Di Pede et al (1991) suggests that urinary furoic acid concentration may not be a sufficiently sensitive marker to detect low exposure to furfural (below a TWA of 8 mg/m³).

REFERENCES

ACGIH. American Conference of Governmental Industrial Hygienists. Threshold Limit Values for Chemical Substances and Physical Agents and Biological Exposure Indices. Cincinnati, 1991-1992.

M. Boeniger. Formate in urine as a biological indicator of formaldehyde exposure: a review. *Am. Ind. Hyg. Assoc. J.* 48:900, 1987.

C. Di Pede, G. Viegi, R. Taddeucci et al. Biological monitoring of work exposure to furfural. *Arch. Environ. Health* 46:125, 1991.

H. Einbrodt, D. Prajsnar and J. Erpenbeck. Der Formaldehyd-und Ameisensäurespiegel im Blut and Urin beim Menschen nach Formaldehydexposition. *Zentralbl. Arbeitsmed. Arbeitsschutz Prophyl.* 26:154, 1976.

J. Flek and V. Sedivec. The absorption metabolism and excretion of furfural in man. *Int. Arch. Occup. Environ. Health* 41:159, 1978.

L. Gottschling, H. Beaulier and W. Melvin. Monitoring of formic acid in urine of humans exposed to low levels of formaldehyde. *Am. Ind. Hyg. Assoc. J.* 45:19, 1984.

B. Heinzow and T. Ellrott. Formic acid in urine. An useful indicator in environmental medicine. *Zentralbl. Hyg. Umweltmed.* 192:455, 1992.

J. Piotrowski. *Exposure Tests for Organic Compounds in Industrial Toxicology.* U.S. Government Printing Office, Washington, D.C., 1977.

V. Sedivec and J. Flek. Biologic monitoring of persons exposed to furfural vapors. *Int. Arch. Occup. Environ. Health* 42:41, 1978.

9. AMIDES AND ANHYDRIDES
9.1. Dimethylformamide

Toxicokinetics

Dimethylformamide (DMF) is a colorless volatile liquid. Its absorption occurs through inhalation of vapors and by direct contact with the liquid form (Kimmerle and Eben 1975; Lauwerys et al 1980; Maxfield et al 1975; Grandjean 1990). Vapors can also be absorbed through the skin (Maxfield et al 1975). However, experiments on volunteers suggest that penetration of liquid DMF through skin usually contributes substantially more to the total intake than percutaneous penetration of DMF vapors. A 10-min immersion of one hand resulted in as much urinary N-hydroxymethyl-N-methylformamide (HMMF) as after an 8-h inhalation exposure to DMF vapor at a TWA level of 30 mg/m³ (Mraz and Nohova 1992).

Dimethylformamide is rapidly metabolized *in vivo*. A negligible fraction of the administered dose is excreted unchanged in urine and by the gastrointestinal tract. The proposed metabolic pathway of DMF in human is summarized in Figure 23. It has been demonstrated that the metabolite measured as *N*-methylformamide (NMF) by gas chromatography (Barnes and Henry 1974) is mainly HMMF, a stable carbinolamine which breaks down in the injector of the gas chromatograph to give NMF (Scailteur and Lauwerys 1984a,b). By analogy, *N*-hydroxymethyl-formamide (HMF) is considered to be the metabolite initially described as formamide (F). Only a very small percentage of the absorbed DMF, however, is transformed into NMF, HMF, and F (probably less than 5%).

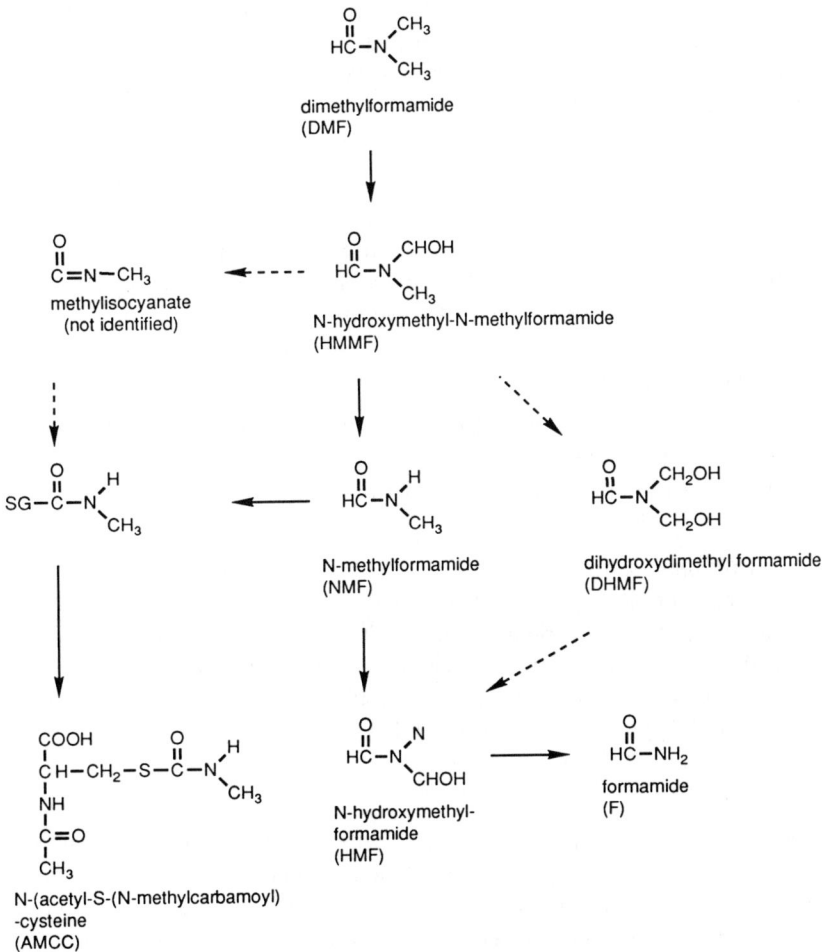

FIGURE 23 Main metabolic pathways of dimethylformamide.

A mercapturic acid derivative N-(acetyl-S-(N-methylcarbamoyl)cysteine (AMCC) has also been identified in the urine of subjects exposed to dimethyl-formamide (Mraz and Turecek 1987; Mraz et al 1989; Kestell et al 1986). The hypothesis has been raised that its precursor might be methylisocyanate, partly responsible for the toxicity of dimethylformamide.

The observations made by Kimmerle and Eben (1975) on volunteers indicate that the dimethylformamide concentration in blood decreases rapidly after the end of the exposure, with a biological half-life of approximately 1 h.They also suggest that the excretion of formamide in urine, like that of N-methylformamide, is dose dependent. However, the excretion of formamide is slightly delayed. The highest formamide concentration in urine was found between 4 and 20 h after the end of the test. Within the first 24 h after the start of exposure, its concentration is much lower than that of N-methylformamide, which constitutes the main urinary me-tabolite in man. Their results also show that N-methylformamide appears rapidly in urine of humans exposed to dimethylformamide; after the end of the exposure, its biological half-life is around 12 h.

Biological monitoring

Several biological parameters can be considered for the evaluation of exposure to dimethylformamide: dimethylformamide in whole blood and in expired air, the sum of HMMF and NMF in whole blood and urine, formamide and N-acetyl-S-(N-methylcarbamoyl)-cysteine (AMCC) in urine.

Blood analyses

Dimethylformamide. At the end of a 4-h exposure of volunteers to average atmospheric concentrations of 21 and 87 ppm, the mean dimethylformamide concentrations in blood amounted to about 0.3 and 1.4 mg/100 ml, respectively (Kimmerle and Eben 1975).

N-methylformamide (sum of HMFF and NMF). The only data available relating N-methylformamide concentration in blood and exposure to dimethylformamide are again those of Kimmerle and Eben (1975) based on four volunteers exposed to dimethylformamide vapors at concentrations ranging from 21 to 87 ppm. At the end of the exposure, N-methylformamide in blood amounted on the average to 0.2 mg/100 ml (21 ppm) and 0.6 mg/100 ml (87 ppm).

Breath analysis

Dimethylformamide. When skin contact with dimethylformamide can be pre-vented, there is during exposure a significant correlation between environmental exposure and dimethylformamide concentration in alveolar air. According to Brugnone and co-workers (Brugnone et al 1980; Brugnone and Perbellini 1985), the alveolar concentration amounts on the average to 3 mg/m³ when the atmo-spheric concentration of dimethylformamide is 10 mg/m³. The BEI corresponding to a TWA of 10 ppm (30 mg/m³) is 8.44 mg/m³ (2.8 ppm).

Urine analyses

No data are available on the urinary excretion of formamide in workers exposed to dimethylformamide.

N-methylformamide (sum of HMMF and NMF). The human volunteer experiments of Kimmerle and Eben (1975) and of Maxfield et al (1975) have demonstrated that N-methylformamide analysis of urine is a useful biological parameter for monitoring exposure to dimethylformamide. Catenacci et al (1980), Yonemoto and Suzuki (1980), and Wicarova and Dadak (1981) have examined workers exposed to dimethylformamide and have found a relationship between the environmental concentration of this solvent and the amount of N-methylformamide excreted in 24 h. Their quantitative results, however, are very different since, for an exposure to 30 mg/m³ (10 ppm) dimethylformamide, the first group of authors found excretion of 12-15 mg N-methylformamide per 24 h, whereas Yonemoto and Suzuki (1980) and Wicarova and Dadak (1981) reported an average excretion of 5 and 6 mg of N-methylformamide per 24 h, respectively, for the same exposure level. Differences in the percutaneous absorption of dimethylformamide and in alcohol consumption, which interferes with dimethylformamide metabolism, may explain this discrepancy.

On the basis of results obtained in workers exposed to dimethylformamide in an acrylic fiber factory, we have suggested that a concentration of N-methylformamide in urine samples collected at the end of the workshift not exceeding 40 to 50 mg/g creatinine indicates an exposure that is probably safe with regard to the acute and long-term action of dimethylformamide on workers' health (Lauwerys et al 1980). Such exposure, however, may still be associated with signs of alcohol intolerance. Our results also demonstrate very clearly that for a substance like dimethylformamide, which can also enter the organism through skin contact, biological monitoring is much superior to airborne concentration monitoring for evaluating the health risk.

Sala et al (1984) found a correlation between urinary NMF levels (4 h after the workshift) and the airborne concentration of DMF in an artificial-leather production. The data corresponding to the different investigated work areas were the following: mean N-methylformamide urinary values of 12 mg/g creatinine, 16 mg/g creatinine, and 188.3 mg/g creatinine in workers occupied at workplaces where TWA exposures were 9.4 mg DMF/m³, 4.5-14 mg/m³, and 86.3 mg/m³, respectively.

The values recommended by Bardodej and Malonova (1987) as tentative biological limits of N-methylformamide in urine are 200 µmol/day (12 mg/d); 100 µmol/shift (6 mg/shift); 19 µmol/mmol creatinine (10 mg/g creatinine) corresponding to a TWA of 10 ppm (30 mg/m³).

The study of Kawai et al (1992) also describes a linear correlation between DMF TWA and concentration of N-methylformamide in the urine collected at the end of the workshift; an 8-h exposure to DMF at 10 and 20 ppm being associated with a urinary N-methylformamide excretion of 18 mg/l (15.8 mg/g creatinine) as a group mean (95% confidence range: 16-21 mg/l) and 35 mg/l (29.6 mg/g creatinine) (30-40 mg/l), respectively.

Mercapturates. Bardodej and Malonova (1987) determined mercapturates in urine samples collected from unexposed and DMF-exposed workers (air concentration not reported) in an artificial-leather factory. The results obtained in control smokers differ significantly from those found in control nonsmokers (mean 6.3 µmol/mmol creatinine; SD: 2.6 and 4.7 ± 1.7 µmol/mmol creatinine, respectively). The authors also observed a good correlation between the urinary concentrations of mercapturates and *N*-methylformamide. According to their results the 19 µmol MMF/mmol creatinine (10 mg/g creatinine) proposed by the same authors as biological limit corresponds to 16 µmol mercapturates/mmol creatinine.

In summary, the determination of the sum of *N*-methyl-*N*-hydroxymethylformamide and *N*-methylformamide in end-of-shift urine (both measured as NMF by gas chromatography) appears to be the most practical method to assess the amount of DMF absorbed. No definite biological threshold limit value can be proposed. It seems, however, that following an 8-h exposure to 10 ppm DMF, the urinary concentration of NMF in urine (end-of-shift) would probably not exceed 30 mg/g creatinine and such level should not be associated with adverse effects.

9.2. Dimethylacetamide

Toxicokinetics

The metabolism of dimethylacetamide in man is similar to that of dimethylformamide. The chemical is absorbed both by the pulmonary and the cutaneous routes. Man excretes monomethylacetamide (*N*-methylacetamide — MMAC) following an exposure to this chemical (Maxfield et al 1975). After an exposure to 10 ppm dimethylacetamide vapor for 6 h, the urinary excretion of MMAC is completed within 30 h.

Biological monitoring

Urine analysis

Kennedy and Pruett (1989) have proposed the determination of urinary MMAC in the end-of-shift sample to assess occupational exposure to dimethylacetamide. At the exposure concentrations encountered (0.5 to 2 ppm), for each 1 ppm dimethylacetamide in air, approximately 10 mg MMAC/l could be measured in the end-of-shift urine sample. Interindividual variation was small and no evidence of build-up in MMAC urinary levels was seen. In contrast to these findings, Borm et al (1987) did not find a correlation between airborne dimethylacetamide and urinary MMAC and found a progressive increase in the MMAC excretion during the week. Difference in skin contact may be responsible for these discrepancies.

9.3. Anhydrides

Biological monitoring

The determination of phthalic acid and hexahydrophthalic acid in urine have been used for biological monitoring of exposure to phthalic anhydride and hexahydrophthalic anhydride. Both are excreted as free acid, phthalic acid with a half-life of around 14 h and hexahydrophthalic acid with a half-life of 2-3 h. Significant correlations were found between the anhydride TWA concentrations and the corresponding acid concentrations in urine.

According to the linear regression line, a TWA exposure to 1 mg/m³ and 6 mg/m³ phthalic anhydride would correspond to postshift mean urinary phthalic acid concentrations of 1.8 µmol/mmol creatinine (2.6 mg/g creatinine) and 5.5 µmol/mmol creatinine (8.15 mg/g creatinine), respectively. The mean control value was 0.34 µmol/mmol creatinine (SD: 0.25) (Pfäffli 1986).

A TWA exposure to 0.1 mg/m³ and 0.6 mg/m³ hexahydrophthalic anhydride would lead to postshift mean urinary acid concentrations of 1 µmol/mmol creatinine (1.53 mg/g creatinine) and 5.7 µmol/mmol creatinine (8.75 mg/g creatinine), respectively. The mean control value was < 0.1 µmol/mmol creatinine (Jönsson et al 1991; Jönsson and Skarping 1991).

REFERENCES

Z. Bardodej and H. Malonova. Exposure tests for dimethylformamide. In: *Biological Monitoring of Exposure to Chemicals. Organic Compounds*. Eds.: M. Ho, M. Dillon. Wiley Interscience, New York, p. 89, 1987.

J. Barnes and N. Henry. The determination of *N*-methylformamide and *N*-methylacetamide in urine. *Am. Ind. Hyg. Assoc. J*. 35:84, 1974.

P. Borm, L. de Jong and A. Vliegen. Environmental and biological monitoring of workers occupationally exposed to dimethylacetamide. *J. Occup. Med.* 29:898, 1987.

F. Brugnone, L. Perbellini and E. Gaffuri. N,N-Dimethylformamide concentration in environmental and alveolar air in an artificial leather factory. *Br. J. Ind. Med.* 37:185, 1980.

G. Brugnone and L. Perbellini. Biological monitoring of occupational exposure to solvents by analysis of alveolar air and blood. Workshop on Chronic Effects of Organic Solvents on the Central Nervous System and Diagnostic Criteria. Report on a joint WHO/Nordic Council of Ministers Working Group. Copenhagen 10-14 June 1985.

G. Catenacci, S. Ghittori, D. Cottica et al. Occupational exposure to dimethylformamide and urinary excretion of monomethylformamide. *G. Ital. Med. Lav.* 2:53, 1980.

Ph. Grandjean. *Skin Penetration: Hazardous Chemicals at Work*. A report prepared for the Commission of the European Communities, Luxembourg. Eds.: Taylor and Francis, London, Eur 12599 En, 1990.

B. Jönsson, H. Welinder and G. Skarping. Hexahydrophthalic acid in urine as an index of exposure to hexahydrophthalic anhydride. *Int. Arch. Occup. Environ. Health* 63:77, 1991.

B. Jönsson and G. Skarping. Method for the biological monitoring of hexahydrophthalic anhydride by the determination of hexahydrophthalic acid in urine using gas chromatography and selected-ion monitoring. *J. Chromatogr. Biomed. Appl.* 572:117, 1991.

T. Kawai, T. Yasugi, K. Mizunuma et al. Occupational dimethylformamide exposure. 2. Monomethylformamide excretion in urine after occupational dimethylformamide exposure. *Int. Arch. Occup. Environ. Health* 63:455, 1992.

G. Kennedy and J. Pruett. Biologic monitoring for dimethylacetamide: measurement for 4 consecutive weeks in a work place. *J. Occup. Med.* 31:47, 1989.

P. Kestell, A. Gledhill, M. Threadgill and A. Gescher. S-(N-methylcarbamoyl)-N-acetylcysteine: a urinary metabolite of the hepatotoxic experimental antitumour agent N-methyl-formamide in mouse, rat and man. *Biochem. Pharmacol.* 35:2283, 1986.

G. Kimmerle and A. Eben. Metabolism studies of N,N-dimethylformamide. II. Studies in persons. *Int. Arch. Arbeitsmed.* 34:127, 1975.

R. Lauwerys, S. Kivits, M. Lhoir et al. Biological surveillance of workers exposed to dimethylformamide and the influence of skin protection on its percutaneous absorption. *Int. Arch. Occup. Environ. Health* 45:189, 1980.

M. Maxfield, J. Barnes, A. Azar and H. Trochimowicz. Urinary excretion of metabolite following experimental human exposure to DMF and DMAC. *J. Occup. Med.* 17:506, 1975.

J. Mraz and F. Turecek. Identification of N-acetyl-S-(N-methylcarbomoyl) cystein, a human metabolite of N,N-dimethylformamide and N-methylformamide. *J. Chromatogr.* 414:399, 1987.

J. Mraz, H. Cross, A. Gescher et al. Differences between rodents and humans in the metabolic toxification of N,N-dimethylformamide. *Toxicol. Appl. Pharmacol.* 98:507, 1989.

J. Mraz and H. Nohova. Percutaneous absorption of N,N-dimethylformamide in humans. *Int. Arch. Occup. Environ. Health* 64:79, 1992.

P. Pfäffli. Phthalic acid excretion as an indicator of exposure to phthalic anhydride in the work atmosphere. *Int. Arch. Occup. Environ. Health* 58:209, 1986.

C. Sala, F. Bernabeo, G. Colombo et al. Dimethylformamide risk. An evaluation in the production of artificial organic leather. *Med. Lav.* 6:143, 1984.

V. Scailteur and R. Lauwerys. In vivo and in vitro oxidative biotransformation of dimethylformamide in rat. *Chem. Biol. Interact.* 50:327, 1984a.

V. Scailteur and R. Lauwerys. In vivo metabolism of dimethylformamide and relationship to toxicity in the male rat. *Arch. Toxicol.* 56:87, 1984b.

O. Wicarova and O. Dadak. N-Methylformamide levels in urine of persons exposed to dimethylformamide vapor. *Prac. Lek.* 33:42, 1981 (in Czech).

J. Yonemoto and S. Suzuki. Relation of exposure to dimethylformamide vapor and the metabolite, methylformamide, in urine of workers. *Int. Arch. Occup. Environ. Health* 46:159, 1980.

10. PHENOL AND DERIVATIVES

Toxicokinetics

Phenol is easily absorbed through the lung and the skin. Due to the low vapor pressure at normal temperature, respiratory exposure at the workplace would tend to be limited, adding to the relative significance of percutaneous absorption

(Grandjean 1990). It is also rapidly excreted in the urine within 24 h, mostly in a conjugated form (Ohtsuji and Ikeda 1972; Piotrowski 1971).

Biological monitoring

Since these substances can be absorbed through the skin, monitoring via urinary level seems the most reliable and comprehensive measure for evaluating personal exposure (Kosaka et al 1989).

Urine analyses

Parent compounds. Levels of phenol in nonexposed adults are generally less than 20 mg/l. However, there is a wide variation in "normal" persons; Ogata and Taguchi (1988) found that the arithmetic mean of urinary phenol concentration from 32 persons with no occupational exposure to phenol was 15 mg/l with a SD of 14 mg/l. Exposure to benzene (>10 ppm) can also result in an increased urinary excretion of phenol. Phenol is a common compound in antiseptic medicines and is also used in disinfectants and other household products.

Piotrowski (1971) estimated that, in workers exposed for 8 h to 1.3 ppm (5 mg/m^3) or 5 ppm (19 mg/m^3) of phenol, the excretion rate at the end of the exposure period would reach about 4.4 and 15.3 mg phenol per hour, respectively. This estimation agrees well with the observations of Ohtsuji and Ikeda (1972). According to their regression line, one should find approximately 300 mg of phenol per gram of creatinine in urine samples collected toward the end of an 8-h exposure to 15 mg phenol/m^3.

The biological limit values recommended by the ACGIH (1991-1992) and the DFG (1991) are 250 mg/g creatinine and 300 mg/l, respectively (end-of-shift sample) corresponding to a TWA of 5 ppm (19 mg/m^3).

p-tert-Butylphenol, *p*-octylphenol, and 4-tert-butylcatechol have been found in urine of workers occupied in the production of these compounds (Ikeda et al 1978).

To prevent the leucodermogenic activity of *p*-tert-butylphenol in workers, Ikeda et al (1978) proposed a biological threshold limit value of 2 mg *p*-tert-butylphenol/l of urine.

Other phenolic compounds (hydroquinone, catechol, resorcinol, cresols, etc.) can be analyzed in urine of exposed persons, but no biological limits have yet been proposed (Fassett 1949). Dinitro-ortho-cresol and pentachlorophenol are dealt with in Section 15.

Phenylsulfate, phenylglucuronide. Ohtsuji and Ikeda (1972) found that neither urinary ethereal sulphate nor urinary ethereal glucuronide was a good index of phenol exposure.

Ogata et al (1986) measured the end-of-shift urinary levels of phenol metabolites phenylsulfate (PhS) and phenylglucuronide (PhG) from subjects (not equipped with protection mask) engaged in the treatment of chemical fibers with phenol. They observed a significant correlation with the environmental concentration of phenol. The urinary concentration of phenol metabolites (sum of PhS and PhG), corresponding to 5 ppm of environmental phenol was 251 mg phenol/g creatinine.

Other analysis

The determination of phenol in saliva might also be used to estimate exposure (Lutogniewska 1980).

Although the data are somewhat limited it can tentatively be proposed that a concentration of 250 mg total phenol/g creatinine in end-of-shift urine sample corresponds to an 8-h TWA exposure of 5 ppm phenol.

REFERENCES

ACGIH. American Conference of Governmental Industrial Hygienists. Threshold Limit Values for Chemical Substances and Physical Agents and Biological Exposure Indices. Cincinnati, 1991-1992.

DFG. Deutsche Forschungsgemeinschaft. *Maximum Concentrations at the Workplace and Biological Tolerance Values for Working Materials.* Report No. XXVII. Commission for the Investigation of Health Hazards of Chemical Compounds in the Work Area. VCH, Weinheim, 1991.

D. Fassett. Method for determining hydroquinone excretion in the urine. *Fed. Proc.* 8:290, 1949.

Ph. Grandjean. *Skin Penetration: Hazardous Chemicals at Work.* A report prepared for the Commission of the European Communities, Luxembourg. Eds.: Taylor and Francis, London, Eur 12599 En, 1990.

M. Ikeda, T. Hirayama and I. Hara. GLC analysis of alkylphenols, alkylleucoderma. *Int. Arch. Occup. Environ. Health* 41:125, 1978.

M. Kosaka, T. Ueda, M. Yoshida and I. Hara. Urinary metabolite levels in workers handling *p*-tert-butylphenol as an index of personal exposure. *Int. Arch. Occup. Environ. Health* 61:451, 1989.

J. Lutogniewska. A simple test for an evaluation of occupational exposure to phenol. *Med. Pr.* 31:399, 1980 (in Polish).

M. Ogata, Y. Yamasaki and T. Kawai. Significance of urinary phenyl sulfate and phenyl glucuronide as indices of exposure to phenol. *Int. Arch. Occup. Environ. Health* 58:197, 1986.

M. Ogata and T. Taguchi. Simultaneous determination of urinary creatinine and metabolites of tolene, xylene, styrene, ethylbenzene and phenol by automated high performance liquid chromatography. *Int. Arch. Occup. Environ. Health* 61:131, 1988.

H. Ohtsuji and M. Ikeda. Quantitative relationship between atmospheric phenol vapour and phenol in the urine of workers in Bakelite factories. *Br. J. Ind. Med.* 29:70, 1972.

J. Piotrowski. Evaluation of exposure to phenol: absorption of phenol vapor in the lungs and through the skin and excretion of phenol in urine. *Br. J. Ind. Med.* 28:172, 1971.

11. CARBON MONOXIDE

Toxicokinetics

Carbon monoxide (CO) is absorbed and eliminated unchanged via the respiratory tract. It easily crosses the alveolar capillary barrier and binds to hemoglobin to form carboxyhemoglobin (HbCO). CO and oxygen compete for hemoglobin, but the affinity of hemoglobin for CO is approximately 220-240 times higher than for oxygen. The health risk of CO is mainly related to the HbCO level. The concentration of COHb rises from the start of exposure to CO and under standard conditions declines with a half-life of 4-5 h after the end of exposure.

There is an endogenous production of CO due to heme catabolism which is increased in several diseases, including hemolytic anemia. CO is also a ubiquitous pollutant and a metabolite of methylene chloride.

Biological monitoring

Three biological methods are being used for monitoring current exposure to CO: determination of the HbCO level in blood and analysis of CO concentration in blood or in expired air.

Blood analyses

Carboxyhemoglobin. In subjects exposed to CO, the HbCO level at equilibrium is proportional to the concentration of CO in inspired air. For moderate exposure (less than 100 ppm), the following approximate relationship has been found in nonsmokers (Antweiler 1973): % HbCO = 0.16 × CO (ppm). The time to reach equilibrium, however, is proportional to the partial pressure of CO in blood, and hence in alveolar air. For the same exposure level, the time to reach equilibrium will therefore vary with the pulmonary ventilation. Equilibrium will be attained sooner during physical activity than at rest. It has been estimated that, during a constant exposure to 30 ppm CO, equilibrium of the HbCO level is reached at rest after approximately 8 h. Before equilibrium is reached, it has been calculated that in nonsmokers the HbCO level is related to the concentration of CO in inspired air, the duration of exposure, and physical work by the following formula: % HbCO = % CO in air × time (in minutes) × K. K is a constant, the value of which is 3 at rest, 5 for light physical work, 8 for moderate physical work, and 11 for heavy physical work. A more complex equation has been developed by Coburn et al (1965), but for practical purposes the above formula is more convenient.

In a sedentary nonsmoker, a TWA exposure of 35 and 50 ppm CO will lead to an HbCO level of about 5% and 8%, respectively, at the end of the shift. Such an HbCO level or even higher values (up to 15%) can be found in smokers nonoccupationally exposed to CO (Jones et al 1972). It has therefore been proposed that thiocyanate in serum or urine be measured to help in interpreting the carboxyhemoglobin level. According to Gilli et al (1979), the simultaneous presence of a serum thiocyanate concentration above 0.38 mg/100 ml and a HbCO level in the range of 2.5 to 6% is a certain indication of cigarette smoking. A serum thiocyanate concentration below 0.38 mg/100 ml and a HbCO level above 5% indicates occupational exposure to CO.

The COHb level (end-of-shift) reflects the exposure level during the shift, provided this exposure level has been more or less constant.

There is conflicting evidence concerning the toxic COHb levels under long-term exposure. The WHO recommends as exposure limit for the protection of the general population (nonsmokers, but including sensitive groups) a carboxyhemoglobin level of 2.5-3% (WHO 1987). The ACGIH (1991-1992) has recently recommended 3.5% COHb as BEI, this level being most likely reached at the end of an 8-h exposure to 25 ppm CO (sample collected immediately at the end of the exposure). This BEI is not applicable to tobacco smokers.

The biological limit value recommended by the DFG (1991) is 5% COHb, corresponding to a TWA of 30 ppm.

Carbon monoxide. Since the amount of CO dissolved in blood represents a negligible fraction (less than 1%) of the amount bound to hemoglobin, the total amount of CO that can be liberated from blood is directly related to the hemoglobin content of blood and its degree of saturation by CO. Under standard conditions, 1 g of hemoglobin can bind 1.49 ml of CO. Therefore, in a nonsmoker with 15 g of hemoglobin per 100 ml of blood, the volume of CO normally present in blood is less than 0.15 ml/100 ml. When the HbCO level is 5%, 10 ml of CO can be released per 100 ml of blood.

Breath analysis

Carbon monoxide. In healthy nonsmokers breathing CO-free air, the level of CO in end-exhaled air is usually less than 2 ppm. This level is increased in the urban environment (6-12 ppm) and even higher on urban highways.

Provided the level of exposure has been more or less constant, CO in exhaled air sample taken at the end of the exposure reflects the exposure level during the shift.

The partial pressure of CO in alveolar air is in equilibrium with the partial pressure of CO in blood, which is itself in equilibrium with the CO bound to hemoglobin. Therefore, the determination of CO in alveolar air can be used to evaluate the HbCO level. Jones et al (1976) showed that the level of CO in the end-expired gas after a breath-hold of 20 s at total lung capacity correlates well with the venous HbCO level. The following equation has been proposed by Stewart (1973) to describe the relationship between CO in the breath and the HbCO level in blood.

$$\text{CO in alveolar air (ppm)} = 4 \, (\% \, HbCO) - 1.9$$

Stewart (1973) recognizes, however, that CO in its more permanent HbCO form is still a more reliable through less accessible means of determining the CO content of the blood.

The ACGIH (1991-1992) recommends 20 ppm as BEI which should not be applied to samples collected in emergency, during the first 3 h of exposure, later than 15 min after the end of exposure, or to tobacco smokers.

In an attempt to protect those workers with latent cardiovascular impairments, we suggest the following biological limit values for nonsmokers: carboxy-hemoglobin level <3.5% and carbon monoxide in end-expired air <12 ppm.

REFERENCES

ACGIH. American Conference of Governmental Industrial Hygienists. Threshold Limit Values for Chemical Substances and Physical Agents and Biological Exposure Indices. Cincinnati, 1991-1992.

H. Antweiler. In European Colloquium on Health Effects of Carbon Monoxide Environmental Pollution, Commission of the European Communities, Luxembourg, p. 197, 1973.

R. Coburn, R. Forster and P. Kane. Considerations of the physiological variables that determine the blood carboxyhemoglobin concentration in man. *J. Clin. Invest.* 44:1899, 1965.

DFG. Deutsche Forschungsgemeinschaft. *Maximum Concentrations at the Workplace and Biological Tolerance Values for Working Materials.* Report No. XXVII. Commission for the Investigation of Health Hazards of Chemical Compounds in the Work Area. VCH, Weinheim, 1991.

G. Gilli, E. Scursatone, G. Vanini and E. Vercellotti. Serum thiocyanate values as a discriminating element in assessment of exposure to carbon monoxide. *Minerva Med.* 70:2803, 1979.

R. Jones, B. Commins and A. Cernik. Blood lead and carboxyhaemoglobin levels in London taxi drivers. *Lancet* 2:302, 1972.

R. Jones, C. Coppin and A. Guz. Carbon monoxide in alveolar air as an index of exposure to cigarette smoke. *Clin. Sci. Mol. Med.* 51:495, 1976.

R. Stewart. Research study to determine the range of carboxyhemoglobin in various segments of the American population. Grant No. EPA-R1-73-004, Office of Research and Monitoring, U.S. Environmental Protection Agency, Washington, D.C., 1973.

WHO. World Health Organization. Air Quality Guidelines for EUROPE. WHO Regional Publications, European Series No. 23, Copenhagen, 1987.

12. CYANIDES AND NITRILES

Toxicokinetics

Cyanide salts are solids which can be absorbed by the respiratory tract, following inhalation of cyanide-containing aerosol, the gastrointestinal tract after ingestion, or possibly the skin after direct contact with cyanide solutions. Hydrogen cyanide being a gas under standard conditions is easily absorbed by the respiratory tract.

Most of the nitriles used in the industry are volatile liquids. They can enter the body by inhalation, ingestion, or cutaneous contact.

Acrylonitrile is absorbed through the lungs, with an average respiratory retention of 52% (Jakubowski et al 1987).

Significant amounts may also be absorbed through the skin (Vogel and Kirkendall 1984; Jakubowski et al 1987; Grandjean 1990).

Biological monitoring

Exposure to cyanides and aliphatic nitriles that can release cyanide *in vivo* causes an increased concentration of thiocyanate in plasma and urine and of cyanide in blood (Chandra et al 1980). Since hydrogen cyanide is present in tobacco smoke, the "normal" level of cyanide and thiocyanate is higher in smokers than in nonsmokers. In nonsmokers, the plasma thiocyanate concentration is usually less than 0.58 mg/100 ml, whereas in smokers the concentration may be as high as 1.5 mg/100 ml (Butts et al 1974; Pettigrew and Fell 1972; Weuffen et al 1981). In nonsmokers, the urinary excretion of thiocyanate does not exceed 2.5 mg/g creatinine. In a person smoking 15 cigarettes per day, the concentration of thiocyanate in urine amounts on the average to 4 mg/l (corrected for SG 1024). In nonsmoking, not occupationally-exposed subjects, normal cyanide blood values are usually below 10 µg/100 ml (Pettigrew and Fell 1973; McAnalley et al 1979; Kato et al 1988; Lundquist and Sörbo 1989; Cailleux et al 1988; Baud et al 1991). However, values up to 26 µg/100 ml have been reported (Symington et al 1978). Cyanide blood concentrations as high as 52 µg/100 ml have been observed in smokers (Symington et al 1978).

A group of nonoccupationally exposed subjects showed a significant increase of acrylonitrile urinary concentration with increasing number of cigarettes smoked (nonsmokers: 2 µg/g creatinine; 1-10 cigarettes/day: 3.6 µg/g creatinine; 11-20 cigarettes/day: 6.7 µg/g creatinine; 21-30 cigarettes/day: 9 µg/g creatinine) (Houthuijs et al 1982).

Urine analyses

Thiocyanate. In 36 nonsmokers exposed to cyanide in electroplating industries, El Ghawabi et al (1975) found that the concentration of thiocyanate in urine increased towards the middle of the working week and became almost stationary during the last 3 d. The relationship between thiocyanate in urine and the concentration of hydrogen cyanide in air is represented by the equation: M = 0.65 C, where M is the thiocyanate in the 24-h urine (in mg) and C is the concentration of cyanide in the air (ppm). Hence, a hydrogen cyanide TWA of 10 ppm would result in the urinary excretion of about 6.5 mg within 24 h (nonsmokers).

In order to better assess the exposure to cyanide-releasing substances *in vivo* without interference by the smoking habits of the workers, Della Fiorentina and De Wiest (1978) have proposed determination of the ratio between urinary thiocyanate concentration (in mg/g creatinine) and the carboxyhemoglobin level (in percent). A ratio above 3 suggests an excessive cyanide exposure.

Sakurai et al (1978) have investigated the relationship between the degree of exposure to acrylonitrile and the urinary excretion of acrylonitrile in 102 workers from acrylic fiber factories. They found that an 8-h average acrylonitrile exposure of 4.2 ppm by personal sampling leads to a mean urinary thiocyanate concentration of 11.4 mg/l (SG 1024) (end-of-shift sample).

Acrylonitrile. In the above-mentioned study, Sakurai et al (1978) observed that a TWA exposure of 4.2 ppm acrylonitrile entails a mean acrylonitrile concentration of 360 µg/l (SG 1024) in urine samples collected at the end of the working period.

The excretion pattern in urine of acrylonitrile exposed workers revealed that the concentration peaked at the end or shortly after the end of the workday and decreased rapidly until the beginning of the next workday (Houthuijs et al 1982).

According to Houthuijs et al (1982) a mean 8-h TWA exposure to 0.13 ppm corresponds to a mean postshift concentration of 39 µg acrylonitrile/l.

Cyanoethyl mercapturic acid. In a study performed on volunteers, 21.8% of the retained acrylonitrile was found to be excreted as N-acetyl-S-(2-cyanoethyl)-L-cysteine (2-cyanoethyl mercapturic acid)(CEMA) in urine with an elimination half-life of about 8 h.

Because of the individual differences in the kinetics this test cannot be applied to monitor a subject exposure. On a group basis, however, there was a good correlation between the uptake of acrylonitrile by lungs and excretion of this metabolite (urine collected between 6 and 8 h after the beginning of exposure). According to Jakubowski et al (1987), a urinary concentration of CEMA lower than 40 µmol/l (SG 1016) indicates that acrylonitrile exposure was below 4.5 mg/m^3. A concentration greater than 130 µmol/l would strongly suggest that this permissible exposure (TLV 4.7 mg/m^3 or 2 ppm) was exceeded.

REFERENCES

B. Ballantyne. In vitro production of cyanide in normal human blood and the influence of thiocyanate and storage temperature. *Clin. Toxicol.* 11:173, 1977.

F. Baud, P. Barriot, V. Toffis et al. Elevated blood cyanide concentrations in victims of smoke inhalation. *N. Engl. J. Med.* 325:1761, 1991.

W. Butts, M. Kuehneman and G. Widdowson. Automated method for determination of serum thiocyanate to distinguish smokers and non smokers. *Clin. Chem.* 20:1344, 1974.

A. Cailleux, J. Subra, P. Riberi et al. Cyanide and thiocyanate blood levels in patients with renal failure or respiratory disease. *J. Med.* 19:345, 1988.

H. Chandra, B. Gupta, S. Bhargava et al. Chronic cyanide exposure — a biochemical and industrial hygiene study. *J. Anal. Toxicol.* 4:1261, 1980.

C. Clark, D. Campbell and W. Reid. Blood carboxyhaemoglobin and cyanide levels in fire survivors. *Lancet* 1:1332, 1981.

H. Della Fiorentina and F. De wiest. Recherche d'un indice biologique d'exposition chronique des travailleurs aux dérivés cyanés. *Arch. Mal. Prof.* 40:699, 1978.

S. El Ghawabi, M. Gaafar, A. El Saharti et al. Chronic cyanide exposure: a clinical, radioisotope and laboratory study. *Br. J. Ind. Med.* 32:215, 1975.

Ph. Grandjean. *Skin Penetration: Hazardous Chemicals at Work.* Report prepared for the Commission of the European Communities, Health and Safety Directorate, Luxembourg, Eur 12599 En. Eds.: Taylor and Francis, London, 1990.

D. Houthuijs, B. Remijn, H. Willems et al. Biological monitoring of acrylonitrile exposure. *Am. J. Ind. Med.* 3:313, 1982.

M. Jakubowski, I. Linhart, G. Pielas and J. Kopecky. 2-Cyanoethylmercapturic acid (CEMA) in the urine as a possible indicator of exposure to acrylonitrile. *Br. J. Ind. Med.* 44:834, 1987.

T. Kato, M. Kameyama, Sh. Nakamura et al. Cyanide metabolism in motor neuron disease. *Acta Neurol. Scand.* 72:151, 1985.

P. Lundquist and B. Sörbo. Rapid determination of toxic cyanide concentrations in blood. *Clin. Chem.* 35:617, 1989.

B. McAnalley, W. Lowry, R. Oliver and J. Garriott. Determination of inorganic sulfide and cyanide in blood using specific ion electrodes application to the investigation of hydrogen sulfide and cyanide poisoning. *J. Anal. Toxicol.* 3:111, 1979.

A. Pettigrew and G. Fell. Microdiffusion method for estimation of cyanide in whole blood and its application to the study of conversion of cyanide to thiocyanate. *Clin. Chem.* 19:466, 1973.

A. Pettigrew and G. Fell. Simplified colorimetric determination of thiocyanate in biological fluids, and its application to investigation of the toxic amblyopias. *Clin. Chem.* 18:996, 1972.

H. Sakurai, M. Onodera, T. Utsunomija et al. Health effects of acrylonitrile in acrylic fibre factories. *Br. J. Ind. Med.* 35: 219, 1978.

I. Symington, R. Anderson, I. Thompson et al. Cyanide exposure in fires. *Lancet* ii:92, 1978.

R. Vogel and W. Kirkendall. Acrylonitrite (vinyl cyanide) poisoning a case report. *Toxicol. Med.* 80:48, 1984.

W. Weuffen, H. Schroeder, A. Kramer et al. Untersuchungen zum Thiocyanat-Serumspiegel bei Blutspendern. Deutch. Gesundheitwes. *Z. Klin. Med.* 36:174, 1981.

13. ISOCYANATES

Toxicokinetics

In volunteers exposed to 1,6-hexamethylene diisocyanates (HDI), 11 to 21% of the estimated inhaled dose was excreted as 1,6-hexamethylene diamine (HDA). The urinary elimination was rapid (half-time about 1.2 h) and more than 90% of the elimination of HDA was completed within 4 h after termination of exposure (Brorson et al 1990). The same authors (Brorson et al 1991; Skarping et al 1991) exposed volunteers to toluene diisocyanate (TDI) and followed the kinetics of the related amines: 2,4 and 2,6-toluene diamine (TDA) (after hydrolysis of plasma and urine). In plasma the amines seem to undergo a two-phase elimination pattern. A correlation was observed between the concentration of 2,4-TDI and 2,6-TDI in air and the levels of 2,4-TDA and 2,6-TDA in plasma. The urinary elimination of 2,4 and 2,6-TDA after inhalation of the related toluene diisocyanates was rapid, a major part (>80%) being excreted within 6 h after termination of exposure. The total amount of TDA excreted in the urine within 24 h from the start of exposure ranged between about 12 and 21% of the estimated inhaled dose of TDI.

The metabolite S-(N-methylcarbamoyl)-N-acetyl cysteine, a mercapturic acid conjugate found in the urine of workers exposed to dimethylformamide, has been identified in the urine of animals treated with methyl isocyanate (Slatter et al 1991).

Biological monitoring

Urine analysis

Related amines. In 5 volunteers exposed to HDI (mean air concentration: 25 µg/ m^3) for 7.5 h at rest, Brorson et al (1990) determined that the cumulated urinary excretion of the related amine HDA during 28 h was 8 to 14 µg (after acid hydrolysis of urine). The amount of HDA (sum of free and conjugated HDA) in samples collected immediately after the end of the exposure was on average 0.02 mmol/mol creatinine (0.01-0.03). No HDA was detected in hydrolyzed plasma samples (<0.5 µg/l).

An exposure for 7.5 h to a mean air concentration of 40 µg TDI/m^3 (approximately 48% 2,4-TDI and 52% 2,6-TDI) led to an average urinary level of 5 µg 2,4-TDA/l (range: 2.8-9.6 µg/l) and 8.6 µg 2,6-TDA/l (range: 5.6-16.6 µg/l) immediately after the end of the exposure (Brorson et al 1991; Skarping et al 1991).

Although data are still limited, several studies (Brorson et al 1990, 1991; Skarping et al 1991; Rosenberg and Savolainen 1986a,b) suggest that biological monitoring of monomeric diisocyanates by the analysis of their related amines (free and conjugated) in urine is possible. However, as their elimination is rapid, the sampling period is critical and the end-of-shift sample only gives information about the diisocyanate air concentration during the last hours of exposure.

REFERENCES

T. Brorson, G. Skarping and J. Nielsen. Biological monitoring of isocyanates and related amines. II. Test chamber exposure of humans to 1-6-heramethylene diisocyanate. *Int. Arch. Occup. Environ. Health* 62:385, 1990.

T. Brorson, G. Skarping and C. Sangö. Biological monitoring of isocyanates and related amines. IV. 2,4- and 2,6-toluenediamine in hydrolysed plasma and urine after test chamber exposure of humans to 2,4- and 2,6-toluene diisocyanate. *Int. Arch. Occup. Environ. Health* 63:253, 1991.

C. Rosenberg and H. Savolainen. Determination of occupational exposure to toluene diisocyanate by biological monitoring. *J. Chromatogr.* 367:385, 1986a.

C. Rosenberg and H. Savolainen. Determination in urine of diisocyanate. Derived amines from occupational exposure by gas chromatography. Mass fragmentography. *Analyst* III:1069, 1986b.

G. Skarping, T. Brorson and C. Sangö. Biological monitoring of isocyanates and related amines. III. Test chamber exposure of humans to toluene diisocyanate. *Int. Arch. Occup. Environ. Health* 63:83, 1991.

J. Slatter, M. Rashed, P. Pearson et al. Biotransformation of methyl isocyanate in the rat. Evidence for glutathione conjugation as a major pathway of metabolism and implications for isocyanate. Mediated toxicities. *Chem. Res. Toxicol.* 4:157, 1991.

14. CYANAMIDE

Confirming the findings of Shirota et al (1984) that acetylcyanamide is a main urinary metabolite of cyanamide (H_2NCN), Mertschenk et al (1991) stated that its determination in the urine of persons exposed to cyanamide is a suitable method to obtain a rough estimate of the total absorption of cyanamide by different routes of exposure. They found, in volunteers, that a mean of 40% of an orally administered dose of cyanamide and a mean of 7.7% of the dose in case of skin contact were excreted as acetylcyanamide in urine. Cyanamide does not seem to be metabolized to cyanide in man (Mertschenk et al 1991).

REFERENCES

B. Mertschenk, W. Bornemann, J. Filser et al. Urinary excretion of acetylcyanamide in rat and human after oral and dermal application of hydrogen cyanamide H_2NCN. *Arch. Toxicol.* 65:268, 1991.
F. Shirota, H. Nagasawa, C. Kwon and E. Desmaster. *N*-Acetylamide, the major urinary metabolite of cyanamide in rat, rabbit, dog and man. *Drug. Metab. Dispos.* 12:337, 1984.

15. PESTICIDES
15.1. Organochlorine Pesticides

Most of the organochlorine pesticides are lipid-soluble chemicals, easily absorbed by all the routes. They are very persistent in the environment and tend to accumulate in the organism. They can be measured in blood (whole blood or plasma) and in adipose tissue of the exposed workers. For some compounds, urinary metabolites have also been identified. The number of compounds for which sufficient information is available to suggest a biological threshold limit value is rather limited.

However, since many of these compounds (endrin, aldrin, dieldrin, chlordane, DDT, etc.) tend to be replaced by other less persistent products, further research is probably not justified.

15.1.1. Hexachlorocyclohexane (Lindane)

Toxicokinetics

Lindane is the gamma isomer of 1,2,3,4,5,6-hexachlorocyclohexane (γ-HCH). HCH may enter the body through the gastrointestinal and respiratory tract and through the skin. Inhalation and skin absorption are the prominent routes of uptake in the occupational setting. It accumulates in fatty tissues, but to a lesser extent than other organochlorine pesticides. The blood half-life is about 20 h. The main metabolites determined in the urine of workers exposed to γ-HCH are 2,3,5-, 2,4,5-, and 2,4,6-trichlorophenols (Angerer et al 1983; Drummond et al 1988). These metabolites accounted for almost 58% of the Lindane metabolites identified in the urine collected during the last 2 h of the workers' shift (Angerer et al 1983). The other metabolites identified include other trichlorophenols, monochlorophenols, tetrachlorophenols, dihydroxychlorobenzene (Angerer et al 1983), and penta-chlorophenol (Engst et al 1979). *In vitro* investigations indicate that human liver microsomes convert γ-HCH to pentachlorocyclohexane (with the formation of an epoxide form), 2,4,6-trichlorophenol, 2,3,4,6-tetrachlorophenol, and pentachloro-benzene (Fitzloff et al 1982; Fitzloff and Pan 1984). Chlorophenols are metabo-lites of all HCH isomers, but also of chlorobenzene and chlorophenols.

During the production of HCH and its purification to Lindane, workers are also exposed to the alpha and beta isomers (Baumann et al 1980). The beta isomer is more cumulative than the other two compounds.

Biological monitoring

Blood analysis

Hexachlorocyclohexane (HCH). In workers exposed to HCH isomers, serum levels of HCH were clearly related to the worker's job-related close contact, the intensity and duration of exposure (Gupta et al 1982; Kashyap 1986; Nigam et al 1986). The β-HCH isomer accounted for 60-70%; the longer the duration or the higher the intensity of exposure, the higher was the concentration of this isomer in serum. The serum level of β-HCH correlates with that of adipose tissue, which suggests that for this isomer the serum level is mainly a reflection of the body burden (Baumann et al 1980). This has been confirmed by Nigam et al (1986), who suggest that as they differ in their toxicity, a regular monitoring of the different isomers in the blood of the exposed workers is essential, but that for routine monitoring the determination of the β-isomer is most useful as it may serve as an index of the body burden of HCH. On the contrary it seems that the blood level of γ-HCH is not influenced by the duration of exposure and could be mainly a marker of recent exposure (Milby et al 1968). Comparing plasma γ-HCH and the urinary trichlorophenols in forestry workers exposed to γ-HCH as poten-tial biological monitoring methods, Drummond et al (1988) considered the mea-surement of plasma γ-HCH to be a valid method for use in routine practice and a specific indicator of γ-HCH absorption. The extensive and variable metabolism of γ-HCH makes trichlorophenol measurement a less suitable approach for bio-logical monitoring of this isomer. These conclusions are in agreement with the study of Angerer et al (1983). Concerning γ-HCH, several authors (Ginsburg et al 1977; Kolmodin-Hedman 1976; Samuels and Milby 1971) have claimed that blood levels of 0.003 mg/100 ml are well tolerated. Baumann et al (1981) examined chronically exposed workers with a mean blood value of about 2.5 µg/

100 ml and found no evidence of impairment of the central or peripheric nervous system. However, according to Czegledi-Janko and Avar (1970), 2 µg/100 ml is the critical blood level with respect to neurological symptoms and signs. Kashyap (1986) and Nigam et al (1986) have reported that the majority of the workers exposed to γ-HCH described toxic symptoms during their work. About 90% of workers with a mean serum level of 6 µg/100 ml and 75% of the workers with a mean serum level of 2.65 µg/100 ml suffered from paresthesia of the face and extremities, headache, giddiness, tremor, loss of sleep. In 1982, the WHO recommended that a biological exposure index limit of 2 µg/100 ml in blood should be set for workers occupationally exposed to Lindane (end-of-shift). The same limit has been proposed by the DFG (1991).

15.1.2. Dichlorodiphenyltrichloroethane (DDT)

The level of total DDT (sum of DDT and its metabolites DDE (dichlorodiphenyl-dichloroethylene) and DDA (dichlorodiphenylacetic acid)) in the blood of non-occupationally exposed persons is usually below 2 µg/100 ml (Radomski et al 1971). Bowman et al (1991) have reported a mean serum level of total DDT of 14.09 µg/100 ml in 71 people living in DDT-treated dwellings, compared to 0.604 µg/100 ml in a control group. According to Poland et al (1970), an average level of 5.7 µg DDT/100 ml blood in exposed workers corresponds to an average daily intake of about 18 mg. For the general population, an acceptable daily intake of 0.25 mg has been recommended by WHO/FAO. DDA represents the major urinary metabolite of DDT in occupationally exposed persons (Durham et al 1965; Wolfe et al 1970). On the contrary to DDT and DDE, which tend to accumulate in the adipose tissue, DDA is excreted in urine (Ramachandran et al 1984). The concentration of DDA in urine is believed to reflect recent exposure to DDT. Under field conditions, there is an average interval of 10.1 h before the excretion of DDA reaches a peak after the exposure to DDT begins (Wolfe et al 1970). Laws et al (1967) found an average urinary DDA concentration of 1.27 mg/l in 20 workers whose average daily intake of DDT was estimated at 18 mg.

15.1.3. Hexachlorobenzene

Toxicokinetics

Hexachlorobenzene is a widespread environmental contaminant (IARC 1986). It can be absorbed via ingestion, inhalation, and skin contact. The low vapor pressure of hexachlorobenzene probably limits the uptake via the respiratory tract. Koizumi (1991) has recently provided experimental evidence for the possible occupational exposure to hexachlorobenzene by skin contamination. Results of animal studies show that hexachlorobenzene is excreted as such in the feces or urine or is slowly metabolized to less chlorinated benzene, chlorinated phenols, other minor metabolites, and glucuronide and glutathione conjugates. In man, the main metabolite which has been identified is pentachlorophenol (Renner and Schuster 1977; Braun et al 1979; Van Ommen et al 1985; Kos et al 1986). Hexachlorobenzene is a very stable and cumulative toxin with a biological half-life of about 2 years in man (Currier et al 1980). However, Burton and Bennett (1987) have estimated that the mean retention time of chlorobenzene in adipose tissue is 15 years.

Biological monitoring

The concentration of hexachlorobenzene in blood is mainly a reflection of the body burden of the pesticide. Bertram et al (1986) have reported median whole blood values of 0.230 µg/100 ml (range: 0.005-1.35 µg/100 ml; n: 118) in 1977 and 0.355 µg/100 ml (range: 0.094-1.63 µg/100 ml; n: 125) in 1982 in two comparable "normal" populations. They found an age-dependent increase of the hexachloro-benzene concentration in the adipose tissue. Rutten et al (1988) found blood concentrations ranging between 0.14 and 0.27 µg/100 ml in six nonoccupationally exposed subjects. According to Burns et al (1974) and Currier et al (1980), blood levels of hexachlorobenzene below 30 µg/100 ml are not associated with distur-bances of porphyrin metabolism or signs of liver dysfunction.

The DFG (1991) recommends a biological tolerance value of 150 µg/l for hexachlorobenzene in blood.

15.1.4. *p*-Dichlorobenzene

Toxicokinetics

p-Dichlorobenzene can be absorbed by the oral, cutaneous, and pulmonary (va-por) routes.

Biological monitoring

Urine analyses

2,5-Dichlorophenol. Experimental studies suggest that 2,5-dichlorophenol (free and conjugated) and sulfur containing metabolites resulting from conjugation with glutathion may be found in urine of exposed subjects (Hawkins et al 1980; Kimura et al 1979).

p-Dichlorobenzene. The parent compound is also partly excreted unchanged in urine and according to Ghittori et al (1985), an 8-h TWA of 450 mg/m^3 will lead to a difference between pre- and postshift urinary concentration of 250 µg/l.

15.1.5. Aldrin and Dieldrin

Toxicokinetics

Aldrin and dieldrin are absorbed from the gastrointestinal tract, through the skin, or by inhalation of the vapor. After absorption, aldrin is rapidly converted to dieldrin and therefore is rarely found in the blood or tissues. Dieldrin is lipophilic and accumulates in body fat. Excretion is primarily in the feces via the bile. The urine constitutes only a minor route of elimination. The major metabolite identified in the feces of workers exposed to aldrin and dieldrin is a hydroxy derivative (Richardson and Robinson 1971; Van Sittert and Tordoir 1987a). According to Jager (1970), the mean biological half-life of dieldrin in blood is approximately 266 d.

Biological monitoring

Blood analysis

The measurement of dieldrin in blood seems to be the most valid test to determine aldrin and dieldrin absorption and is an indicator of dieldrin body burden.

Hope-Sandifer et al (1981) found a mean "normal" blood value of 4.8 μg/100 ml. Brown et al (1964) have estimated that a blood concentration exceeding 15 μg/ 100 ml may be associated with neurological symptoms. Jager (1970) has proposed the removal from exposure of workers with blood concentrations of dieldrin above 20 μg/100 ml. A case-control study of persons with blood levels of dieldrin ranging between 0.04 and 5.4 μg/100 ml has not detected any neurological deficit (Hope-Sandifer et al 1981). Van Sittert and Tordoir (1981) suggest a biological limit value for dieldrin in blood for the prevention of acute health effects of 20 μg/ 100 ml. We propose a limit of 15 μg/100 ml to prevent chronic effects.

Urine analysis

Hayes and Curley (1968) have suggested that urine analysis may be useful for monitoring the absorption of dieldrin. However, the data are insufficient to indicate which concentration of dieldrin in urine corresponds to a dangerous level of absorption.

15.1.6. Endrin

Toxicokinetics

The dermal route of exposure appears to be very important (Kummer and Van Sittert 1986; Van Sittert and Tordoir 1987b). The half-life of endrin in blood has been estimated at 24 h (Jager 1970), and the half-life of urinary anti-12-hydroxyendrin excretion has been calculated to be 55-75 h (Van Sittert 1985).

Endrin, an isomer of dieldrin, is metabolized by mammals (including man) much more rapidly than dieldrin. Anti-12-hydroxyendrin (conjugated with glucuronic acid) is a urinary metabolite of endrin (Baldwin and Hutson 1980; Rummer and Van Sittert 1986).

Biological monitoring

Because of the rapid metabolism of endrin, blood concentration of endrin is an indicator of short-term exposure, while urinary anti-12-hydroxyendrin is a more valid parameter to assess long-term exposure.

According to Ottevanger and Van Sittert (1979), a urinary level of this metabolite of 0.130 μg/g creatinine corresponds to an exposure level, below which it is likely that microsomal enzyme induction will not occur in exposed workers. Jager (1970) has indicated that the level of endrin concentration in blood, below which no signs or symptoms of intoxication ever occurs, is 5 to 10 μg/100 ml.

15.1.7. Chlordane and Heptachlor

Toxicokinetics

Chlordane and heptachlor can be absorbed by the gastrointestinal and respiratory tracts and probably by skin contact. They are stored in the adipose tissue.

Biological monitoring

Data obtained from humans exposed via inhalation to chlordane in termite-treated homes in Japan (Taguchi and Yakushiji 1988) or as the result of pesticide spraying (Kawano and Tatsukawa 1982; Saito et al 1986; Takamiya 1987) indicate that blood or tissue levels of chlordane and/or its metabolites can be useful indicators of chlordane exposure. Saito et al (1986) studied the relationship between chlordane and its metabolites in blood of pest-control operators occupationally exposed to chlordane. The concentration of chlordane and its metabolites in blood (trans-nonachlor, oxychlordane, heptachlor, heptachlor epoxide) was significantly correlated with the number of spraying days and the amount of chlordane sprayed. Chlordane was not detected in blood of the nonexposed subjects.

Serum concentration of heptachlor epoxide in Finnish plywood workers varied from below the detection limit of 0.1 ng/g (9.8 ng/100 ml) to 19.2 ng/g (1880 ng/100 ml)(mean: 3.2 ng/g or 313 ng/100 ml; SD: 3.9 ng/g or 380 ng/100 ml). The control values ranged from below the detection limit to 1.2 ng/g (117.5 ng/100 ml). The duration of exposure to heptachlor correlated with the levels of heptachlor epoxide in serum (Mussalo-Rauhamaa et al 1991).

15.1.8. Various Organochlorine Pesticides

Monitoring of occupational exposure to chlordimeform has been carried out by urinary determination of its metabolite: 4-chloro-2-toluidine (Lemesch et al 1987).

Since organochlorine pesticides are usually potent microsomal enzyme inducers, the determination of urinary D-glucaric acid has been suggested as a method of detecting exposure to these chemicals (Hunter et al 1972; Notten and Henderson 1975; Vrij-Standhardt et al 1979). However, Morgan and Roan (1974) did not find a stimulation of D-glucaric acid synthesis in workers exposed to pesticides in the chemical industry. Furthermore, the specificity and the sensitivity of this test are low (Notten and Henderson 1977). Knowledge about the influence of sex, age, and smoking habits on the urinary glucaric acid excretion in the normal population is a necessary prerequisite for its use as an index of microsomal enzyme induction in occupational and environmental toxicology (Colombi et al 1987).

15.2. Organophosphorus Pesticides

Toxicokinetics

Organophosphorus pesticides (OP) may enter the body through ingestion, inhalation, and percutaneous absorption, the absorption rate depending upon the chemical structure and formulation of the compound. Their biotransformation usually occurs as a complex multistep process (for review see Dillon and Ho 1987). Elimination of the organophosphorus compounds and their metabolites occurs via the urine and to a lesser extent the feces (via the bile) and is rapid. With most compounds 80-90% of the absorbed dose is eliminated within 48 h, although small amounts can be recovered in urine for a few days (Maroni 1986). However, some compounds (e.g., fenthion) may have a longer biological half-life. Organophosphorus pesticides are cholinesterase inhibitors.

Biological monitoring

Monitoring of exposure to these compounds is usually carried out by determining the cholinesterase activity of plasma (pseudocholinesterase) or red blood cells (true cholinesterase) or by analyzing the urine of the workers for the presence of metabolites, in particular alkylphosphates that are *in vivo* hydrolytic products of some organophosphorus pesticides. In some cases, monitoring of the pesticide itself in plasma (Roan et al 1969) or in urine (Maroni et al 1990) has also been recommended.

Blood analysis

Cholinesterase activity. The World Health Organization (WHO 1975a) has proposed that a reduction of plasma or erythrocyte cholinesterase activity to 50 or 70%, respectively, of the pre-exposure level justifies removal of the workers from exposure to organophosphorus pesticide. Plasma cholinesterase (pseudocholinesterase, or PchE) may be depressed in situations other than exposure to organophosphorus or carbamate insecticides, i.e., in early hepatitis, in alcoholic cirrhosis, and in other liver diseases as well as by various drugs (Zavon 1976; Dillon and Ho 1987). As acetylcholinesterase is found in nervous system tissues (which is the target organ of OP) its measurement in erythrocytes is a better indicator of the health risk than plasma cholinesterase. However, plasma cholinesterase is usually more susceptible to inhibition by organophosphorus pesticides than true acetylcholinesterase (e.g., erythrocyte acetylcholinesterase, or AchE). After a single OP exposure, serum PchE activity recovers more quickly than AchE activity in red blood cells. After a severe intoxication, the reduction of enzyme activity may last up to 30 d in plasma and 100 d in erythrocytes, corresponding to the time of liver PchE resynthesis and red cell replacement, respectively (Maroni 1986). Acetyl-cholinesterase inhibition is usually related to the severity of the acute intoxication by OP. In case of chronic or repeated exposure, the correlation with the toxic effects may be lower or even nonexistent. After prolonged exposure clinical signs of intoxication may appear for a higher reduction of the pre-exposure AchE level than after acute exposure. The appearance of symptoms depends more upon the rate of fall in cholinesterase activity than the absolute level of activity reached (Maroni 1986; Coye et al 1986a). At a workshop on biological monitoring of persons working with cholinesterase inhibitors held in 1975 (Zavon 1976), it was concluded that "there is nevertheless a place for plasma cholinesterase monitor-ing, as in a well controlled manufacturing situation where minor breaks in control procedures may be detected by this means. However, the erythrocyte cholinest-erase is the better measure in most situations and, when possible, plasma and erythrocyte determinations should both be done. Whole blood cholinesterase could also be used, however the contribution of each enzyme to the total activity should be known under the given conditions of the assay". It has also been stressed that, because the natural cholinesterase level varies considerably from one indi-vidual to another, it is desirable before handling or coming into contact occupa-tionally with organophosphorus compounds that everyone should have at least one estimation of both plasma and erythrocyte cholinesterase activities. These figures then provide a baseline against which any subsequent exposure estima-tions can be compared for their significance. It is proposed that action must be taken to minimize exposure when the depression from the baseline for erythrocyte (or whole blood) cholinesterase and plasma cholinesterase amounts to 30 and 50%, respectively.

Urine analyses

Alkylphosphates. Exposure to certain OP pesticides may be detected by measure-ment of urinary alkylphosphates (Morgan et al 1977; Shafik et al 1973). Depend-ing on the chemical structure of the pesticide, various alkylphosphates may be detected (Brokopp et al 1981; Dillon and Ho 1987; Coye et al 1986b) (see Table 14).

Usually increased urinary excretion of alkylphosphates is detectable at an exposure level insufficient to depress plasma or red blood cell cholinesterase (Franklin et al 1981). Morgan et al (1977) showed that total excretion of alkylphosphate metabolites is dose dependent in humans in the case of a single ingested dose. Franklin et al (1981) reported a high correlation between alkylphosphate excretion and amount of active ingredient sprayed in the course of occupational exposure to gluthion. However, since numerous OP pesticides yield similar metabolites, the use of this method alone to identify the type of OP exposure is limited. Moreover, the measurement of these metabolites requires complex analytical methods and no relationship between these metabolite urinary levels and the amount absorbed has been established yet.

Table 14
Alkylphosphates Derived From Various OP Pesticides

Metabolite	Principal Parent Compounds
Monomethylphosphate	Malathion
Dimethylphosphate	Dimethylparathion, dichlorvos, trichlorfon, mevinphos, malaoxon, dimethoate, fenchlorphos, dicrotophos, guthion, malathion, phosphamidon
Diethylphosphate	Parathion, disulfoton, phorate, paraoxon, tetraethyl, pyrophosphate, demeton-oxon, diazinon-oxo, dichlofenthion, chlorpyrifos, ethion, systox
Dimethylthiophosphate	Fenitrothion, fenchlorphos, malathion, dimethoate, guthion, methylparathion
Diethylthiophosphate	Disulfoton, phorate, diazinon, demeton, parathion, fenchlorphos, chlorpyrifos, ethion, systox
Dimethyldithiophosphate	Malathion, dimethoate, azinphos-methyl, guthion
Diethyldithiophosphate	Disulfoton, phorate, ethion
Phenylphosphoric acid	Leptophos, EPN

Other metabolites. Absorption of parathion is rapidly followed by an increased urinary excretion of *p*-nitrophenol (Arterberry et al 1961; Durham et al 1971; Roan et al 1969; Wolfe et al 1970). Under field conditions, there is an average interval of 8.7 h before the excretion of *p*-nitrophenol reaches a peak after exposure to parathion begins (Hayes 1971). It has been reported that absorption of parathion is tolerated without illness and with little or no reduction in cholinesterase activity so long as the concentration of *p*-nitrophenol in the urine does not rise much above 2 mg/l (Hayes 1971; WHO 1975a). Higher rates of excretion, reflecting higher tissue levels and, eventually higher absorption are accompanied by cholinesterase inhibition and illness. It should be noted that *p*-nitrophenol is also the phenolic metabolite of methyl parathion and *O*-ethyl *O*-4-nitrophenyl phenyl phosphonothioate (EPN). According to Dillon and Ho (1987), considerable potential exists in the biological monitoring of exposure to OP esters by monitoring metabolites other than those discussed above (for example, see Table 15). However, this needs further research.

Fenske and Elkner (1990) have demonstrated that the urinary levels of the principal metabolite of chlorpyrifos 3,5,6-trichloro-2-pyridinol, collected 24-48 h postexposure, were highly correlated with total absorbed dose estimates (approximately 73% contributed by the dermal route).

As Hayes (1971) has pointed out, both approaches, measurement of cholinesterase activity and pesticide metabolite determination in urine, give complementary information because metabolite excretion occurs rather rapidly, while enzyme activity recovers slowly. The cholinesterase determination integrates

Table 15
Examples of Organophosphorous Metabolites

Metabolite	Parent Compound
Glucuronide of:	
Dichlorovinyl alcohol	Dichlorovos, trichlorfon
Trichloroethanol	Trichlorfon
Sulfoxides, sufones	p-S-demeton, p-O-demeton, disulfoton, phorate
Mono-dicarboxylic acids	Malathion, malaoxon
Thiodiazole leaving group	Methidathion, its methyl sulfide, methyl sulfoxide, methyl sulfone
Conjugated enols and carboxylic acids	Dicrotophos, monocrotophos, dimethoate

the effects of exposure during several days, while urinary metabolite measurement gives information on very recent exposure. A reduction of red cell acetylcholinesterase activity to 70% (inhibition 30%) of the individual reference value is suggested as an indicator of risk of overexposure. This level has been adopted by the ACGIH (1991-1992) and the DFG (1991). Pseudocholinesterase activity in plasma is generally more sensitive, but less specific; an inhibition level of 50% is proposed as a biological limit value. The alkylphosphates and other urinary metabolites can be used as indicators of exposure to OP, but data are too limited to establish BEI. A maximum permissible concentration of 0.5 mg p-nitrophenol/g creatinine is recommended to monitor occupational exposure to parathion.

15.3. Carbamate Insecticides

Several N-methyl and N-dimethyl carbamate esters are used as insecticides. They are cholinesterase inhibitors, but cholinesterase when inhibited by these compounds is reactivated rapidly. Because of this, cholinesterase activity at the time of blood analysis may not necessarily be a true indication of cholinesterase activity in the circulating blood at the time of blood withdrawal. The determination of urinary metabolites may be a more practical method of monitoring exposure, but insufficient human data exist for the proposal of meaningful biological monitoring methods for the majority of carbamate insecticides.

15.3.1. Carbaryl

Toxicokinetics

Carbaryl is 1-naphthyl-N-methylcarbamate. Carbaryl enters the organism by all the routes. It is rapidly metabolized and the metabolites are excreted mainly through the kidneys. Free and conjugated 1-naphthol are the main urinary metabolites (NIOSH 1976).

Biological monitoring

Exposure to carbaryl can be monitored by the measurement of pseudocholines-terase activity in plasma and by the determination of total 1-naphthol in the urine (Best and Murray 1962; Comer et al 1975; Vandekar 1965). The relationship between the external exposure to carbaryl and the concentration of 1-naphthol in urine has not yet been clearly established, but it seems that when the 1-naphthol concentration in urine does not exceed 10 mg/l at the end of the exposure period, the risk of occurrence of symptoms or signs of clinical intoxication is low (WHO 1975b). Because of the rapid reactivation of cholinesterase inhibited by carbaryl, blood samples should be drawn and analyzed within 4 h after exposure.

15.3.2. Isopropoxyphenyl-N-Methylcarbamate

2-Isopropoxyphenol can be detected in urine of persons exposed to this carbamate insecticide (Dawson et al 1964).

15.4. Phenoxy Acid Herbicides

Toxicokinetics

The phenoxy acid herbicides are absorbed through the skin, lungs, and gas-trointestinal tract. Dermal absorption is the major route of uptake in the occupa-tional setting. They are rapidly eliminated mainly unchanged in urine. However, Knopp and Glass (1991) found a relatively slow elimination of 2,4-dichloro-phenoxyacetic acid in urine, its clearance occurring with a 12- to 22-h half-life.

Fjeldstad and Wannag (1977) followed the urinary excretion of 2-methyl-4-chlorophenoxyacetic acid (MCPA) after administration of 7 mg of the herbicide to volunteers. About 50% of the ingested dose was detected in the urine within 48 h.

Biological monitoring

Urine analysis

Kolmodin-Hedman and Erne (1980) have studied occupational exposure to 2,4-dichlorophenoxyacetic acid (2,4-D) and 2,4,5-trichlorophenoxyacetic acid (2,4,5-T) in 4 men spraying a 2% emulsion in kerosene. The mean airborne concentration in the breathing zone was 0.1 mg/m^3 for 2,4-D and 0.2 mg/m^3 for 2,4,5-T. The mean 24-h excretion in urine was 9 mg for 2,4-D and approximately 1 mg for 2,4,5-T.

Smith et al (1981) have also detected 2,4-D in urine of forestry workers. They indicate that since the pharmacokinetic behavior of this compound is independent of the route of administration and since more than 95% of the absorbed dose is excreted in urine, the amount excreted is a direct reflection of the internal dose.

Kolmodin-Hedman et al (1983a,b) and Knopp and Glass (1991) have also observed that the urinary concentration of MCPA, 2,4-D, dichlorprop, mecaprop, at the end of the shift or the following morning reflects the exposure intensity. No toxic manifestation was detected in subjects with urinary MCPA concentrations up to 12 mg/l (Kolmodin-Hedman et al 1983a).

Many other authors (Linnainmaa 1983; Frank et al 1985; Yeary 1986; Nigg and Stamper 1983; Libich et al 1984; Vural and Burgaz 1984) have measured the concentration of chlorophenoxy acid herbicides (2,4-D; 2,4,5-T; MCPA) in the urine of occupationally exposed workers, the concentrations varying from nondetectable to about 30 mg/l (IARC 1986). Nevertheless, data are still lacking to propose a maximum permissible urinary concentration of phenoxy acid derivatives.

Blood analysis

In the past, technical preparations of phenoxy acid herbicide, particularly 2,4,5-T, were frequently contaminated by various polychlorinated dibenzo-*p*-dioxins and dibenzofurans. Serum levels of polychlorinated dibenzo-*p*-dioxins and dibenzofurans, which were substituted with chlorine at the 2,3,7,8 position, were measured in nine workers who had sprayed pesticides, although not necessarily 2,4,5-T, for a range of 83-372 months and in nine matched control subjects. The average level of 2,3,7,8-tetrachlorodibenzo-*p*-dioxin (TCDD) for applicators was almost 10 times that of the control value. The average levels of all other congeners and isomers did not differ substantially. The variation in TCDD levels among the applicators was related to their duration of exposure to 2,4,5-T (Smith et al 1992).

15.5. 4,6-Dinitro-Ortho-Cresol (DNOC)

DNOC can be absorbed by all routes. Its concentration in blood (90% in plasma) seems to be related to the severity of adverse effects. Blood DNOC concentrations higher than 2 mg/100 ml indicate a health hazard and levels greater than 4 mg/100 ml may be associated with clinical poisoning (Bidstrup et al 1952). The lowest effect level (change in biochemical indicators) was found at 1.1 mg/100 ml serum (Jastroch et al 1978; Thiele 1983).

NIOSH (1978) has proposed the following biological surveillance program. If the blood DNOC level on Monday morning (assuming end of exposure on Friday afternoon) is equal to or greater than 1 mg/100 ml of whole blood (the "warning" level), continued work is permitted but an industrial hygiene survey should be conducted to ascertain whether or not prescribed control procedures are adequate and are being followed by the workers. If the blood DNOC level is 2 mg/100 ml of whole blood or greater, the worker should be removed from further contact with DNOC until the blood level falls below this concentration.

The DFG (1991) recommends 1 mg/100 ml whole blood as biological tolerance value.

On the basis of Thiele's assessment (1983), we propose a biological exposure limit of 1 mg/100 ml in serum or 0.5 mg/100 ml in whole blood (sample taken at the end of the workshift).

15.6. Pentachlorophenol

Toxicokinetics

The chlorophenols can be absorbed by all routes, but in occupational settings, the main routes of exposure are the respiratory tract and the skin (Kauppinen and Lindroos 1985; Lindroos et al 1987; Embree et al 1984; Fenske et al 1987; Grandjean 1990). The fungicide pentachlorophenol (PCP) is mostly excreted in urine, as the parent compound and conjugates. Metabolic transformation by human liver to tetrachlorohydroquinone has been demonstrated in *in vitro* studies (Juhl et al 1985) (Figure 24). The relative amount of PCP excreted by humans as conjugates and as the free form has not yet been clearly established. Braun et al (1979) found that 74% of a single dose (0.1 mg/kg) of PCP administered orally to volunteers was excreted as free PCP with a half-time of 33 h, while 12% was excreted as a glucuronide conjugate with a half-time of 13 h. On the other hand, Edgerton and Moseman (1979), who compared methods with and without acid hydrolysis for PCP determination in urine, found that hydrolyzed samples gave results generally 4 to 6 times higher than untreated samples. The mean percentage of PCP conjugated to glucuronic acid in the urine of nonspecifically exposed persons examined by Uhl et al (1986) was 65%. Gomez-Catalan et al (1987) also found that PCP is excreted mainly as a conjugate; free PCP in urine amounted to only about 13% (13.2 ± 1.3) of the total PCP excreted in a normal population and about 9.5% (9.3 ± 3.8) in occupationally exposed workers. In the study of Pekari et al (1991) conducted on workers exposed to chlorophenols, urinary PCP was conjugated to a smaller extent (about 70-75%) than either tetra- or trichlorophenols

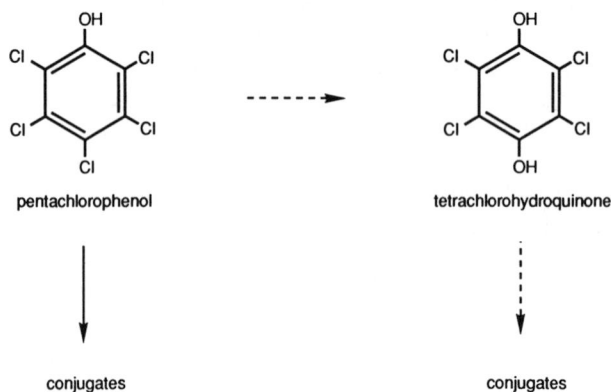

FIGURE 24 Main metabolic pathways of pentachlorophenol.

(about 80-98%). Morover, it seems that sulfate conjugation dominates at low urinary concentrations, but when chlorophenol concentration increases glucuronic acid conjugation becomes more important (Pekari et al 1991). The absorption resulting from a single workday exposure to 0.5 mg PCP/m^3 has been shown to require about 1 week (168 h) for the body to eliminate 86% of the original dose (Braun et al 1979). From investigations on volunteers, Uhl et al (1986) derived blood and urine elimination half-lives of 16-20 d. They explain this slow elimination of PCP by the low urinary clearance due to high plasma protein binding (>96%) and tubular reabsorption. The time lag necessary to attain a steady state after a change in the exposure level is estimated to be about 3 months. The half-times reported by Pekari et al (1991) in the urine of workers exposed to chlorophenols were 18 h for trichlorophenol, 4.3 d for tetrachlorophenol, and 16 d for pentachlorophenol. The discrepancy between those results and the study of Braun et al (1979) might be explained by differences in the kinetics of PCP following a single oral dose and the kinetics during chronic exposure to lower doses. It should also be noted that PCP is a metabolite of hexachlorobenzene.

Biological monitoring

Blood and urine analyses

Pentachlorophenol. PCP is widespread in the environment and so can be detected in the urine and plasma of most nonoccupationally exposed persons (mean levels probably <40 ng/ml). Higher concentrations are found in residents of log homes or subjects living in homes where the wood has been treated by PCP. Within families, Cline et al (1989) have observed that the ratio of the children's PCP levels to the average of their parents PCP levels was close to 1.8. Workers occupationally exposed to PCP show the highest levels. However, the concentrations reported are very different among the authors (Bevenue et al 1967; Wood et al 1983; Klemmer et al 1980; Bomhard et al 1984; Embree et al 1984; Butte 1985; Kleinman et al 1986; Maroni et al 1986; Jones et al 1986, Currie and McDonald 1986; Uhl et al 1986; Cline et al 1989; Seiler 1991).

Repeated workday exposure to 0.5 mg/m^3 would result in a maximum steady-state plasma concentration of approximately 50 µg/100 ml (Braun et al 1979; Wood et al 1983). However, the importance of skin absorption must not be neglected. According to Zober et al (1981), a concentration of free pentachlorophenol in urine below 2 mg/g creatinine can still be associated with biological signs of liver dysfunction. The data of Begley et al (1977) suggest that PCP affects renal function if the mean free PCP urinary and serum concentrations are about 9 mg/l urine and 0.26 mg/100 ml serum. Klemmer et al (1980) could not evidence specific long-term effects, except irritation of the skin and mucous membranes, at a mean serum concentration of 0.378 mg PCP/100 ml. Workers having mean urinary and serum levels of 0.932 µmol chlorophenate (tetra, penta)/l urine and 3.580 µmol chlorophenate/l serum had a lower hematocrit and microscopic hematuria (Enarson et al 1986). Jones et al (1986) did not detect chloracne, lipid metabolism, or liver function disturbances at a mean urinary concentration of 650 µg/g creatinine. No respiratory symptoms were observed at preshift value of 105 µg PCP/l urine and 71.4 µg PCP/100 ml serum (Embree et al 1984).

The BEI adopted by ACGIH (1991-1992) are 2 mg/g creatinine (total PCP in urine sample taken prior to the last shift of the workweek) and 5 mg/l in plasma (free PCP in blood sample taken at the end of the shift).

The biological tolerance values recommended used as guidelines by DFG (1991) are 1 mg pentachlorophenol/l plasma or serum and 300 µg pentachlorophenol/l in urine (sampling time not fixed).

PCP concentration in blood and urine can be used as an indicator of PCP exposure and probably reflects integrated exposure over several weeks. The no-observed-effect-level in case of long-term occupational exposure is not yet clearly established. On the basis of the available data, maximum permissible concentrations of 1 mg PCP/g urinary creatinine and 0.05 mg PCP/100 ml serum are proposed.

15.7. 2-Ethylhexanoic Acid

Exposure to 2-ethylhexanoic acid (EHA) (wood preservative agent) has been determined by urinalysis of the compound in workers exposed in Finnish saw-mills. The main route for entrance into the body was by breathing. The mean concentrations of 2-EHA in the breathing zone correlated linearly with urinary levels of 2-EHA: a mean air concentration of 0.5 mg 2-EHA/m^3, leading to a mean urinary concentration of 0.68 mmol 2-EHA/mol creatinine (sample taken immediately after a workshift)(Kröger et al 1990; Pennanen and Manninen 1990).

15.8. Pyrethroids
15.8.1. Cypermethrin

Cypermethrin is absorbed by all routes. Single oral doses of cypermethrin given to male volunteers has resulted in a rapid urinary excretion of its metabolites: on average, 78% of the dose of the trans isomer and 49% of the cis isomer within 24 h. The ester cleavage was a major route of biotransformation (Eadsforth and Baldwin 1983). A similar experiment conducted on five consecutive days did not reveal any accumulation in the body (Van Sittert et al 1985a). Cyclopropane carboxylic acid has been identified in the urine of subjects exposed to cypermethrine. A urinary level reaching 0.4 mg/l did not cause clinical manifestations (Coveney and Eadsforth 1982; Van Sittert et al 1985b; Eadsforth and Baldwin 1983).

15.8.2. Deltamethrin and Fenvalerate

Oral exposure of volunteers to labelled deltamethrin has shown that the apparent plasma and urinary elimination half-lives were consistent: between 10 and 11.5 h, and 10 and 13.5 h, respectively. Urinary excretion accounted for 51-59% of the administered radioactivity, 90% of this radioactivity was excreted during the 24 h following absorption (Papalexiou et al, cited by the WHO 1990). Zhang et al (1991) have assessed the occupational exposure to deltamethrin and fenvalerate by measuring their concentrations in urine. Dermal exposure was found to constitute the main route of absorption: the exposure through the pulmonary route was less than 1% of that through the skin. The symptoms described by the workers (headache, dizziness, nausea) showed no significant correlation with the urinary pyrethroid excretion.

15.9. 1,3-Dichloropropene

Toxicokinetics

The respiratory tract is the predominant route of uptake of 1,3-dichloropropene (DCP). However, cutaneous absorption might play a significant role in case of spills.

The mercapturic acid metabolites *N*-acetyl-*S*-(cis and trans 3-chloropropenyl-2)-L-cysteine (DCP-MA) are excreted in urine (Osterloh et al 1984, 1989; Van Welie et al 1989, 1991a,b,c). Urinary half-lives are about 5 (SD: 1.2) h for cis DCP-MA and 4.7 (SD: 1.3) h for trans DCP-MA. Urinary half-lives of elimination of thioethers are almost twofold higher.

Biological monitoring

Occupational exposure to 1,3-dichloropropene has been determined by measuring the urinary excretion of the mercapturic acid, *N*-acetyl-*S*-(3-chloropropenyl-2)-L-cysteine.

Urine analyses

Mercapturic acid. Mercapturic acids of DCP are not detected in urine of subjects nonexposed to DCP. However, nonoccupational exposure to DCP of bystanders living in the neighborhood of fields that have been treated with DCP can be detected by urinary excretion of DCP-MA (Van Welie et al 1991b).

A first field study of Osterloh et al (1984) showed a good correlation between integrated exposure to airborne DCP (mg.min/m^3) and the 24-h DCP-MA urinary excretion (r: 0.854; $p < 0.001$). A further study described an even more precise correlation with DCP-MA excretion in the urine of the next morning (r: 0.914; $p < 0.001$) (Osterloh et al 1989). According to Osterloh et al (1989), an integrated exposure to DCP >700 mg.min/m^3 correspond to a urinary excretion of >1.5 mg C DCP-MA/d, and such exposure intensity may lead to renal injury.

The regression equations drawn from the relationship observed in DCP applicators between respiratory 8-h TWA exposure to DCP and the cumulative urinary excretion of the corresponding mercapturic acids showed that an exposure to 5 mg DCP/m^3 (ratio of cis versus trans DCP: 1.36 ± 0.07) would lead to a urinary excretion of about 24 mg for the cis-DCP isomer and 9 mg for the trans-DCP isomer (Van Welie et al 1991c).

Thioether. It is well known that thioethers are normally present in the urine and their background level depends on lifestyle.

Van Welie et al (1991a) studied the excretion of thioethers in urine of applicators occupationally exposed to DCP. In nonexposed subjects the mean background level of urinary thioethers was 6.05 mmol SH/mol creatinine. Post-minus preshift thioether concentration and cumulative thioether excretion showed good linear relations with exposure to DCP. An 8-h TWA exposure to 5 mg/m^3 DCP resulted in a post-minus preshift thioether concentration of 9.6 mmol SH/mol creatinine (95% confidence interval 7.4-11.8 mmol SH/mol creatinine) and in a cumulative thioether excretion of 139 μmol SH (95% confidence interval 120-157 μmol SH).

15.10. Fungicides

Some studies have suggested potential biological indicators to monitor exposure of workers to several fungicides: urinary ethylenethiourea for exposure to ethylenebisdithiocarbamate (Kurttio and Savolainen 1990), urinary tetrahydrophtalimide and 2-thiothiazolidine-4-carboxylic acid for exposure to captan (Van Welie et al 1991d), 5-ethoxy-1,2,4-thiodiazole-3-carboxylic acid, and to a lesser extent N-acetyl-S-(5-ethoxy-1,2,4-thiodiazole-3-yl-methyl)-L-cysteine) for exposure to etridiazole (Van Welie et al 1991e).

15.11. Ethylene Oxide

Toxicokinetics

Exposure to ethylene oxide mainly occurs via inhalation. The alveolar retention of ethylene oxide lies around 75-80% of the inhaled dose (Brugnone et al 1985, 1986). Two groups of subjects with significant differences in the metabolism of ethylene oxide in blood have been identified: the "conjugators" (with a rapid decrease of the ethylene oxide) and the "nonconjugators". This enzyme polymorphism may influence the cytogenetic toxicity of ethylene oxide (Föst et al 1991; Leutbecher et al 1991).

Biological monitoring

Blood and breath analyses

Ethylene glycol. Wolfs et al (1983) observed in workers exposed to ethylene oxide (mean atmospheric concentration: 4.2 ppm; range: 0.3-52) an increase in ethylene glycol concentration in blood at the end of the shift (mean concentration: 9.03 mg/100 ml; SE: 1.26; controls: 4.5 mg/100 ml; SE: 1.31).

Ethylene oxide. On the basis of a field study conducted on workers employed in a hospital sterilizer unit, Brugnone et al (1985, 1986) have estimated that a TWA exposure to 2 mg/m^3 would lead at the end of the shift to mean alveolar and blood concentrations of ethylene oxide of 0.5 µg/l and 0.8 µg/100 ml, respectively. This is in agreement with the estimation of the DFG (1991): exposure to mean air concentrations of 0.92, 1.83, 3.66, 5.49, 7.32, and 9.15 mg/m^3 leading to mean alveolar air and whole blood ethylene oxide concentrations of 0.22, 0.44, 0.88, 1.32, 1.76, and 2.2 mg/m^3 and 3, 6.1, 12.1, 18.1, 24.3, and 30.3 µg/l, respectively (sampling time: during exposure >4 h).

Hemoglobin adduct. Some authors (Sarto et al 1991; Hagmar et al 1991; Van Sittert et al 1985a; Osterman-Golkar and Bergmark 1988; Farmer et al 1986; Kautiainen and Törnqvist 1991) have assessed occupational exposure to ethylene oxide by measuring the amount of 2-hydroxyethyl adduct to hemoglobin. According to Kautiainen and Törnqvist (1991), a hemoglobin adduct level of 100 pmol/g globin can be expected in a person exposed to about 50 ppb ethylene oxide during working hours (40 h/week).

Comparing the relative sensitivity of endpoints used for detection of ethylene oxide exposure, Tates and co-workers (1991) established that *N*-(2-hydroxyethyl)-valine adducts in hemoglobin were the most sensitive indicator for relatively recent exposure to ethylene oxide, i.e., exposure that occurred during about 4 months (life span of erythrocytes) prior to blood sampling. It should be possible to detect a TWA exposure (40 h/week) of about 0.01 ppm in nonsmoking individuals and still lower exposures at the group level. The relative sensitivity for the other tests was in the following order: SCEs > chromosomal aberrations > micronuclei > HPRT mutants.

Attempts have also been made to measure adducts of ethylene oxide on lymphocyte DNA.

REFERENCES

ACGIH. American Conference of Governmental Industrial Hygienists. Threshold Limit Values for Chemical Substances and Physical Agents and Biological Exposure Indices. Cincinnati, 1991-1992.

J. Angerer, R. Maas and R. Heinrick. Occupational exposure to mexachlorocyclohexane. VI. Metabolism of γ-hexachlorocyclohexane in man. *Int. Arch. Occup. Environ. Health* 52:59, 1983.

J. Arterberry, W. Durham, J. Elliott and H. Wolfe. Exposure to parathion: measurement by blood cholinesterase level and urinary *p*-nitrophenol excretion. *Arch. Environ. Health* 3:476, 1961.

M. Baldwin and D. Hutson. Analysis of human urine for a metabolite of endrin by chemical oxidation and gas-liquid chromatography as an indicator of exposure to endrin. *Analyst* 105:60, 1980.

K. Baumann, J. Angerer, R. Heinrich and G. Lehnert. Occupational exposure to hexachlorocyclohexane. 1. Body burden of HCH-isomers. *Int. Arch. Occup. Environ. Health* 47:119, 1980.

K. Baumann, K. Behling, H. Brassow and K. Stapel. Occupational exposure to hexachlorocyclohexane. III. Neurophysiological findings and neuromuscular function in chronically exposed workers. *Int. Arch. Occup. Environ. Health* 48:165, 1981.

J. Begley, E. Reichert, M. Rasmad and H. Klemmer. Association between renal function tests and pentachlorophenol exposure. *Clin. Toxicol.* 11:97, 1977.

H. Bertram, F. Kemper and C. Müller. Hexachlorobenzene content in human whole blood and adipose tissue: experiences in environmental specimen banking. In: *"Hexachlorobenzene: Proceedings of an International Symposium"*. Ed.: C. Morris, J. Cabral. IARC Scientific Publication No. 77, Lyon 1986.

E. Best, Jr. and B. Murray. Observation on workers exposed to Sevin insecticide. A preliminary report. *J. Occup. Med.* 4:507, 1962.

A. Bevenue, J. Wilson, L. Casarett and H. Klemmer. A survey of pentachlorophenol content in human urine. *Bull. Environ. Contam. Toxicol.* 2:319, 1967.

P. Bidstrup, J. Bonnel and D. Harvey. Prevention of acute dinitro-ortho-cresol (DNOC) poisoning. *Lancet* 1:794, 1952.

A. Bomhard, K. Schaller and G. Triebig. Quantitative determination of pentachlorophenol in human plasma and urine by capillary gas chromatography with ECD and MS detection. *Presenius. Anal. Chem.* 319:516, 1984.

H. Bowman, R. Cooppan, P. Becker and S. Ngxongo. Malaria control and levels of DDT in serum of two populations in Kwazulu. *J. Toxicol. Environ. Health* 33:141, 1991.

W. Braun, G. Blau and M. Chenoweth. The metabolism/pharmacokinetics of pentachlorophenol in man, and a comparison with the rat and monkey. *Dev. Toxicol. Environ. Sci.* 4:289, 1979.

C. Brokopp, J. Wyatt and J. Gabica. Dialkyl phosphates in urine samples pesticide formulators exposed to disulfoton and phorate. *Bull. Environ. Contam. Toxicol.* 26:524, 1981.

V. Brown, C. Hunter and A. Richardson. A blood test diagnostic to exposure to aldrin and dieldrin. *Br. J. Ind. Med.* 21:283, 1964.

F. Brugnone, L. Perbellini, G. Faccini and F. Pasini. Concentration of ethylene in the alveolar air of occupationally exposed workers. *Am. J. Ind. Med.* 8:67, 1985.

F. Brugnone, L. Perbellini, G. Faccini et al. Ethylene oxide exposure. Biological monitoring by analysis of alveolar air and blood. *Int. Arch. Occup. Environ. Health* 58:105, 1986.

J. Burns, F. Miller, E. Gomes and R. Albert. Hexachlorobenzene exposure from contaminated DCPA in vegetable spraymen. *Arch. Environ. Health* 29:192, 1974.

M. Burton and B. Bennett. Exposure of man to environmental hexachlorobenzene. An exposure commitment assessment. *Sci. Total Environ.* 66:137, 1987.

W. Butte. Pentachlorophenol concentrations in serum and urine of persons that have used wood preserving agents. *J. Clin. Chem. Clin. Biochem.* 23:599, 1985.

R. Cline, R. Hill, D. Phillips and L. Needham. Pentachlorophenol measurements in body fluids of people in log homes and workplaces. *Bull. Environ. Contam. Toxicol.* 18:475, 1989.

A. Colombi, M. Maroni, C. Antonini et al. Reference values of D-glucaric acid excretion. In: *Biological Monitoring of Exposure to Chemicals. Organic Compounds*, p. 297. Eds.: M. Ho, H. Dillon. Wiley Interscience, New York, 1987.

S. Comer, D. Staiff, S. Armstrong and H. Wolfe. Exposure of workers to carbaryl. *Bull. Environ. Contam. Toxicol.* 13:385, 1975.

P. Coveney and C. Eadsforth. The metabolism of cypermethrin in man. Urinary excretion following a single dermal dose of cypermethrin. Sittingbourne, Shell Research (SBGR 82.290), 1982.

M. Coye, J. Lowe and K. Maddy. Biological monitoring of agricultural workers exposed to pesticides. I. Cholinesterase activity determinations. *J. Occup. Med.* 28:619, 1986a.

M. Coye, J. Lowe and K. Maddy. Biological monitoring of agricultural workers exposed to pesticides. II. Monitoring of intact pesticides and their metabolites. *J. Occup. Med.* 28:628, 1986b.

K. Currie and E. Mc Donald. Uptake/excretion of chlorophenols by sawmill employees. Phase II. Ottawa, Health and Welfare Canada (B.C. Research Report to Pesticides Division; contract No. 1424), 1986.

M. Currier, C. McClimans and G. Barna-Lloyd. Hexachlorobenzene blood levels and the health status of men employed in the manufacture of chlorinated solvents. *J. Toxicol. Environ. Health* 6:367, 1980.

G. Czegledi-Janko and P. Avar. Occupational exposure to lindane: clinical and laboratory findings. *Br. J. Ind. Med.* 27:283, 1970.

J. Dawson, D. Heath, J. Rose et al. The excretion by humans of the phenol derived in vivo from 2-isopropoxyphenyl N-methylcarbamate. *Bull. WHO* 30:127, 1964.

DFG. Deutsche Forschungsgemeinschaft. *Maximum Concentrations at the Workplace and Biological Tolerance Values for Working Materials.* Report No. XXVII. Commission for the Investigation of Health Hazards of Chemical Compounds in the Work Area. VCH, Weinheim, 1991.

H. Dillon and M. Ho. Review of the biotransformation of organophosphorus pesticides. In: *Biological Monitoring of Exposure to Chemicals. Organic Compounds,* p. 227. Eds.: M. Ho, H. Dillon. Wiley-Interscience, New York, 1987.

L. Drummond, E. Gillanders and H. Wilson. Plasma γ-hexachlorocyclohexane concentrations in forestry workers exposed to lindane. *Br. J. Ind. Med.* 45:493, 1988.

W. Durham, J. Armstrong and G. Quinby. DDA excretion levels: studies in persons with different degrees of exposure to DDT. *Arch. Environ. Health* 11:76, 1965.

W. Durham, H. Wolfe and J. Elliot. Absorption and excretion of parathion by spraymen. *Arch. Environ. Health* 24:381, 1971.

C. Eadsforth and M. Baldwin. Human-dose excretion studies with the pyrethroid insecticide cypermethrin. *Xenobiotica* 13:67, 1983.

T. Edgerton and R. Moseman. Determination of pentachlorophenol in urine: the importance of hydrolysis. *J. Agr. Food Chem.* 27:197, 1979.

V. Embree, D. Enarson, M. Chan-Yeung et al. Occupational exposure to chlorophenates: toxicology and respiratory effects. *Clin. Toxicol.* 22:317, 1984.

D. Enarson, M. Chang-Yeung, V. Embree et al. Occupational exposure to chlorophenates— renal, hepatic and other health effects. *Scand. J. Work Environ. Health* 12:144, 1986.

R. Engst, R. Macholz and M. Kujawa. Metabolism of lindane in microbial organisms, warmblooded animals and humans. *Gig. Sanit.* 10:64, 1979.

P. Farmer, E. Bailey, S. Gorf et al. Monitoring human exposure to ethylene oxide by the determination of haemoglobin adducts using gas chromatography-mass spectrometry. *Carcinogenesis* 7:637, 1986.

R. Fenske and K. Elkner. Multi-route exposure assessment and biological monitoring of urban pesticide applicators during structural control treatment with chlorpyrifos. *Toxicol. Ind. Health* 6:349, 1990.

R. Fenske, S. Horstman and R. Bentley. Assessment of dermal exposure to chlorophenols in lumber mills. *Appl. Ind. Hyg.* 2:143, 1987.

J. Fitzloff, J. Portig and K. Stein. Lindane metabolism by human and rat liver microsomes. *Xenobiotica* 12:197, 1982.

J. Fitzloff and J. Pan. Epoxidation of the lindane metabolite β-PCCH by human and rat liver microsomes. *Xenobiotica* 14:599, 1984.

P. Fjeldstad and A. Wannag. Human urinary excretion of the herbicide 2-methyl-4-chlorophenoxyacetic acid. *Scand. J. Work Environ. Health* 3:100, 1977.

V. Föst, E. Hallier, H. Ottenwalder et al. Distribution of ethylene oxide in human blood and its implication for biomonitoring. *Hum. Exp. Toxicol.* 10:25, 1991.

R. Frank, R. Campbell and G. Sirons. Forestry workers involved in aerial application of 2,4-dichlorophenoxyacetic acid (2,4-D) exposure and urinary excretion. *Arch. Environ. Contam. Toxicol.* 14:427, 1985.

C. Franklin, R. Fenske, R. Greenhalfgh et al. Correlation of urinary pesticide metabolite excretion with estimated dermal contact in the course of occupational exposure to guthion. *J. Toxicol. Environ. Health* 7:715, 1981.

C. Ginsburg, W. Lowry and J. Reisch. Absorption of lindane (gamma benzene hexachloride) in infants and children. *J. Pediatr.* 91:998, 1977.

S. Ghittori, M. Imbriani, G. Pezzagno and E. Capodaglio. Urinary elimination of *p*-dichlorobenzene (p-dcn) and weighted exposure concentration. *G. Ital. Med. Lav.* 7:59, 1985.

J. Gomez-Catalan, J. To-Fiqueras, J. Planas et al. Pentachlorophenol and hexachlorobenzene in serum and urine of the population of Barcelona. *Hum. Toxicol.* 6:397, 1987.

Ph. Grandjean. *Skin Penetration: Hazardous Chemicals at Work.* Report prepared for the Commission of the European Communities, Health and Safety Directorate, Luxembourg, Eur 12599. Eds.: Taylor and Francis, London, 1990.

S. Gupta, J. Patrikh, M. Shah et al. Changes in serum hexachlorocyclohexane residues in malaria spraymen after short term occupational exposure. *Arch. Environ. Health* 37:41, 1982.

L. Hagmar, H. Welinder, K. Linden et al. An epidemiological study of cancer risk among workers exposed to ethylene oxide using haemoglobin adducts to validate environmental exposure assessments. *Int. Arch. Occup. Environ. Health* 63:271, 1991.

D. Hawkins, L. Chasseaud, R. Woodhouse and D. Cresswell. The distribution, excretion and biotransformation of p-dichloro(^{14}C)benzene in rats after repeated inhalation, oral and subcutaneous doses. *Xenobiotica* 10:81, 1980.

W. Hayes, Jr. and A. Curley. Storage and excretion of dieldrin and related compounds. Effect of occupational exposure. *Arch. Environ. Health* 16:155, 1968.

W. Hayes, Jr. Studies on exposure during the use of anticholinesterase pesticides. *Bull. OMS* 44:277, 1971.

S. Hope-Sandifer, C. Cupp, T. Wilkins et al. A case control study of persons with elevated blood levels of Dieldrin. *Arch. Environ. Contam. Toxicol.* 10:35, 1981.

J. Hunter, J. Maxwell, D. Stewart and R. Williams. Increased hepatic enzyme activity from occupational exposure to certain organochlorine pesticides. *Nature* 237:399, 1972.

IARC. International Agency for Research on Cancer. *Some Halogenated Hydrocarbons and Pesticides Exposures.* Monographs on the evaluation of the carcinogenic risk of chemicals to humans. Vol. 41, IARC, Lyon, 1986.

K. Jager. *Aldrin, Dieldrin, Endrin and Telodrin: an Epidemiological and Toxicological Study of Long-Term Occupational Exposure,* p. 121. Elsevier, New York, 1970.

S. Jastroch, W. Knoll, B. Lange et al. Studies on exposure to dinitro-o-cresol in agricultural chemists. *Z. Gesamte Hyg. Ihre Grenzgeb.* 24:340, 1978.

R. Jones, D. Winter and A. Cooper. Absorption study of pentachlorophenol in persons working with wood preservatives. *Hum. Toxicol.* 5:189, 1986.

V. Juhl, I. Witte and W. Butte. Metabolism of pentachlorophenol to tetrachlorohydroquinone by human liver homogenate. *Bull. Environ. Contam. Toxicol.* 35:596, 1985.

S. Kashyap. Health surveillance and biological monitoring of pesticide formulators in India. *Toxicol. Lett.* 33:107, 1986.

T. Kauppinen and L. Lindroos. Chlorophenol exposure in sawmills. *Am. Ind. Hyg. Assoc. J.* 46:34, 1985.

A. Kautiainen and M. Törnqvist. Monitoring exposure to simple epoxides and alkenes through gas chromatographic determination of hemoglobin adducts. *Int. Arch. Occup. Environ. Health* 63:27, 1991.

M. Kawano and R. Tatsukawa. Chlordanes and related compounds in blood of pest control operators. *J. Agric. Chem. Soc. Jpn.* 56:923, 1982.

R. Kimura, T. Hayashi, M. Sato et al. Identification of sulfur-containing metabolites of p-dichlorobenzene and their disposition in rats. *J. Pharmacol. Dyn.* 2:237, 1979.

G. Kleinman, S. Hortsman, D. Kalman et al. Industrial hygiene, chemical and biologcial assessments of exposures to a chlorinated phenolic sapstain control agent. *Am. Ind. Hyg. Assoc. J.* 47:731, 1986.

H. Klemmer, L. Wong, M. Sato et al. Clinical findings in workers exposed to pentachlorophenol. *Arch. Environ. Contam. Toxicol.* 9:715, 1980.

D. Knopp and S. Glass. Biological monitoring of 2,4-dichlorophenoxy acetic acid exposed workers in agriculture and forestry. *Int. Arch. Occup. Environ. Health* 63:329, 1991.

A. Koizumi. Experimental evidence for the possible exposure of workers to hexachlorobenzene by skin contamination. *Br. J. Ind. Med.* 48:622, 1991.

B. Kolmodin-Hedman. Exposure to lindane: plasma levels and health aspects in occupationally exposed Swedes. Proceedings of the Symposium on Lindane in Lyon-Chazay, 1976.

B. Kolmodin-Hedman and K. Erne. Estimation of occupational exposure to phenoxy acids (2.4-D and 2.4.5-T). *Arch. Toxicol. Suppl.* 4:318, 1980.

B. Kolmodin-Hedman, S. Höglund and M. Akerblom. Studies on phenoxy acid herbicides: I. Field study, occupational exposure to phenoxy acid herbicides (MCPA, Dichloroprop, Mecoprop and 2,4-D) in agriculture. *Arch. Toxicol.* 54:257, 1983a.

B. Kolmodin-Hedman, S. Höglund, A. Swensson and M. Akerblom. Studies on phenoxy acid herbicides. II. Oral and dermal uptake and elimination in urine of MCPA in humans. *Arch. Toxicol.* 54:267, 1983b.

G. Koss, A. Reuter and W. Koransky. Excretion of metabolites of hexachlorobenzene in the rat and in man. In: *Hexachlorobenzene: Proceedings of an International Symposium.* Eds.: C. Morris, J. Cabral. IARC Scientific Publication No. 77, Lyon, 1986.

S. Kröger, J. Liesivuori and A. Manninen. Evaluation of workers' exposure to 2-ethylhexanoic acid (2-EHA) in Finnish sawmills. *Int. Arch. Occup. Environ. Health* 62:213, 1990.

A. Kummer and W. Van Sittert. Field studies on health effects from the application of two organophosphorus insecticide formulations by hand-held ULV to cotton. *Toxicol. Lett.* 33:7, 1986.

P. Kurttio and K. Savolainen. Ethylenethiourea in air and in urine as an indicator of exposure to ethylenebis dithio carbamate fungicides. *Scand. J. Work Environ. Health* 16:203, 1990.

E. Laws, A. Curley and F. Biros. Men with intensive occupational exposure to DDT. A clinical and chemical study. *Arch. Environ. Health* 15:76, 1967.

C. Lemesch, Y. Wolf and E. Gabai. Occupational exposure to chlordimeform in Israel. *Isr. J. Med. Sci.* 23:1261, 1987.

M. Leutbecher, Th. Langhof, H. Peter and U. Föst. Ethylene oxide: metabolism in human blood and its implication to biological monitoring. The Eurotox Congress, Book of Abstracts, Maastricht, The Netherlands, Sept. 1991.

S. Libich, J. To, R. Frank and G. Sirons. Occupational exposure of herbicide applicators to herbicides used along electric power transmission line right of way. *Am. Ind. Hyg. Assoc. J.* 45:56, 1984.

L. Lindroos, H. Koskinen, P. Mutanen and J. Järvisalo. Urinary chlorophenols in sawmill workers. *Int. Arch. Occup. Environ. Health* 59:463, 1987.

K. Linnainmaa. Sister chromatid exchanges among workers occupationally exposed to phenoxy acid herbicides 2,4-D and MCPA. *Teratog. Carcinog. Mutag.* 3:269, 1983.

M. Maroni. Organophosphorus pesticides. In: *Biological Indicators for the Assessment of Human Exposure to Industrial Chemicals.* Eds.: L. Alessio, A. Berlin, M. Boni, R. Roi. Commission of the European Communities, Luxembourg. Eur 10704 En, 1986.

M. Maroni, G. Catenacci, D. Galli et al. Biological monitoring of human exposure to acephate. *Arch. Environ. Contam. Toxicol.* 19:782, 1990.

M. Maroni, V. Foa, A. Colombi et al. Urinary pentachlorophenol elimination after occupational and environmental exposure. *Toxicol. Lett.* 31:232, 1986.

T. Milby, A. Samuels and F. Ottoboni. Human exposure to lindane: blood lindane levels as a function of exposure. *J. Occup. Med.* 10:84, 1968.

D. Morgan and C. Roan. Liver function of workers having high tissue stores of chlorinated hydrocarbon pesticides. *Arch. Environ. Health* 29:14, 1974.

D. Morgan, H. Hetzler, E. Sclach and L. Lin. Urinary excretion of paranitrophenol and alkylphosphates following ingestion of methyl or ethyl parathion by human subjects. *Arch. Environ. Contam. Toxicol.* 6:153, 1977.

H. Mussalo-Rauhamaa, H. Pyysalo and K. Antervo. Heptachlor epoxide, and other chlordane compounds in Finnish plywood workers. *Arch. Environ. Health* 46:340, 1991.

S. Nigam, A. Karnic, S. Majunder et al. Serum hexachlorocyclohexane residues in workers engaged at a hexachlorocyclohexane manufacturing plant. *Int. Arch. Occup. Environ. Health* 57:315, 1986.

H. Nigg and J. Stamper. Exposure of Florida air boat aquatic weed applicators to 2,4-dichlorophenoxyacetic acid (2,4-D). *Chemosphere* 12:209, 1983.

NIOSH. Criteria for a Recommended Standard — Occupational Exposure to Carbaryl, National Institute for Occupational Safety and Health, Cincinnati, 1976.

NIOSH. Criteria for a Recommended Standard — Occupational Exposure to Dinitroorthocresol. National Institute of Occupational Safety and Health, Publication No. 78-131, U.S. Government Printing Office, Washington, D.C., 1978.

W. Notten and P. Henderson. Alterations in urinary D-glutaric acid excretion as an indication of exposition to xenobiotics. In: *Proceedings of the International Symposium — Environment and Health,* CEC-EPA-WHO, Paris, 1975.

W. Notten and P. Henderson. The interaction of chemical compounds with the functional stage of the liver II. Estimation of changes in D-glucaric acid synthesis as a method for diagnosing exposure to xenobiotics. *Int. Arch. Occup. Environ. Health* 38:209, 1977.

J. Osterloh, B. Cohen, W. Popendorf and S. Pond. Urinary excretion of the *N*-acetyl cysteine conjugate of cis-1,3-dichloropropene by exposed individuals. *Arch. Environ. Health* 39:271, 1984.

J. Osterloh, R. Wang, F. Schneider and K. Maddy. Biological monitoring dichloropropene: air concentrations, urinary metabolite and renal enzyme excretion. *Arch. Environ. Health* 44:207, 1989.

S. Osterman-Golkar and E. Bergmark. Occupational exposure to ethylene oxide: relation between in-vivo dose and exposure dose. *Scand. J. Work Environ. Health* 14:372, 1988.

C. Ottevanger and N. Van Sittert. Relation between anti-12-hydroxyendrin excretion and enzyme induction in workers involved in the manufacture of endrin. In: *Chemical Porphyria in Man,* p. 23. Eds.: J. Strik, J. Koeman. Elsevier, Amsterdam, 1979.

K. Pekari, M. Luotamo, J. Järvisalo et al. Urinary excretion of chlorinated phenols in saw-mill workers. *Int. Arch. Occup. Environ. Health* 63:57, 1991.

S. Pennanen and A. Manninen. Urinary arginine and ornithine in occupational exposure to 2-ethylhexanoic acid. *Arch. Toxicol.* 64:426, 1990.

A. Poland, D. Smith, R. Kuntzman et al. Effect of intensive occupational exposure to DDT on phenylbutazone and cortisol metabolism in human subjects. *Clin. Pharmacol. Ther.* 11:724, 1970.

I. Radomski, E. Astolfi, W. Deickmann and A. Alberto. Blood levels of organochlorine pesticide in Argentina. Occupationally and non-occupationally exposed adults, children and newborn infants. *Toxicol. Appl. Pharmacol.* 20:186, 1971.

M. Ramachandran, S. Zaidis, B. Barnejee and Q. Hussein. Urinary excretion of DDA: 2,2-Bis(4-chlorophenyl) acetic acid as an index of DDT exposure in men. *Indian J. Med. Res.* 80:483, 1984.

G. Renner and K. Schuster. 2,4,5-Trichlorophenol a new urinary metabolite of hexachlorobenzene. *Toxicol. Appl. Pharmacol.* 39:355, 1977.

A. Richardon and J. Robinson. The identification of a major metabolite of HEOD (Dieldrin) in human feces. *Xenobiotica* 1:213, 1971.

C. Roan, D. Morgan and E. Paschal. Blood cholinesterase, serum parathion concentration and urine p-nitrophenol concentration in exposed individuals. *Bull. Environ. Contam. Toxicol.* 4:362, 1969.

G. Rutten, A. Schoots, R. Van Holder et al. Hexachlorobenzene and 1,1-di(4-chlorophenyl)-2,2-dichloroethene in serum of uremic patients and healthy persons: determination by capillary gas chromatogrpahy and electron capture detection. *Nephron* 48:217, 1988.

I. Saito, N. Kawamura, K. Uno et al. Relationship between chlordane and its metabolites in blood of pest control operators under spraying conditions. *Int. Arch. Occup. Environ. Health* 58:91, 1986.

A. Samuels and H. Milby. Human exposure to lindane: comparison of an exposed and unexposed population. *J. Occup. Med.* 13:147, 1971.

F. Sarto, M. Törnqvist, R. Tomanin et al. Studies of biological and chemicals monitoring of low level exposure to ethylene oxide. *Scand. J. Work Environ. Health* 17:60, 1991.

J. Seiler. Pentachlorophenol. *Mutat. Res.* 257:27, 1991.

T. Shafik, E. Bradway, H. Enos and A. Yobs. Human exposure to organophosphate pesticides. A modified procedure for the gas liquid chromatographic analysis of alkyl phosphate metabolites in urine. *J. Agr. Food Chem.* 21:625, 1973.

F. Smith, J. Ramsey, M. Dryzga and W. Braun. Calculated dose levels of 2,4-D in exposed forest workers based on urinary excretion data and dermal pharmacokinetics of 2,4-D in the rat. *Toxicologist* 1:9, 1981.

A. Smith, D. Patterson, M. Warner et al. Serum 2,3,7,8-tetrachlorodibenzo-p-dioxin levels of New Zealand pesticide applicators and their implication for cancer hypotheses. *J. Natl. Cancer Inst.* 84:104, 1992.

S. Taguchi and T. Yakushiji. Influence of termite treatment in the home on the chlordane concentration in human milk. *Arch. Environ. Contam. Toxicol.* 17:65, 1988.

K. Takamiya. Residual levels of plasma oxychlordane and trans-nonachlor in pest control operators and some characteristics of these accumulations. *Bull. Environ. Contam. Toxicol.* 39:750, 1987.

A. Tates, T. Grummt, M. Törnqvist et al. Biological and chemical monitoring of occupational exposure to ethylene oxide. *Mutat. Res.* 250:483, 1991.

E. Thiele. Dose-effect relations in the occupational handling of DNOC. *Wissenscha. Zeitschr. Ernst-Moritz-Arndt-Univers. Greifsw. Medizin. Reihe* 32:7, 1983.

S. Uhl, P. Schmid and C. Schlatter. Pharmacokinetics of pentachlorophenol in man. *Arch. Toxicol.* 58:182, 1986.

M. Vandekar. Observations on the toxicity of carbaryl, folithion and 3-isopropylphenyl-N-methylcarbamate in a village-scale trial in Southern Nigeria. *Bull. WHO* 33:107, 1965.

B. Van Ommen, P. Van Bladeren, J. Tenmink and F. Müller. Formation of pentachlorophenol as the major product of microsomal oxidation of hexachlorobenzene. *Biochem. Biophys. Res. Commun.* 126:25, 1985.

N. Van Sittert, G. De Jong, M. Clare et al. Cytogenetic immunological and haematological effects in workers in an ethylene oxide manufacturing plant. *Br. J. Ind. Med.* 42:19, 1985a.

N. Van Sittert, C. Eadsforth and P. Bragt. Human oral dose-excretion study with Ripcord. The Hague, Shell Internationale Petroleum Maatschappij. (HSE 85.008), 1985a. HSE 85.009, 1985b.

N. Van Sittert. Biologische monitoring in de praktijk van de bedrijfsgezondheidszorg. De mens als meetinstrument op het werk. Coronel PAOG Nascholingssymposium. Amsterdam, Vrije Universiteit, 1985.

N. Van Sittert and W. Tordoir. Aldrin and Dieldrin. In: *Biological Indicators for the Assessment of Human Exposure to Industrial Chemicals.* Eds.: L. Alessio, A. Berlin, M. Boni, R. Roi. Commission of the European Communities, Luxembourg. Eur 11135 En, 1987a.

N. Van Sittert and W. Tordoir. Endrin. In: *Biological Indicators for the Assessment of Human Exposure to Industrial Chemicals.* Eds.: L. Alessio, A. Berlin, M. Boni, R. Roi. Commission of the European Communities, Luxembourg. Eur 11135 En, 1987b.

R. Van Welie, P. Van Duyn and N. Vermeulen. Determination of two mercapturic acid metabolites of 1,3-dichloropropene in human urine with gas chromatography and sulphur selective detection. *J. Chromatog.* 496:463, 1989.

R. Van Welie, C. Van Marrewijk, F. de Wolff and N. Vermeulen. Thioether excretion in urine of applicators exposed to 1,3-dichloropropene: a comparison with urinary mercapturic acid excretion. *Br. J. Ind. Med.* 48:492, 1991a.

R. Van Welie, P. Van Duyn and N. Vermeulen. Environmental and biological monitoring of non-occupational exposure to 1,3-dichloropropene. *Int. Arch. Occup. Environ. Health* 63:169, 1991b.

R. Van Welie, P. Van Duyn, D. Brouwer et al. Inhalation exposure to 1,3-dichloropropene in the dutch flower-bulb capture. Part II. Biological monitoring by measurement of urinary excretion of two mercapturic acid metabolites. *Arch. Environ. Contam. Toxicol.* 20:6, 1991c.

R. Van Welie, P. Van Duyn, E. Lamme et al. Determination of tetrahydrothalimide and 2-thiothiazolidine-4-carboxylic acid, urinary metabolites of the fungicide captan in rats and humans. *Int. Arch. Occup. Environ. Health* 63:181, 1991d.

R. Van Welie, R. Mensert, P. Van Duyn and N. Vermeulen. Identification and quantitative determination of a carboxylic and a mercapturic acid metabolite of etridiazole in urine of rat and man. Potential tools for biological monitoring. *Arch. Toxicol.* 65:625, 1991e.

M. Veutbecher, Th. Langhof, H. Peter and U. Fröst. Ethylene oxide: metabolism in human blood and its implication to biological monitoring. The 1991 Eurotox Congress, Book of Abstracts, Maastricht, The Netherlands, Sept. 1991.

W. Vrij-Standhardt, J. Strick, C. Ottevanger and N. Van Sittert. Urinary D-glucaric acid and urinary total porphyrin excretion in workers exposed to endrin. In: *Chemical Porphyria in Man.* Eds.: J. Strick, J. Koeman. Elsevier, Amsterdam, 1979.

N. Vural and S. Burgaz. A gas chromatographic method for determination of 2,4-D residues in urine after occupational exposure. *Bull. Environ. Contam. Toxicol.* 33:518, 1984.

WHO. World Health Organization. Data Sheets on Pesticides No. 6. Parathion, Geneva, 1975a.

WHO. World Health Organization. Data Sheets on Pesticides No. 3. Carbaryl, Geneva, 1975b.

WHO. World Health Organization. Recommended Health-Based Limits in Occupational Exposure to Pesticides. Techn. Rep. Series No. 677, Geneva, 1982.

WHO. IPCS, International Program on Chemical Safety. Environmental Health Criteria 97, Deltamethrin. World Health Organization, Geneva 1990.

H. Wolfe, W. Durham and J. Armstrong. Urinary excretion of insecticide metabolites. Excretion of para-nitrophenol and DDA as indicators of exposure to parathion and DDT. *Arch. Environ. Health* 21:711, 1970.

P. Wolfs, M. Dutrieux, V. Scailteur et al. Surveillance des travailleurs exposés à l'oxyde d'éthylène dans une entreprise de distribution de gaz stérilisants et dans des unités de stérilisation de matériel médical. *Arch. Mal. Prof.* 44:321, 1983.

S. Wood, W. Rom, G. White and D. Logan. Pentachlorophenol poisoning. *J. Occup. Med.* 25:527, 1983.

R. Yeary. Urinary excretion of 2,4-D in commercial lawn specialists. *Appl. Ind. Hyg.* 1:119, 1986.

Z. Zhang, J. Sun, Sh. Chen et al. Levels of exposure and biological monitoring of pyrethroid in spraymen. *Br. J. Ind. Med.* 48:82, 1991.

M. Zavon. Biological monitoring in exposure to cholinesterase inhibitors. *Int. Arch. Occup. Environ. Health* 37:67, 1976.

A. Zober, K. Schaller, K. Gobler and H. Krekeker. Pentachlorphenol und Leberfunktion: eine Untersuchung an beruflich belasteten Kollectiven. *Int. Arch. Occup. Environ. Health* 48: 347, 1981.

16. HORMONES

Diethylstilbestrol production in pharmaceutical or chemical plants may lead to exposure of male workers. The amount of the compound excreted in urine appears to be related to the risk of breast tenderness and enlargement and impotence (Shmunes and Burton 1981; Watrous and Olsen 1959). Shmunes and Burton have proposed the removal of individuals with diethylstilbestrol levels that exceed 30 μg/l of urine in a 24-h urine collection.

REFERENCES

E. Shmunes and D. Burton. Urinary monitoring for diethylstilbestrol in male chemical workers. *J. Occup. Med.* 23:179, 1981.

R. Watrous and B. Olsen. Diethylstilbestrol absorption in industry, a test for early detection as an aid in prevention. *Am. Ind. Hyg. Assoc. J.* 20:469, 1959.

17. MUTAGENIC AND CARCINOGENIC SUBSTANCES

In addition to the detection of mutagenic or carcinogenic substances or their metabolites in blood and urine, six types of biological analyses have been considered for detecting human exposure to carcinogens (for review see: IARC 1988; Aitio et al 1988). They are:

1. the analysis of the mutagenic activity of the urine of exposed workers
2. the determination of thioether concentration in urine
3. the determination of alkylated nucleic acids in urine, alkylated amino acids in circulating proteins, or DNA adducts
4. chromosomal analysis
5. the analysis of spermatozoa, morphologically or genetically
6. the detection of oncogen proteins

17.1. Analysis of the Mutagenic Activity of the Urine

It is sometimes possible to detect mutagenic activity in body fluids, especially urine, of workers exposed to mutagenic (and, hence, potentially carcinogenic) substances. Falk et al (1979, 1980) and Thiringer et al (1991) found that rubber workers and nurses handling cytostatic drugs exhibit significantly higher mutagenic activity in their urine than the unexposed workers. Increased urinary mutagenicity has also been found in anesthesiologists (McCoy et al 1978), epichlorhydrin-exposed workers (Kilian et al 1978), coke plant workers (De Meo et al 1987; Kriebel et al 1983), carbon electrode workers (Pasquini et al 1982), anode plant workers (Heussner et al 1985), workers exposed to bitumen fumes (Pasquini et al 1989), in subjects exposed to potentially genotoxic substances in a chemical plant producing a large variety of pharmaceuticals and explosives (Ahlborg et al 1985), and in subjects treated topically with coal tar (Wheeler et al 1981, Clonfero et al 1986).

Other studies, however, failed to detect an increased mutagenic activity of urine following occupational exposure in an anode plant (Clonfero et al 1984), a coke plant (Moller and Dybing 1980), a coal liquefaction plant (Recio et al 1984), a coal tar distillation plant (Jongeleenen et al 1986), coke ovens (Reuterwall et al 1991) and during handling of petroleum products (Nylander and Berg 1991) or wood treatment with creosote (Bos et al 1984).

Since smokers show mutagenicity in their urine (Van Doorn et al 1979; Yamasaki and Ames 1977), one must account for this variable in the interpretation of the results. The importance of the smoking factor has been confirmed in many of the above cited studies, and also by Bartsch et al (1990). It seems, however, that smoking induces mutation preponderantly in bacterial strains that are sensitive to frameshift mutation (Vainio et al 1981). The mutagenicity of urine is also increased in some liver diseases (Gelbart and Sontag 1980). In view of the large individual variability of urine mutagenicity, this analysis is mainly useful for identifying groups at risk of exposure to certain mutagenic substances or their active metabolites eliminated via the urine.

17.2. Analysis of Thioether Detoxication Products in Urine

Many electrophilic compounds (a class of chemicals that include the genotoxic compounds) react *in vivo* with glutathione. The glutathione conjugates formed in this reaction are partly excreted in the urine as premercapturic acids or other thioethers. Several authors have evaluated the possibility of measuring urinary concentration of thioethers as a method for detecting exposure to electrophilic agents and their precursors (Kilpikari 1981; Kilpikari and Sarolainen 1982; Seutter-Berlage et al 1977, 1978; Vainio et al 1978; Van Doorn et al 1979, 1981; Jagun et al 1982). Many studies have been carried out on workers exposed to chemicals with potential genotoxic activity such as petroleum retailers, workers occupied in rubber industry, asphalt producing or using plants and coke ovens, subjects exposed to pesticides, styrene, bitumen fumes, polyurethane foams or various environmental pollutants, and nurses handling cytostatic drugs (Stock and Priestly 1986; Burgaz et al 1988; Hagmar et al 1988; Que Hee et al 1987; Ahlborg et al 1985; Pasquini et al 1989; Holmen et al 1988; Norström et al 1988; Dehnen 1990; Thiringer et al 1991; Aringer et al 1991; Reuterwall et al 1991; Bayhan et al 1987; Mallol and Nogues 1991; Van Welie et al 1991).

In general, the increased excretion of thioethers over baseline level was small and frequently nonstatistically significant. There was also large inter- or intraindividual variations, probably because the main factor determining the amount of thioethers excreted in urine is the diet which can, in extreme situations enhance the excretion over 20-fold. The influence of cigarette smoke which contains several compounds that are excreted in urine as thioethers has been confirmed in several studies (Aringer and Lidums 1988; Heinonen et al 1983; Lafuente and Mallol 1986; Burgaz et al 1992).

Although most mercapturic acids that have been identified as urinary metabolites originate from xenobiotic compounds, some endogenous electrophiles are also excreted as mercapturic acids. Hence, there is a difference in urinary baseline thioether excretion between men and women because of female steroids, specially estrogen conjugates. In 196 nonexposed persons, including smokers, Van Doorn et al (1981) found a mean and a 95 percentile value for the urinary thioether excretion of 3.8 and 5.0 mmol SH/mol creatinine.

In conclusion, the primary value of the thioether assay is its signal function, which means that when an increase in the urinary thioether excretion is found (and possible interference by exogenous thio compounds can be excluded) it may be concluded that exposure to and absorption of one or more electrophilic substances have occurred. A negative result (i.e., a value within the normal range) unfortunately does not permit a conclusion that no or negligible exposure to electrophilic compounds has occurred (Van Doorn et al 1981).

17.3. Determination of Alkylated Nucleic Acids in Urine or Alkylated Amino Acids in Circulating Proteins or DNA Adducts

Many carcinogens are electrophilic reagents or are metabolized to such species *in vivo*. These electrophilic compounds can bind to nucleophilic sites on biological macromolecules such as DNA or proteins. The binding of carcinogens to DNA and the resulting formation of DNA adducts can be directly demonstrated by identifying these adducts either in cellular DNA or in degradation products of DNA or RNA excreted in urine or indirectly by measuring the adducts formed with nontarget macromolecules such as proteins.

Protein adducts. It has been suggested that monitoring of the reaction products formed with amino acids in proteins might provide an indication of the extent of DNA alkylation. Among the amino acids likely to be alkylated following exposure are cysteine, histidine, lysine, and the *N*-terminal amino acid of the protein. Because of its long lifetime and ready availability, hemoglobin was suggested as a suitable dose-monitoring compound (Osterman-Golkar et al 1976). Calleman et al (1978) have found some dose-response relationship between the intensity of exposure of workers to ethylene oxide and the degree of *N*-3-hydroxyethylation of histidine of hemoglobin. Several other studies have reported the occurrence of hemoglobin adducts in humans exposed to ethylene oxide (Van Sittert et al 1985; Farmer et al 1986; Bolt et al 1988; Törnqvist et al 1986a; Kautiainen and Törnqvist 1991; Tates et al 1991) and other genotoxic chemicals such as propylene oxide (Pero et al 1985; Osterman-Golkar et al 1984; Kautiainen and Törnqvist 1991), ethene (Törnqvist et al 1989), 1,3-butadiene (Müller et al 1991), 4-aminobiphenyl (Bryant et al 1987; Weston et al 1991; Skipper et al 1986) and other aromatic amines (Lewalter and Korallus 1985; Neumann 1988; Birner and Neumann 1988; Birner et al 1990; Braselton et al 1988), polycyclic aromatic hydrocarbon epoxides (Day et al 1990), methylbromide (Iwasaki 1988), and styrene (Goergens et al 1991; Brenner et al 1991). Kautiainen and Törnqvist (1991) have estimated that 100 pmol hydroxyethyl/g globin is the adduct level corresponding to the exposure to ethene present in inhaled tobacco smoke in a smoker of about ten cigarettes/day or to an average exposure to about 50 ppb ethylene oxide. The amount of *N*-(2-hydroxyethyl)-valine adducts in hemoglobin could permit the detection of the ethylene oxide that are formed from the ethene in the smoke of one cigarette per day. Bartsch et al (1990) have also reported that the levels of 4-aminobiphenyl-hemoglobin-adduct is higher in cigarette smokers than in nonsmokers.

As carcinogens may vary in their ability to cross the erythrocyte membrane, the degree of alkylation of serum albumin might constitute a better index of exposure, at least to some electrophylic compounds (Bailey et al 1981). It has been used to assess exposure to aflatoxin B1 (Gan et al 1988; Sabbioni 1990; Wild et al 1990; Astrup et al 1991) and to polycyclic aromatic hydrocarbons (PAH) (Lee et al 1991).

According to Sherson et al (1990), measurement of benzo(a)pyrene (BP) serum protein adduct concentrations appears to be a useful method by which groups exposed to benzopyrene may be biologically monitored. The mean adduct concentrations for both smoking (28 BP equivalents/100 μg protein; SD: 18.2; n: 26) and nonsmoking (24 BP equivalents/100 μg protein; SD: 21; n: 19) foundry workers were significantly higher than in control smokers (14.2; SD: 24.4; n: 26) and nonsmokers (7.3; SD: 8.72; n: 19).

DNA adducts. Two complementary approaches are available to estimate DNA adducts. Firstly, DNA adducts can be quantified in cells of accessible tissues. DNA adducts in white blood cells or lymphocytes have been identified in coke oven workers, roofers, foundry workers, styrene-exposed workers, PAH-exposed workers in an aluminium plant (Perera et al 1988; Phillips et al 1988; Shamsuddin et al 1985; Vahakangas et al 1985; Herbert et al 1990; Reddy et al 1991; Liu et al 1988; Shuker 1989; Hemminki et al 1990; Van Schooten et al 1991).

Another approach for monitoring DNA adducts takes advantage of the fact that adducts removed from cellular nucleic acids are excreted in urine. The determination in urine of methylated nucleic acids (3-methyladenine, 7-methylguanine) and 1-methylnicotinamide has been suggested as a biological method for detecting exposure to methylating carcinogens (Chu and Lawley 1975; Craddock and Magee 1967; Shaikh et al 1980). Excretion of 3-methyladenine, which is not a normal constituent of urine, has been found after treatment of animals with direct-acting methylating agents, dichlorvos and dimethylsulfate (Löfroth and Wennerberg 1974; Löfroth et al 1974). It has been shown that 3- methyladenine is normally present at low levels in human urine (5-15 μg/24 h) (Shuker et al 1987).

17.4. Chromosomal Analysis

The application of somatic cell monitoring for detection of exposure to various mutagenic or carcinogenic agents (such as heavy metals, benzene, toluene, vinylchloride, epichlorhydrin, cytostatic drugs, propylene oxide, styrene, isocyanates, polyester resins, ethylene oxide, PAH, etc.) has raised great interest (Burgdorf et al 1977; Forni et al 1971; Garry et al 1979; Kucerova et al 1979; Norppa et al 1980; Picciano 1980; Purchase et al 1978; Sram et al 1980; Hogstedt et al 1981; Miner et al 1983; Garner 1985; Stich and Rosin 1985; Van Sittert et al 1985; Bender et al 1988; Walles et al 1988; Holmen et al 1988; Yager et al 1988; Thiringer et al 1991; Reuterwall et al 1991; Sarto et al 1991; Sardas and Karakaya 1991; Popp et al 1991; Hageman et al 1991; Tates et al 1991).

The role of smoking habits is not well defined but probably constitute a confounding factor.

The different suggested analyses are chromosomal aberrations, sister chromatid exchanges, unscheduled DNA synthesis, single strand breaks, point mutations (e.g., hypoxanthine-guanine-phosphoribosyltransferase gene [HGPRT]) in peripheral blood lymphocytes, micronuclei in peripheral blood lymphocytes, or exfoliated cells from the buccal cavity or the bladder. According to Fabricant and Legator (1981), chromosomal studies are the best methods for examining genetic damage in routine industrial medical surveillance. Sister chromatid exchanges and micronuclei also offer some promise. Like mutagenicity of urine, these tests can only be used to detect groups at risk.

17.5. Analysis of Spermatozoa

Sperm studies (morphologic and genetic) may also be used to detect exposure to mutagenic substances (Evenson 1986; Wyrobek 1982). Morphologic analysis of the spermhead and the study of Y chromosome dysjunction during spermatogenesis are examples of such tests (Kapp et al 1979).

17.6. Detection of Oncogen Proteins

New techniques are used to assay for assumed mutant proteins (Hemminki 1992). Polycyclic aromatic hydrocarbons have been shown to cause specific mutational lesions that can lead to the activation of the ras oncogene and expression of its p21 protein product (Brandt-Rauf and Niman 1988; Brandt-Rauf et al 1989; Brandt-Rauf 1991a,b). These authors also measured the presence of oncogene proteins in serum (fes oncogene-related proteins, H-ras oncogene-related proteins) to ascertain the potential carcinogenic risk from exposure to PCB (Brandt-Rauf and Niman 1988).

17.7. Conclusions

All the biological tests for occupational genetic monitoring discussed above are not appropriate for risk assessment at the present time, because their clinical relevance must still be assessed through appropriate prospective studies. As opposed to the majority of biological tests, the currently available genetic tests cannot presently be used for evaluating the risk of excessive exposure for individual workers. Furthermore, multiple exposure and various confounding factors influencing the background levels of cytogenetic alteration or adduct production (such as cigarette smoke, drug, environmental pollution, etc.) make the interpretation of the results difficult. However, when applied under well-defined conditions (e.g., inclusion of a properly selected control population in each survey), they may be valuable to identify groups at risk of exposure to genetically active substances and hence to justify the implementation of better preventive measures.

The findings of a Nordic cohort data base, comprising 3,190 subjects and covering the time period 1970 through 1985 indicate a possible association between rate of chromosome aberrations in peripheral lymphocytes and subsequent cancer morbidity. The SMR (standardized morbidity ratio) for cancer were

90, 92, and 180 in the low (≤P33), medium (P34-P66) and high (≥P67) chromosome aberration rates and 119, 109, and 169 in the low, medium, and high sister chromatid exchange rates. A further follow-up period of 5 years is estimated to result in statistical powers of about 80% for chromosome aberrations, and 70% for sister chromatid exchange, to significantly demonstrate a "true" twofold or more risk increase for cancer in subjects with "high" rates of chromosomal changes compared to those with "medium" rates (Nordic Council 1990).

REFERENCES

G. Ahlborg, B. Bergström, C. Hogstedt et al. Urinary screening for potentially genotoxic exposures in a chemical industry. *Br. J. Ind. Med.* 12:691, 1985.

L. Aringer and V. Lidums. Influence of diet and other factors on urinary levels of thioethers. *Int. Arch. Occup. Environ. Health* 61:123, 1988.

L. Aringer, A. Löf and C. Elinder. The applicability of the measurement of urinary thioethers. A study of humans exposed to styrene during diet standardization. *Int. Arch. Occup. Environ. Health* 63:341, 1991.

J. Astrup, J. Schmidt, T. Seremet and H. Autrup. Determination of exposure to aflatoxins among Danish workers in animal-feed production through the analysis of aflatoxin B adducts to serum albumin. *Scand. J. Work Environ. Health* 17:436, 1991.

E. Bailey, T. Conners, P. Farmer et al. Methylation of cysteine in hemoglobin following exposure to methylating agents. *Cancer Res.* 41:2514, 1981.

H. Bartsch, N. Caporaso, M. Coda et al. Carcinogen hemoglobin adducts, urinary mutagenicity, and metabolic phenotype in active and passive cigarette smokers. *J. Natl. Cancer. Inst.* 82:1826, 1990.

A. Bayhan, S. Burgaz and A. Karakaya. Urinary thioether excretion in nurses at an oncologic department. *J. Clin. Pharmacol. Ther.* 12:303, 1987.

M. Bender, R. Leonard, O. White et al. Chromosomal aberrations and sister chromatic exchanges in lymphocytes from coke oven workers. *Mutat. Res.* 206:11, 1988.

G. Birner and H. Neumann. Biomonitoring of aromatic amines. Hemoglobin binding of some monocyclic aromatic amines. *Arch. Toxicol.* 62:110, 1988.

G. Birner, W. Albrecht and H. Neumann. Biomonitoring of aromatic amines. Hemoglobin binding of benzidine and some benzidine congeners. *Arch. Toxicol.* 64:97, 1990.

H. Bolt, H. Peter and U. Föst. Analysis of macromolecular ethylene oxide adducts. *Int. Arch. Occup. Environ. Health* 60:141, 1988.

R. Bos, T. Hulshoff, J. Theuws and P. Henderson. Genotoxic exposure of workers creosoting wood. *Br. J. Ind. Med.* 41:260, 1984.

P. Brandt-Rauf and H. Niman. Serum screening for oncogene proteins in workers exposed to PCBs. *Br. J. Ind. Med.* 45:689, 1988.

P. Brandt-Rauf, S. Smith and H. Niman. Serum oncogene proteins in hazardous waste workers. *J. Soc. Occup. Med.* 39:141, 1989.

P. Brandt-Rauf. Oncogene proteins as biomarkers in the molecular epidemiology of occupational carcinogenesis. The example of the ras oncogene-encoded p21 protein. *Int. Arch. Occup. Environ. Health* 63:1, 1991a.

P. Brandt-Rauf. Advances in cancer biomarkers as applied to chemical exposures: the ras oncogene and p21 protein and pulmonary carcinogenesis. *J. Occup. Med.* 33:951, 1991b.

W. Braselton, T. Chen and B. Kuslikis. Dose-monitoring of exposure to 4,4-methylene bis (2-chloroaniline) MBOCA by determination of hemoglobin adducts. *Toxicologist* 8:183, 1988.

D. Brenner, A. Jeffrey, L. Latriano et al. Biomarkers in styrene-exposed boatbuilders. *Mutat. Res.* 261:225, 1991.

M. Bryant, P. Skipper and S. Tannenbaum. Hemoglobin adducts of 4-aminobiphenyl in smokers and non-smokers. *Cancer Res.* 47:602, 1987.

S. Burgaz, A. Bayman and A. Karakaya. Thioether excretion of workers exposed to bitumen fumes. *Int. Arch. Occup. Environ. Health* 60:347, 1988.

S. Burqaz, P. Borm and F. Jonqeneelen. Evaluation of urinary excretion of 1-hydroxypyrene and thioethers in workers exposed to bitumen fumes. *Int. Arch. Occup. Environ Health* 63:397, 1992.

W. Burgdorf, K. Kurvink and J. Cervenka. Elevated sister chromatid exchange rate in lymphocytes of subjects treated with arsenic. *Hum. Genet.* 36:69, 1977.

C. Calleman, L. Ehrenberg, B. Jansson et al. Monitoring and risk assessment by means of alkyl groups in hemoglobin in persons occupationally exposed to ethylene oxide. *J. Environ. Pathol. Toxicol.* 2:427, 1978.

B. Chu and P. Lawley. Increased urinary excretion of nucleic acid and nicotinamide derivatives by rats after treatment with alkylating agents. *Chem. Biol. Interact.* 10:333, 1975.

E. Clonfero, P. Venier, D. Toffolo et al. Mutagenesis test on urine of workers exposed to polycyclic aromatic hydrocarbon in an anode plant. *Med. Lav.* 75:275, 1984.

E. Clonfero, M. Zordan, D. Cottica et al. Mutagenic activity and polycyclic aromatic hydrocarbon levels in urine of humans exposed to therapeutical coal tar. *Carcinogenesis* 7:819, 1986.

Commission of the European Communities — International Programme on Chemical Safety (UNEP-ILO-WHO). Indicators for assessing exposure and biological effects of genotoxic chemicals. Eds.: A. Aitio, G. Becking. A. Berlin et al. Eur 11642 En, Luxembourg 1988.

V. Craddock and P. Magee. Effect of administration of the carcinogen dimethylnitrosamine on urine 7-methylguanine. *Biochem. J.* 104:435, 1967.

B. Day, S. Naylor, L. Gan et al. Molecular dosimetry of polycyclic aromatic hydrocarbon epoxides and diol epoxides via hemoglobin adducts. *Cancer Res.* 50:4611, 1990.

W. Dehnen. A study on urinary thioethers by detecting n-acetylcysteine and thiophenol after alkaline hydrolysis. *Zentralbl. Hyg. Umweltmed.* 189:441, 1990.

M. De Meo, G. Dumenil, A. Botta et al. Urine mutagenicity of steel workers exposed to coke oven emissions. *Carcinogenesis* 8:363, 1987.

D. Evenson. Flow cytometry of acridine orange stained sperm is a rapid and practical method for monitoring occupational exposure to genotoxicants. *Prog. Clin. Biol. Res.* 207:121, 1986.

J. Fabricant and M. Legator. Etiology, role and detection of chromosomal aberrations in man. *J. Occup. Med.* 23:617, 1981.

K. Falk, P. Grohn, M. Sorsa et al. Mutagenicity in urine of nurses handling cytostatic drugs. *Lancet* 1:1250, 1979.

K. Falk, M. Sorsa, H. Vainio and I. Kilpikari. Mutagenicity in urine of workers in rubber industry. *Mutat. Res.* 79:45, 1980.

P. Farmer, E. Bailey, S. Gorf et al. Monitoring human exposure to ethylene oxide by the determination of haemoglobin adducts using gas chromatography-mass spectrometry. *Carcinogenesis* 7:637, 1986.

A. Forni, A. Cappellini, E. Pacifido and E. Vigliani. Chromosome changes and their evolution in subjects with past exposure to benzene. *Arch. Environ. Health* 23:385, 1971.

L. Gan, P. Skipper, X. Peng et al. Serum albumin adducts in the molecular epidemiology of aflatoxin carcinogenesis: correlation with aflatoxin b1 intake and urinary excretion of aflatoxin m1. *Carcinogenesis* 9:1323, 1988.

R. Garner. Assessment of carcinogen exposure in man. *Carcinogenesis* 6:1071, 1985.

V. Garry, J. Hozier, D. Jacobs et al. Ethylene oxide evidence of human chromosomal effects. *Environ. Mut.* 1:375, 1979.

S. Gelbart and S. Sontag. Mutagenic urine in cirrhosis. *Lancet* 1:894, 1980.

H. Goergens, M. Linscheid, K. Golkar et al. Evaluation of different biomonitors for occupational exposure to styrene. The 1991 Eurotox Congress. Book of abstracts. Maastricht, The Netherlands, Sept 1991.

G. Hageman, I. Welle and J. Kleinjans. Validation of cytogenetic biomarkers used in genotoxic risk assessment: relation with exposure to tobacco smoke and life style factors. The 1991 Eurotox Congress. Book of abstracts. Maastricht, The Netherlands, Sept 1991.

L. Hagmar, T. Bellander, L. Persson et al. Biological effects in a chemical factory with mutagenic exposure. III. Urinary mutagenicity and thioether excretion. *Int. Arch. Occup. Environ. Health* 60:453, 1988.

T. Heinonen, V. Kytöniemi, M. Sorsa and H. Vainio. Urinary excretion of thioether among low-tar and medium-tar cigarette smokers. *Int. Arch. Occup. Environ. Health* 52:11, 1983.

K. Hemminki, K. Randerath, M. Reddy et al. Postlabeling and immunoassay analysis of polycyclic aromatic hydrocarbons — adducts of deoxyribonucleic acid in white blood cells of foundry workers. *Scand. J. Work Environ. Health* 16:158, 1990.

K. Hemminki. Use of chemical biochemical and genetic markers in cancer epidemiology and risk assessment. *Am. J. Ind. Med.* 21:65, 1992.

R. Herbert, M. Marcus, M. Wolff et al. Detection of adducts of deoxuribonucleic acid in white blood cells of roofers by 32-p-post labeling. Relationships of adducts levels to measures of exposure to polycyclic aromatic hydrocarbons. *Scand. J. Work Environ. Health* 16:135, 1990.

J. Heussner, J. Ward and M. Legator. Genetic monitoring of aluminium workers exposed to coal tar pitch volatiles. *Mutat. Res.* 155:143, 1985.

B. Hogstedt, B. Gullberg, E. Mark-Vendel et al. Micronuclei and chromosome aberrations in bone marrow cells and lymphocytes of humans exposed mainly to petroleum vapors. *Hereditas* 94:179, 1981.

A. Holmen, B. Akesson, L. Hansen et al. Comparison among assays in workers producing polyurethane foams. *Int. Arch. Occup. Environ. Health* 60:175, 1988.

IARC. International Agency for Research on Cancer. *Methods for Detecting DNA Damaging Agents in Humans: Applications in Cancer Epidemiology and Prevention.* Eds.: H. Bartsch, K. Hemminki, I. O'Neill. IARC Scientific Publication No. 89, Lyon 1988.

K. Iwasaki. Individual differences in the formation of hemoglobin adducts following exposure to methyl bromide. *Ind. Health* 26:257, 1988.

O. Jagun, M. Ryan and H. Waldron. Urinary thioether excretion in nurses handling cytotoxic drugs. *Lancet* ii:443, 1982.

F. Jongeneelen, R. Bos, R. Anzion et al. Biological monitoring of polycyclic aromatic hydrocarbons. Metabolites in urine. *Scand. J. Work Environ. Health* 2:137, 1986.

R. Kapp, D. Picciano and C. Jacobson. Y chromosome nondisjunction in DBCP exposed workmen. *Mutat. Res.* 64:47, 1979.

A. Kautiainen and M. Törnqvist. Monitoring exposure to simple epoxides and alkenes through gas chromatographic determination of hemoglobin adducts. *Int. Arch. Occup. Environ. Health* 63:27, 1991.

D. Kilian, T. Pullin, T. Conner et al. Mutagenicity of epichlorhydrin in the bacterial assay system: evaluation by direct in vitro activity and in vivo activity of urine from exposed humans and mice. *Mutat. Res.* 53:72, 1978.

I. Kilpikari. Correlation of urinary thioethers with chemical exposure in a rubber plant. *Br. J. Ind. Med.* 38:98, 1981.

I. Kilpikari and Savolainen. Increased urinary excretion of thioether in new rubber workers. *Br. J. Ind. Med.* 39:401, 1982.

D. Kriebel, B. Commoner, D. Bollinger et al. Detection of occupational exposure to genotoxic agents with urinary mutagen assay. *Mutat. Res.* 108:67, 1983.

M. Kucerova, Z. Polikova and J. Batora. Comparative evaluation of the frequency of chromosomal aberrations and the sister chromatid exchange numbers in peripheral lymphocytes of workers occupationally exposed to vinyl chloride monomer. *Mutat. Res.* 67:97, 1979.

A. Lafuente and L. Mallol. Urinary thioethers in low tar cigarette smokers. *Publ. Health* 100:392, 1986.

B. Lee, Y. Baoyun, R. Herbert et al. Immunologic measurement of polycyclic aromatic hydrocarbon-albumin adducts in foundry workers and roofers. *Scand. J. Work Environ. Health* 17:190, 1991.

J. Lewalter and U. Korallus. Blood protein conjugates and acetylation of aromatic amines. *Int. Arch. Occup. Environ. Health* 56:179, 1985.

S. Liu, S. Rappaport, K. Pongracz and W. Bodell. Detection of styrene oxide — DNA adducts in lymphocytes of a worker exposed to styrene. In: *Methods for Detecting DNA Damaging Agents in Humans.* Eds.: H. Bartsch, K. Hemminki and I. O'Neill. IARC Scientific Publication No. 89, p. 217, Lyon, 1988.

G. Löfroth and R. Wennerberg. Methylation of purines and nicotinamide in the rat by dichlorvos. *Z. Naturforsch.* 29:651, 1974.

G. Löfroth, S. Osterman-Golkar and R. Wennerberg. Urinary excretion of methylated purines following inhalation of dimethylsulfate. *Experientia* 15:641, 1974.

E. McCoy, R. Kankel, K. Robbins et al. Presence of mutagenic substances in the urine of anesthesiologists. *Mutat. Res.* 53:71, 1978.

J. Mallol and M. Nogues. Air pollution and urinary thioether excretion in children of Barcelona. *J. Toxicol. Environ. Health* 33:189, 1991.

J. Miner, W. Rom, G. Livingston and J. Lyon. Lymphocytes sister chromatid exchange frequences in coke oven workers. *J. Occup. Med.* 25:30, 1983.

M. Moller and E. Dybing. Mutagenicity studies with urine concentrates from coke plant workers. *Scand. J. Work Environ. Health* 6:216, 1980.

A. Müller, M. Linscheid, Ch. Siethoff et al. Human blood protein adducts as biomonitor for exposure to 1,2-epoxybutene-3, the first reactive intermediate of 1.3-butadiene. The 1991 Eurotox Congress. Book of abstracts. Maastricht, The Netherlands, Sept 1991.

H. Neumann. Biomonitoring of aromatic amines and alkylating agents by measuring hemoglobin adducts. *Int. Arch. Occup. Health* 60:151, 1988.

Nordic Study Group on the Health Risk of Chromosome Damage. Presented by Ch. Reuterwall. A Nordic prospective on the relationship between peripheral lymphocyte chromosome damage and cancer morbidity in occupational groups. Mutation and The Environment part C:357, 1990. Eds.: Mendelson and Albertini. Wiley-Liss, New York.

H. Norppa, M. Sorsa, H. Vainio et al. Increased sister chromatid exchange frequencies in lymphocytes of nurses handling cytostatic drugs. *Scand. J. Work Environ. Health* 6:299, 1980.

A. Norström, B. Anderson, L. Aringer et al. Determination of specific mercapturic acids in human urine after experimental exposure to toluene or o-xylene. In: *Methods for Detecting DNA Damaging Agents in Humans: Application in Cancer Epidemiology and Prevention.* Eds.: H. Bartsch, K. Hemminki, I. O'Neil. IARC Scientific Publication No. 89, Lyon, 1988.

G. Nylander and K. Berg. Mutagenicity study of urine from smoking and non-smoking road tanker drivers. *Int. Arch. Occup. Environ. Health* 63:229, 1991.

S. Osterman-Golkar, L. Ehrenberg, D. Segerback and I. Hallstrom. Evaluation of genetic risks of alkylating agents. II. Haemoglobin as a dose monitor. *Mutat. Res.* 34:1, 1976.

S. Osterman-Golkar, E. Bailey, P. Farmer et al. Monitoring exposure to propylene oxide through the determination of hemoglobin alkylation. *Scand. J. Work Environ. Health* 10:99, 1984.

R. Pasquini, S. Monarca, G. Scasselatti et al. Mutagens in urine of carbon electrode workers. *Int. Arch. Occup. Environ. Health* 50:387, 1982.

R. Pasquini, S. Monarca, G. Scassellati et al. Urinary excretion of mutagens, thioethers and D-glucaric acid in workers exposed to bitumen fumes. *Int. Arch. Occup. Environ. Health* 61:335, 1989.

F. Perera, K. Hemminki, T. Young et al. Detection of polycyclic aromatic hydrocarbon DNA adducts in white blood cells of foundry workers. *Cancer Res.* 48:2288, 1988.

R. Pero, S. Osterman-Golkar and B. Hogstedt. Unscheduled DNA synthesis correlated to alkylation of hemoglobin in individuals occupationally exposed to propylene oxide. *Cell. Biol. Toxicol.* 1:309, 1985.

D. Phillips, K. Hemminki, A. Alhonen et al. Monitoring occupational exposure to carcinogens: detection by 32-P-postlabeling of aromatic DNA adducts in white blood cells from iron foundry workers. *Mutat. Res.* 204:531, 1988.

D. Picciano. Cytogenetic investigation of occupational exposure to epichlorhydrin. *Mutat. Res.* 70:115, 1980.

W. Popp, C. Vahrenholz, K. Norpoth et al. Measuring DNA damage in lymphocytes of workers by the alkaline filter elution in some occupationally exposed groups. The 1991 Eurotox Congress. Book of abstracts. Maastricht, The Netherlands, Sept. 1991.

I. Purchase, C. Richardson, D. Anderson et al. Chromosomal analysis in vinyl chloride exposed workers. *Mutat. Res.* 57:325, 1978.

S. Que Hee, O. Igwe and C. Clark. Thioether excretion of workers in a waste water facility receiving pesticide wastes. In: *Biological Monitoring of Exposure to Chemicals Organic Compounds.* Eds.: M. Ho and H. Dillon. Wiley-Interscience, New York, 1987.

L. Recio, M. Enoch, M. Hannan and R. Hill. Application of urine mutagenicity to monitor coal liquefaction workers. *Mutat. Res.* 136:201, 1984.

M. Reddy, K. Hemminki and K. Randerath. Postlabelling analysis of polycyclic aromatic hydrocarbon — DNA adducts in white blood cells of foundry workers. *J. Toxicol. Environ. Health* 34:177, 1991.

Ch. Reuterwall, L. Aringer, C. Elinder et al. Assessment of genotoxic exposure in Swedish coke oven work by different methods of biological monitoring. *Scand. J. Work Environ. Health* 17:123, 1991.

G. Sabbioni. Chemical and physical properties of the major serum albumin adducts of aflatoxin-b1 and their applications for the quantitation in biological samples. *Chem. Biol. Interact.* 75:1, 1990.

S. Sardas and A. Karakaya. The significance of sister chromatid exchanges for biomonitoring occupational exposure. The 1991 Eurotox Congress. Book of Abstracts. Maastricht, The Netherlands, Sept 1991.

F. Sarto, M. Törnqvist, R. Tomanin et al. Studies of biological and chemical monitoring of low level exposure to ethylene oxide. *Scand. J. Work Environ. Health* 17:60, 1991.

F. Seutter-Berlage, H. Van Dorp, H. Kosse and P. Henderson. Urinary mercapturic acid excretion as a biological parameter of exposure to alkylating agents. *Int. Arch. Occup. Environ. Health* 39:45, 1977.

F. Seutter-Berlage, L. Delbressine, F. Smeets and H. Ketelaars. Identification of three sulphur-containing urinary metabolites of styrene in the rat. *Xenobiotica* 8:413, 1978.

B. Shaikh, S. Huang and N. Pontzer. Urinary excretion of methylated purines and 1-methylnicotinamide following administration of methylating carcinogens. *Chem. Biol. Interact.* 30:253, 1980.

A. Shamsuddin, N. Sinopoli, K. Hemminki et al. Detection of Benzo(a)pyrene — DNA adducts in human white blood cells. *Cancer Res.* 45:66, 1985.

D. Sherson, P. Sabro, T. Sigscaard et al. Biological monitoring of foundry workers exposed to polycyclic aromatic hydrocarbons. *Br. J. Ind. Med.* 47:448, 1990.

D. Shuker. Detection of adducts arising from human exposure to *n*-nitroso compounds. *Cancer Surv.* 8:475, 1989.

D. Shuker, E. Bailey, A. Parry et al. The determination of urinary 3-methyladenine in humans as a potential monitor of exposure to methylating agents. *Carcinogenesis* 8:959, 1987.

P. Skipper, M. Bryant, S. Tannenbaum and J. Groopman. Analytical methods for assessing exposure to 4-aminobiphenyl based on protein adduct formation. *J. Occup. Med.* 28:643, 1986.

R. Sram, Z. Zudova and N. Kuleshov. Cytogenic analysis of peripheral lymphocytes in workers occupationally exposed to epichlorhydrin. *Mutat. Res.* 70:115, 1980.

H. Stich and M. Rosin. Towards a more comprehensive evaluation of a genotoxic hazard in man. *Mutat. Res.* 150:43, 1985.

J. Stock and B. Priestly. Urinary thioether output as an index of occupational chemical exposure in petroleum retailers. *Br. J. Ind. Med.* 43:718, 1986.

A. Tates, T. Grummt, M. Törnqvist et al. Biological and chemical monitoring of occupational exposure to ethylene oxide. *Mutat. Res.* 250:483, 1991.

G. Thiringer, G. Granung, A. Holmen et al. Comparison of methods for the biomonitoring of nurses handling antitumor drugs. *Scand. J. Work Environ. Health* 17:133, 1991.

M. Törnqvist, S. Osterman-Golkar, S. Kautiainen et al. Tissue doses of ethylene oxide in cigarette smokers determined from adduct levels in hemoglobin. *Carcinogenesis* 7:1519, 1986a.

M. Törnqvist, J. Almberg, S. Nilsson and S. Osterman-Golkar. Tissue dose of ethylene oxide from occupational exposure to ethene in fruit stores. *Scand. J. Work Environ. Health* 15:436, 1989.

K. Vahakangas, G. Trivers, H. Rowe and C. Harris. Benzo(a)pyrene diolepoxide — DNA adducts detected by synchronous fluorescence spectrophotometry. *Environ. Health Perspect.* 62:101, 1985.

H. Vainio, H. Savolainen and I. Kilpikari. Urinary thioether of employees of a chemical plant. *Br. J. Ind. Med.* 35:232, 1978.

H. Vainio, M. Sorsa, J. Rantanen et al. Biological monitoring in the identification of the cancer risk of individuals exposed to chemical carcinogens. *Scand. J. Work Environ. Health* 7:241, 1981.

R. Van Doorn, R. Bos, C. Leudekkers et al. Thioether concentration and mutagenicity of urine from cigarette smokers. *Int. Arch. Occup. Environ. Health* 43:159, 1979.

R. Van Doorn, C. Leijdekkers, R. Bos et al. Detection of human exposure to electrophilic compounds by assay of thioether detoxication products in urine. *Ann. Occup. Hyg.* 24:77, 1981.

F. Van Schooten, M. Hillebrand, J. Van Engelen et al. Detection of polycyclic hydrocarbon — DNA adducts in mouse tissues and human white blood cells. The 1991 Eurotox Congress. Book of abstracts, Maastricht, The Netherlands, Sept 1991.

N. Van Sittert, G. De Jong, M. Clare et al. Cytogenetic, immunological and haematological effects in workers in an ethylene oxide manufacturing plant. *Br. J. Ind. Med.* 42:19, 1985.

R. Van Welie, C. Van Marrewijk, F. De Wolff and N. Vermeulen. Thioether excretion in urine of applicators exposed to 1,3-dichloropropene: a comparison with urinary mercapturic excretion. *Br. J. Ind. Med.* 48:492, 1991.

S. Walles, H. Norppa, S. Osterman-Golkar et al. Single-strand breaks in DNA of peripheral lymphocytes of styrene exposed workers. In: *Methods for Detecting DNA Damaging Agents in Humans.* Eds.: H. Bartsch, K. Hemminki and I. O'Neill. IARC Scientific Publication No. 89, p.223, Lyon, 1988.

A. Weston, N. Caporaso, K. Taglizadeh et al. Measurement of 4-aminobiphenyl-hemoglobin adducts in lung cancer cases and controls. *Cancer Res.* 51:5219, 1991.

L. Wheeler, M. Saperstein and N. Lowe. Mutagenicity of urine from psoriatic patients undergoing treatment with coal tar and ultraviolet light. *J. Invest. Dermatol.* 77:181, 1981.

C. Wild, Y. Jiang, G. Sabbioni et al. Evaluation of methods for quantitation of aflatoxin-albumin and their application to human exposure assessment. *Cancer Res.* 50:245, 1990.

A. Wyrobek. Sperm assay as indicators of chemically induced germ-cell damage. In: *Mutagenicity, New Horizons in Genetic Toxicology,* p. 337. Ed.: J. Heddle. Academic Press, New York, 1982.

J. Yager, M. Sorsa and S. Selvin. Micronuclei in cytokinesis-blocked lymphocytes as an index of occupational exposure to alkylating cytostatic drugs. In: *Methods for Detecting DNA Damaging Agents in Humans: Applications in Cancer Epidemiology and Prevention.* Eds.: H. Bartsch, K. Hemminki and I. O'Neill. IARC Scientific Publication No. 89, Lyon, 1988.

E. Yamasaki and B. Ames. Concentration of mutagens from urine with the nonpolar resin XAD-2: cigarette smokers have mutagenic urine. *Proc. Natl. Acad. Sci.* 74:3555, 1977.

4 SUMMARY OF RECOMMENDATIONS

We have summarized in tabular form the principal biological monitoring methods presently available for detecting individual workers or groups of workers exposed to industrial chemicals.

For each chemical agent, we have listed the proposed biological parameters and, when data were available, the reference values and maximum permissible values. The significance of the latter proposals must be kept clearly in mind. They are more frequently an index of exposure than of the health risk because they often derived from studies on external exposure-internal dose relationships and rarely from an assessment of their relationship with adverse effects (see Chapter 1). They are simply tentative guidelines based on the currently available scientific knowledge. They cannot be considered as rigid reference values, against which any observed value may be compared, to reach a conclusion regarding the presence or the absence of a health risk. Furthermore, we have not indicated in the table whether the biological parameters mainly reflect the body burden or the intensity of recent exposure. The reader must refer to the sections of the book dealing with the chemicals listed in the table to interpret the significance of the proposed permissible values. Except when indicated otherwise, the tentative permissible values refer to analyses performed on biological samples collected at the end of the exposure period.

BIOLOGICAL MONITORING OF EXPOSURE TO INDUSTRIAL CHEMICALS

A. Inorganic and Organometallic Substances

Chemical Agents	Parameter	Biological Material	Reference Value	Tentative Maximum Permissible Concentration	Remarks
Aluminium	aluminium	serum	< 1 µg/100 ml		
	aluminium	urine	< 50 µg/g creat.	150 µg/g creat.	
Antimony	antimony	urine	< 1 µg/g creat.	35 µg/g creat.	
Arsenic	total arsenic	urine	< 40 µg/g creat.		interference of arsenic from marine origin
	total arsenic	blood			
	total arsenic	hair	< 1 µg/g		
	sum of inorganic arsenic and methylated metabolites	urine	< 10 µg/g creat.	50 µg/g creat. if TWA: 50 µg/m³ 30 µg/g creat. if TWA: 10 µg/m³	no interference of arsenic from marine origin
Barium	barium	urine	< 15 µg/g creat.		
	barium	blood	< 0.8 µg/100 ml		
Beryllium	beryllium	urine	< 2 µg/g creat.		nonsmokers
Cadmium	cadmium	urine	< 2 µg/g creat.	5 µg/g creat.	
	cadmium	blood	< 0.5 µg/100 ml	0.5 µg/100 ml	
	metallothionein	urine			
Carbon disulfide	iodine-azide test	urine		> 6.5 (Vasak index)	to detect exposure > 100 mg/m³
	2-thiothiazolidine-4-carboxylic acid (TTCA)	urine		5 mg/g creat.	

Substance	Analyte	Specimen			
Chromium VI (soluble compounds)	chromium	urine	< 5 µg/g creat.	30 µg/g creat.	postshift minus preshift value
	chromium	red blood cells			
Cobalt	cobalt	urine	< 2 µg/g creat.	30 µg/g creat.	
	cobalt	blood	< 0.2 µg/100 ml		
	cobalt	serum	< 0.05 µg/100 ml		
Copper	copper	urine	< 50 µg/g creat.		
	copper	serum	< 0.14 mg/100 ml		
Fluoride	fluoride	serum			
	fluoride	urine	< 0.5 mg/g creat.	3-4 mg/g creat.	
Germanium	germanium	urine	< 1 µg/g creat.		
Lead	lead	blood	< 25 µg/100 ml	40 µg/100 ml	
	lead	urine	< 50 µg/g creat.	50 µg/g creat.	
	lead (after 1 g EDTA IV or 2 g DMSA PO)	urine	< 600 µg/24 h	600 µg/24 h	
	free porphyrin	red blood cells	< 75 µg/100 ml RBC	80 µg/100 ml RBC	
	zinc protoporphyrin	blood	< 40 µg/100 ml	40 µg/100 ml	
			< 2.5 µg/g Hb	3 µg/g Hb	
	δ-aminolevulinic acid (ALA)	urine	< 4.5 mg/g creat.	5 mg/g creat.	
	coproporphyrins	urine	< 100 µg/g creat.	100 µg/g creat.	
	ALA dehydratase	red blood cells			
	pyrimidine-5'-nucleotidase	red blood cells			

BIOLOGICAL MONITORING OF EXPOSURE TO INDUSTRIAL CHEMICALS (continued)

A. Inorganic and Organometallic Substances (continued)

Chemical Agents	Parameter	Biological Material	Reference Value	Tentative Maximum Permissible Concentration	Remarks
Lead tetraethyl	lead	urine	< 50 µg/g creat.	100 µg/g creat.	
Manganese	manganese	urine	< 3 µg/g creat.		
	manganese	blood	< 1 µg/100 ml		
Mercury inorganic	mercury	urine	< 5 µg/g creat.	50 µg/g creat.	
	mercury	blood	< 1 µg/100 ml	2 µg/100 ml	
	mercury	saliva			
Methylmercury	mercury	blood	< 1 µg/100 ml	10 µg/100 ml	
	mercury	hair			
Nickel (soluble compounds)	nickel	urine	< 2 µg/g creat.	30 µg/g creat.	
	nickel	plasma	< 0.05 µg/100 ml		
Nickel carbonyl	nickel	urine			
Nitrous oxide	N_2O	urine		60 µg/g creat.	
	N_2O	expired air			
Selenium	selenium	serum	< 15 µg/100 ml		
	selenium	urine	< 25 µg/g creat.		
Silver	silver	urine	< 1 µg/g creat.		
	silver	serum	< 0.5 µg/100 ml		
Tellurium	tellurium	urine	< 1µg/g creat.		

Substance	Determinant	Sample	Value		Sampling time
Thallium	thallium	urine	< 1 µg/g creat.		
	thallium	blood	< 0.1 µg/100 ml		
Uranium	uranium	urine	< 0.1 µg/g creat.		
	uranium	blood	< 0.01 µg/100 ml		
Vanadium	vanadium	urine	< 1 µg/g creat.	50 µg/g creat.	
	vanadium	blood	< 0.1 µg/100 ml		
Zinc	zinc	urine	< 0.9 mg/g creat.		
	zinc	serum	< 170 µg/100 ml		

B. Organic Substances

1. Nonsubstituted aliphatic and alicyclic hydrocarbons

Substance	Determinant	Sample	Value	Sampling time
n-Hexane	2-hexanol	urine	0.2 mg/g creat.	
	2,5-hexanedione	urine	2 mg/g creat.	end first day of work
			4 mg/g creat.	end of workweek
	n-hexane	blood	15 µg/100 ml	during exposure
	n-hexane	expired air	50 ppm	during exposure
2-Methyl-pentane	2-methyl-2-pentanol	urine		
	2-methyl-pentane	expired air	1500 µg/l	
	2-methyl-pentane	blood	35 µg/100 ml	
3-Methyl-pentane	3-methyl-2-pentanol	urine		
	3-methyl-pentane	expired air	1500 µg/l	
	3-methyl-pentane	blood	35 µg/100 ml	

BIOLOGICAL MONITORING OF EXPOSURE TO INDUSTRIAL CHEMICALS (continued)

Chemical Agents	Parameter	Biological Material	Reference Value	Tentative Maximum Permissible Concentration	Remarks
B. Organic Substances (continued)					
Cyclohexane	cyclohexanol	urine		3.2 mg/g creat.	
	cyclohexane	blood		45 µg/100 ml	during exposure
	cyclohexane	expired air		220 ppm	during exposure
2. Nonsubstituted aromatic hydrocarbons					
Benzene	phenol	urine	< 20 mg/g creat.	45 mg/g creat.	if TWA: 10 ppm
				< 20 mg/g creat.	if TWA: 1 ppm
	muconic acid	urine		1.4 mg/g creat.	if TWA : 1 ppm
	benzene	expired air		0.022 ppm	if TWA : 1 ppm (during exposure)
	benzene	blood		< 5 µg/100 ml	if TWA : 1 ppm (during exposure)
Toluene	hippuric acid	urine	< 1.5 g/g creat.	2.5 g/g creat.	
	O-cresol	urine	< 0.3 mg/g creat.	1 mg/g creat.	
	toluene	expired air		20 ppm	during exposure
	toluene	blood		0.1 mg/100 ml	during exposure
Ethylbenzene	mandelic acid	urine		1 g/g creat.	
	ethylbenzene	blood		0.15 mg/100 ml	during exposure
	ethylbenzene	expired air			
Cumene (isopropylbenzene)	2-phenylpropanol	urine		10 mg/hr	last 2 h of shift

Substance	Parameter	Material	Value	Sampling condition
	cumene	expired air		
	cumene	blood		
Mesitylene	3,5-dimethyl-hippuric acid	urine		
Styrene	mandelic acid	urine	800 mg/g creat.	
	phenylglyoxylic acid	urine	250 mg/g creat.	
	styrene	blood	0.1 mg/100 ml	
			0.002 mg/100 ml	16 h after end of exposure
	styrene	expired air	9 ppm	
α-Methylstyrene	atrolactic acid	urine		
Xylene	methylhippuric acid	urine	1.5 g/g creat.	
	xylene	expired air		during exposure
	xylene	blood	0.3 mg/100 ml	
Biphenyl	2,4-hydroxy-biphenyl	urine	1.5 mg/g creat.	
Polycyclic hydrocarbons	1-hydroxypyrene	urine	< 2 µg/g creat.	
	hemoglobin adducts	blood		

3. Halogenated hydrocarbons

Substance	Parameter	Material	Value	Sampling condition
Monochloromethane (methylchloride)	S-methylcysteine	urine		
Monobromomethane (methylbromide)	S-methylcysteine	urine		
	bromine	blood		
Dichloromethane	HbCO	blood	2%	
			< 1%	nonsmokers
	dichloromethane	blood	0.05 mg/100 ml	
	dichloromethane	expired air	15 ppm	

BIOLOGICAL MONITORING OF EXPOSURE TO INDUSTRIAL CHEMICALS (continued)

B. Organic Substances (continued)

Chemical Agents	Parameter	Biological Material	Reference Value	Tentative Maximum Permissible Concentration	Remarks
1,2-Dibromoethane	N-acetyl-S-(2-hydroxy-ethyl)cysteine	urine			
Vinyl chloride	thiodiglycolic acid	urine	< 2 mg/g creat.		
Trichlorethylene	trichloroethanol	urine		150 mg/g creat.	
	trichloroacetic acid	urine		75 mg/g creat.	
	trichloroethanol	plasma		0.25 mg/100 ml	after 5-d exposure
	trichloroethylene	expired air		0.5 ppm	16 h after the end of exposure
				10 ppm	during exposure
	trichloroacetic acid	plasma		5 mg/100 ml	after 5-d exposure
	trichloroethylene	blood		0.06 mg/100 ml	during exposure
1,1,1-Trichloroethane (methylchloroform)	trichloroethanol + trichloroacetic acid	urine		40 mg/g creat.	end of workweek
	trichloroacetic acid	urine		10 mg/g creat.	
	trichloroethanol	urine		30 mg/g creat.	end of workweek
	trichloroethanol	blood		0.1 mg/100 ml	
	trichloroethane	blood		100 µg/100 ml	
	trichloroethane	urine		800 µg/g creat.	
	trichloroethane	expired air		30 ppm	16 h after the end of exposure

Substance	Determinant	Medium	Value	Sampling time / condition
Tetrachloroethylene	tetrachloroethylene	expired air	60 ppm 8 ppm	during exposure 16 h after the end of exposure
	tetrachloroethylene	blood	100 µg/100 ml	16 h after the end of exposure
	tetrachloroethylene	urine	70 µg/g creat.	16 h after the end of exposure
	trichloroacetic acid	urine	5 mg/g creat.	end-of-week
Hexachlorobutadiene	hexachlorobutadiene	blood		
Monochlorobenzene	4-chlorocatechol	urine		
	4-chlorophenol	urine		
p-Dichlorobenzene	p-dichlorobenzene	urine	250 µg/g creat.	
	2,5-dichlorophenol	urine		
Halothane	trifluoroacetic acid	urine	10 mg/g creat.	after 5-d exposure if TWA: 5 ppm
	trifluoroacetic acid	blood	0.25 mg/100 ml	after 5-d exposure if TWA: 5 ppm
	halothane	urine	90 µg/g creat. 10 µg/g creat.	if TWA: 50 ppm if TWA: 5 ppm
	halothane	expired air	0.5 ppm	if TWA : 5 ppm
2,3,7,8-Tetrachloro-dibenzo-p-dioxin (TCDD)	TCDD TCDD	serum blood		
Polychlorinated biphenyl	polychlorinated biphenyl	serum adipose tissue		
Other volatile halogenated hydrocarbons (carbon tetrachloride, chloroform, halogenated anaesthetics, etc.)	substances	expired air blood		

BIOLOGICAL MONITORING OF EXPOSURE TO INDUSTRIAL CHEMICALS (continued)

Chemical Agents	Parameter	Biological Material	Reference Value	Tentative Maximum Permissible Concentration	Remarks
		B. Organic Substances (continued)			
4. Amino- and nitroderivatives					
Triethylamine(TEA)	TEA + triethylamine-N-oxide	urine		60 mg/g creat.	if TWA: 2.5 ppm
Dimethylethylamine (DMEA)	DMEA + dimethylethyl-amine-N-oxide	urine		90 mg/g creat.	if TWA: 5 ppm
Aniline	Aniline	urine			
	p-aminophenol	urine		30 mg/g creat.	
	methemoglobin	blood		2%	
	aniline released from hemoglobin adducts	blood		10 µg/100 ml	
Nitroglycerine	nitroglycerine	blood			
Ethyleneglycol dinitrate	ethyleneglycol dinitrate	urine			
	ethyleneglycol dinitrate	blood			
Isopropylnitrate	isopropylnitrate	blood			
	isopropylnitrate	urine			
	isopropylnitrate	expired air			

Compound	Determinant	Medium		
Several aromatic amino- and nitro compounds	methemoglobin	blood	5%	< 2%
	diazo-positive metabolite	urine		
	parent compound, e.g., benzidine, β-naphthylamine	urine		
	hemoglobin adducts	blood		
Nitrobenzene	p-nitrophenol	urine	5 mg/g creat.	
	methemoglobin	blood	5%	< 2%
4,4'-Methylene bis (2-chloroaniline)or	MOCA	urine		
	MOCA			
Methylene dianiline ou MDA	MDA	urine		
Benzidine-derived azo compounds	benzidine	urine		
Monoacetylbenzidine derived azo compounds	monoacetylbenzidine	urine		
2,4-Dinitrotoluene	2,4-dinitrobenzoic acid	urine		
Trinitrotoluene	2,4- and 2,6-dinitro-aminotoluene	urine		
5. Alcohols				
Methanol	methanol	urine	25 mg/g creat.	< 2.5 mg/g creat.
	methanol	blood		
	formic acid	urine		< 60 mg/g creat.
	formic acid	blood		

BIOLOGICAL MONITORING OF EXPOSURE TO INDUSTRIAL CHEMICALS (continued)

B. Organic Substances (continued)

Chemical Agents	Parameter	Biological Material	Reference Value	Tentative Maximum Permissible Concentration	Remarks
Isopropanol	acetone	urine	< 2 mg/g creat.		
	isopropanol	expired air		500 mg/m^3	
Furfuryl alcohol	furoic acid	urine	< 65 mg/g creat.		
6. Glycols and derivatives					
Ethyleneglycol	oxalic acid	urine	< 50 mg/g creat.		
	glycolic acid	urine			
	ethyleneglycol	serum			
Ethyleneglycol monomethylether (methylcellosolve)	methoxyacetic acid	urine			
Ethyleneglycol monoethylether (ethylcellosolve or 2-ethoxyethanol)	ethoxyacetic acid	urine		150 mg/g creat.	if TWA: 5 ppm
Ethyleneglycol monoethylether acetate (2-ethoxyethanol acetate)	ethoxyacetic acid	urine		150 mg/g creat.	if TWA: 5 ppm
Ethyleneglycol monobutylether (butylcellosolve)	butoxyacetic acid	urine			

Substance	Determinant	Medium			
Ethyleneglycol phenylether (phenylcellosolve)	phenoxyacetic acid	urine			
Propyleneglycol monomethylether γ-isomer (1-methoxy-2-propanol)	propyleneglycol	urine			
Propyleneglycol monomethylether β-isomer (2-methoxy-1-propanol)	methoxypropionic acid	urine			
Dioxane	β-hydroxy-ethoxyacetic acid	urine			
7. Ketones					
Acetone	acetone	urine	< 2 mg/g creat.	30 mg/g creat.	
	acetone	blood	< 0.2 mg/100 ml	5 mg/100 ml	
	acetone	expired air			
Cyclohexanone	cyclohexanol	urine		20 mg/g creat.	
Methylethylketone	methylethylketone	urine		2.5 mg/g creat.	
	methylethylketone	blood			
	methylethylketone	expired air			
	3-hydroxy-2-butanone	urine			
Methyl-n-butylketone	2,5-hexanedione	urine		4 mg/g creat.	end of the workweek
Methylisobutylketone	methylisobutylketone	urine		0.5 mg/g creat.	
8. Aldehydes					
Furfural	furoic acid	urine	< 65 mg/g creat.	80 mg/g creat.	

BIOLOGICAL MONITORING OF EXPOSURE TO INDUSTRIAL CHEMICALS (continued)

B. Organic Substances (continued)

Chemical Agents	Parameter	Biological Material	Reference Value	Tentative Maximum Permissible Concentration	Remarks
9. Amides and anhydrides					
Dimethylformamide	N-methylformamide**	urine		30 mg/g creat.	
	dimethylformamide	blood		0.15 mg/100 ml	
	N-methylformamide**	blood		0.1 mg/100 ml	
	dimethylformamide	expired air		2.5 ppm	during exposure
	N-acetyl-S-(N-methyl-carbamoyl) cysteine	urine			
Dimethylacetamide	N-methylacetamide	urine			
Anhydride phtalic	phtalic acid	urine		8 mg/g creat.	if TWA: 1 ppm
Anhydride hexa-hydrophtalic	hexahydrophtalic acid	urine		8 mg/g creat.	if TWA: 0.1 ppm
10. Phenols					
Phenol	phenol	urine	< 20 mg/g creat.	250 mg/g creat.	
p-tert-Butylphenol	p-tert-butylphenol	urine		2 mg/g creat.	
11. Carbon monoxide					
Carbon monoxide	carboxyhemoglobin	blood	< 1%	3.5%	nonsmokers
	carbon monoxide	blood	< 0.15 ml/100 ml	7 ml/100 ml	nonsmokers
	carbon monoxide	expired air	< 2 ppm	12 ppm	nonsmokers

12. Cyanides and nitriles

Cyanides and aliphatic nitriles	thiocyanate	urine	< 2.5 mg/g creat.	6 mg/g creat.	nonsmokers
	thiocyanate	plasma	< 0.6 mg/100 ml		nonsmokers
	cyanide	blood	< 10 µg/100 ml		nonsmokers
	SCN (mg/g creat.) HBCO (%)	urine + blood	< 50 µg/100 ml	3	smokers
Acrylonitrile	acrylonitrile	urine			
	thiocyanate	urine	< 2.5 mg/g creat.		nonsmokers
Methemoglobin forming agents except for specific compounds mentioned elsewhere	methemoglobin	blood	< 2 %	5%	

13. Isocyanates

Toluene diisocyanate	toluenediamine	urine	

14. Cyanamides

Cyanamide	acetylcyanamide	urine	

15. Pesticides

Organophosphorus	cholinesterase	red blood cells	30% inhibition
	cholinesterase	plasma	50% inhibition
	cholinesterase	whole blood	30% inhibition
	dialkylphosphates	urine	
Parathion	p-nitrophenol	urine	0.5 mg/g creat.

BIOLOGICAL MONITORING OF EXPOSURE TO INDUSTRIAL CHEMICALS (continued)

B. Organic Substances (continued)

Chemical Agents	Parameter	Biological Material	Reference Value	Tentative Maximum Permissible Concentration	Remarks
Carbamates insecticides	cholinesterase	red blood cells		30% inhibition	
	cholinesterase	plasma		50% inhibition	
	cholinesterase	whole blood		30% inhibition	
Carbaryl	1-naphtol	urine		10 mg/g creat.	
Baygon	2-isopropoxyphenol	urine			
DDT	DDT	serum			
	DDT + DDE + DDD	blood			
	DDA	urine			
Dieldrin	dieldrin	blood		15 µg/100 ml	
	dieldrin	urine			
Lindane	lindane	blood		2 µg/100 ml	
Endrin	endrin	blood		5 µg/100 ml	
	anti-12-hydroxy-endrin	urine		0.13 mg/g creat.	
Hexachlorobenzene	hexachlorobenzene	blood	< 0.3 µg/100 ml	30 µg/100 ml	
	2,4,5-trichlorophenol	urine			
	pentachlorophenol	urine	< 30 µg/g creat.		
Pentachlorophenol	pentachlorophenol	urine	< 30 µg/g creat.	1 mg/g creat.	
	pentachlorophenol	plasma		0.05 mg/100 ml	

Substance	Analyte	Biological material	Value	Timing / Notes
Chlorophenoxyacetic acid derivatives (2,4-D; 2,4,5-T; MCPA)	2,4-D	urine		
	2,4,5-T	urine		
	MCPA	urine		
1,3-Dichloropropene	mercapturic acid	urine		
	thioethers	urine		
Dinitroorthocresol	dinitroorthocresol	blood	1 mg/100 ml	
	amino-4-nitro-orthocresol	urine		
Ethylene oxide	ethylene oxide	expired air	0.5 mg/m^3	during exposure
	ethylene oxide	blood	0.8 µg/100 ml	during exposure
	N-acetyl-S(-2-hydroxy-ethyl)cysteine	urine		
16. Hormones				
Diethylstilbestrol	diethylstilbestrol		30 mg/g creat.	24-h urine collection
17. Mutagenic and carcinogenic substances				
	• mutagenic activity	urine		comparison with a control group
	• thioethers	urine		
	• chromosome analysis	lymphocytes		
	• spermatozoa analysis	sperm		
	• protein adducts	blood		
	• DNA adducts	lymphocytes		
	• nucleic acid adducts	urine		
	• oncogen proteins	serum		

* Analyses performed on biologic materials collected at the end of the workday unless otherwise indicated.

** The metabolites measured as N-methylformamide by gas chromatography is mainly N-hydroxymethyl-N-methylformamide.

INDEX